# The Handbook of Environmental Chemistry

Volume 4   Part B

Edited by O. Hutzinger

# Air Pollution

With Contributions by
H. Brauer, J. S. Gaffney, R. Harkov,
M. A. K. Khalil, F. W. Lipfert,
N. A. Marley, E. W. Prestbo, G. E. Shaw

With 93 Figures

Springer-Verlag
Berlin Heidelberg New York
London Paris Tokyo
Hong Kong

Professor Dr. Otto Hutzinger
University of Bayreuth
Chair of Ecological Chemistry and Geochemistry
Postfach 101251, D-8580 Bayreuth
Federal Republic of Germany

ISBN 3-540-50915-1 Springer-Verlag Berlin Heidelberg New York
ISBN 0-387-50915-1 Springer-Verlag New York Berlin Heidelberg

Library of Congress Cataloging-in-Publication Data
Air pollution. (The Handbook of environmental chemistry; v. 4, pt A–B)
Includes bibliographies and index.
1. Air – Pollution – Handbook, manual, etc. 2. Environmental chemistry – Handbooks, Manuals, etc.
I. Dop, H. van (Han van), 1944–. II. Series: Handbook of environmental chemistry; v. 4, etc.
QD31.H335 vol. 4, etc. [TD883.1] 623.5'3 86-219604
ISBN 3-540-15041-2 (v. A)

This work is subject to copyright. All rights are reserved, whether the whole or part of the material is concerned, specifically the rights of translation, reprinting, reuse of illustrations, recitation, broadcasting, reproduction on microfilms or in other ways, and storage in data banks. Duplication of this publication or parts thereof is only permitted under the provisions of the German Copyright Law of September 9, 1965, in its version of June 24, 1985, and a copyright fee must always be paid. Violations fall under the prosecution act of the German Copyright Law.
© Springer-Verlag Berlin Heidelberg 1989
Printed in Germany

The use of registered names, trademarks etc. in this publication does not imply, even in the absence of a specific statement, that such names are exempt from the relevant protective laws and regulations and therefore free for general use.

Typesetting, printing and bookbinding: Brühlsche Universitätsdruckerei, Giessen
2152/3140-543210

# Preface

Environmental Chemistry is a relatively young science. Interest in this subject, however, is growing very rapidly and, although no agreement has been reached as yet about the exact content and limits of this interdisciplinary subject, there appears to be increasing interest in seeing environmental topics which are based on chemistry embodied in this subject. One of the first objectives of Environmental Chemistry must be the study of the environment and of natural chemical processes which occur in the environment. A major purpose of this series on Environmental Chemistry, therefore, is to present a reasonably uniform view of various aspects of the chemistry of the environment and chemical reactions occurring in the environment.

The industrial activities of man have given a new dimension to Environmental Chemistry. We have now synthesized and described over five million chemical compounds and chemical industry produces about one hundred and fifty million tons of synthetic chemicals annually. We ship billions of tons of oil per year and through mining operations and other geophysical modifications, large quantities of inorganic and organic materials are released from their natural deposits. Cities and metropolitan areas of up to 15 million inhabitants produce large quantities of waste in relatively small and confined areas. Much of the chemical products and waste products of modern society are released into the environment either during production, storage, transport, use or ultimate disposal. These released materials participate in natural cycles and reactions and frequently lead to interference and disturbance of natural systems.

Environmental Chemistry is concerned with *reactions in the environment*. It is about distribution and equilibria between environmental compartments. It is about reactions, pathways, thermodynamics and kinetics. An important purpose of this Handbook is to aid understanding of the basic distribution and chemical reaction processes which occur in the environment.

Laws regulating toxic substances in various countries are designed to assess and control risk of chemicals to man and his environment. Science can contribute in two areas to this assessment; firstly in the area of toxicology and secondly in the area of chemical exposure. The available concentration ("environmental exposure concentration") depends on the fate of chemical compounds in the environment and thus their distribution and reaction behaviour in the environment. One very important contribution of Environmental Chemistry to the above mentioned toxic substances laws is to develop laboratory test methods, or mathematical correlations and models that predict the environmental fate of new chemical compounds. The third purpose of this Handbook is to help in the basic understanding and development of such test methods and models.

The last explicit purpose of the handbook is to present, in a concise form, the most important properties relating to environmental chemistry and hazard assessment for the most important series of chemical compounds.

At the moment three volumes of the Handbook are planned. Volume 1 deals with the natural environment and the biogeochemical cycles therein, including some background information such as energetics and ecology. Volume 2 is concerned with reactions and processes in the environment and deals with physical factors such as transport and adsorption, and chemical, photochemical and biochemical reactions in the environment, as well as some aspects of pharmacokinetics and metabolism within organisms. Volume 3 deals with anthropogenic compounds, their chemical backgrounds, production methods and information about their use, their environmental behaviour, analytical methodology and some important aspects of their toxic effects. The material for volumes 1, 2, and 3 was more than could easily be fitted into a single volume, and for this reason, as well as for the purpose of rapid publication of available manuscripts, all three volumes are published as a volume series (e.g. Vol. 1; A, B, C). Publisher and editor hope to keep the material of the volumes 1 to 3 up to date and to extend coverage in the subject areas by publishing further parts in the future. Readers are encouraged to offer suggestions and advice as to future editions of "The Handbook of Experimental Chemistry".

Most chapters in the Handbook are written to a fairly advanced level and should be of interest to the graduate student and practising scientist. I also hope that the subject matter treated will be of interest to people outside chemistry and to scientists in industry as well as government and regulatory bodies. It would be very satisfying for me to see the books used as a basis for developing graduate courses on Environmental Chemistry.

Due to the breadth of the subject matter, it was not easy to edit this Handbook. Specialists had to be found in quite different areas of science who were willing to contribute a chapter within the prescribed schedule. It is with great satisfaction that I thank all authors for their understanding and for devoting their time to this effort. Special thanks are due to the Springer publishing house and finally I would like to thank my family, students and colleagues for being so patient with me during several critical phases of preparation for the Handbook, and also to some colleagues and the secretaries for their technical help.

I consider it a privilege to see my chosen subject grow. My interest in Environmental Chemistry dates back to my early college days in Vienna. I received significant impulses during my postdoctoral period at the University of California and my interest slowly developed during my time with the National Research Council of Canada, before I was able to devote my full time to Environmental Chemistry in Amsterdam. I hope this Handbook will help deepen the interest of other scientists in this subject.

Otto Hutzinger

This preface was written in 1980. Since then publisher and editor have agreed to expand the Handbook by two new open-ended volume series: Air Pollution and Water Pollution. These broad topics could not be fitted easily into the headings of the first three volumes.

All five volume series will be integrated through the choice of topics covered and by a system of cross referencing.

The outline of the Handbook is thus as follows:
1. The Natural Environment and the Biogeochemical Cycles
2. Reactions and Processes
3. Anthropogenic Compounds
4. Air Pollution
5. Water Pollution

Bayreuth, February 1989                                           Otto Hutzinger

# Contents

*J.S. Gaffney, N.A. Marley, E.W. Prestbo*  Peroxyacyl Nitrates (Pans): Their Physical and Chemical Properties  *1*

*R. Harkov*  Semivolatile Organic Compounds in the Atmosphere  *39*

*G.E. Shaw, M.A.K. Khalil*  Arctic Haze  *69*

*F.W. Lipfert*  Air Pollution and Materials Damage  *113*

*H. Brauer*  Air Pollution Control Equipment  *187*

Subject Index  *255*

# List of Contributors

Prof. Dr.-Ing. Heinz Brauer
Technische Universität Berlin
Institut für Chemieingenieurtechnik
Ernst-Reuter-Platz 3–5
D-1000 Berlin 10

Dr. Jeffrey S. Gaffney
Center for Environmental Research
Argonne National Laboratory
Argonne, IL 60439, USA

Dr. Ronald Harkov
ENSR
Somerset Executive Square 1
One Executive Drive
Somerset, NJ 08873, USA

Dr. M.A.K. Khalil
Oregon Graduate Center
Institute of Atmospheric Sciences
19600 NW Von Neumann Drive
Beaverton, OR 97006-1999, USA

Dr. Frederick W. Lipfert
Brookhaven National Laboratory
Upton, NY 11973, USA

Dr. Nancy A. Marley
Center for Environmental Research
Argonne National Laboratory
Argonne, IL 60439, USA

Eric W. Prestbo
Chemistry Department
University of Washington
Seattle, WA 98195, USA

Prof. Glenn E. Shaw
University of Alaska
Geophysical Institute
Fairbanks, AK 99775-0800, USA

# Peroxyacyl Nitrates (PANs): Their Physical and Chemical Properties

*Jeffrey S. Gaffney, Nancy A. Marley*
Center for Environmental Research
Biological, Environmental and Medical Research Division
Argonne National Laboratory, Argonne, Illinois, 60439, USA

*Eric W. Prestbo*
Chemistry Department
University of Washington, Seattle
Seattle, Washington, 98195, USA

Introduction . . . . . . . . . . . . . . . . . . . . . . . . . . . . 2
    Discovery of Peroxyacetyl Nitrate (PAN) . . . . . . . . . . . . . 3
    Relationship to Ozone, Nitrogen Oxides, and Organics . . . . . . . 5
Formation of Peroxyacyl Nitrates . . . . . . . . . . . . . . . . . . 5
    Natural Formation in the Troposphere . . . . . . . . . . . . . . . 5
    Photochemical, Gas-Phase Synthesis . . . . . . . . . . . . . . . . 6
    Liquid-Phase Synthesis . . . . . . . . . . . . . . . . . . . . . . 7
Physical Properties . . . . . . . . . . . . . . . . . . . . . . . . 8
    Spectral Properties . . . . . . . . . . . . . . . . . . . . . . . 8
    Solubility . . . . . . . . . . . . . . . . . . . . . . . . . . . 11
Chemical Properties and Stability . . . . . . . . . . . . . . . . . 11
    Photolysis . . . . . . . . . . . . . . . . . . . . . . . . . . . 12
    Chemical Reactions . . . . . . . . . . . . . . . . . . . . . . . 12
        Unimolecular Decompositions . . . . . . . . . . . . . . . . 12
        Other Reported Reactions . . . . . . . . . . . . . . . . . . 13
    Structure . . . . . . . . . . . . . . . . . . . . . . . . . . . 16
Sampling and Analysis Methods . . . . . . . . . . . . . . . . . . . 17
    Long-Path Infrared Spectroscopy . . . . . . . . . . . . . . . . . 17
    Gas Chromatography with Electron Capture Detection . . . . . . . . 19
    Chemiluminescent Detection . . . . . . . . . . . . . . . . . . . 22
    Chemical Methods . . . . . . . . . . . . . . . . . . . . . . . . 23
    Mass Spectrometry . . . . . . . . . . . . . . . . . . . . . . . 24
    Nuclear Magnetic Resonance . . . . . . . . . . . . . . . . . . . 24
Importance of Peroxyacyl Nitrates in the Global Troposphere . . . . 24
    Nitrogen Oxide Cycle . . . . . . . . . . . . . . . . . . . . . . 24
    Nighttime Chemistry . . . . . . . . . . . . . . . . . . . . . . 27

---

\* This work was supported by the United States Department of Energy's Office of Health and Environmental Research and performed at Los Alamos National Laboratory

    Non-methane Hydrocarbon Cycles . . . . . . . . . . . . . . . . 28
    Relation to Pertoxides, Peracids and Organic Nitrates . . . . . . . . 28
Toxicity . . . . . . . . . . . . . . . . . . . . . . . . . . . . . . . 29
    Contribution to Urban Oxidant Burden . . . . . . . . . . . . . 29
    Toxicity to Flora and Fauna . . . . . . . . . . . . . . . . . . 29
Summary and Recommendations for Future Study . . . . . . . . . . . 30
Acknowledgements . . . . . . . . . . . . . . . . . . . . . . . . . . 30
References . . . . . . . . . . . . . . . . . . . . . . . . . . . . . . 31

## Summary

Peroxyacyl nitrates (PANs) are important organic indicators of tropospheric chemical processes. In this chapter, our current understanding of the chemical and physical properties of PANs are reviewed.

    After a brief discussion of the discovery of peroxyacetyl nitrate (PAN), their synthesis and physical and chemical properties are examined. Measurement techniques and sampling methodologies of PANs are presented. Their formation and importance in atmospheric chemistry is discussed in terms of urban, regional, and global scales. Our knowledge of the toxicity of PANs is examined, followed by some suggestions for further study of PANs and associated organic peroxides, peracids, and organonitrates.

## Introduction

The air we breathe is a most important resource. Our atmosphere is a complex physical and chemical media, which we are just now beginning to understand [1]. Increased emissions of reactive trace gases from man's energy related and industrial activities have led to a wide variety of air pollution problems on urban, regional, and global scales. Both primary emissions and their secondary products produced in the oxidative atmosphere can lead to health and environmental consequences. It is increasingly apparent, that numerous trace gas species can play an important role in the chemistry and the physics (climate) of the troposphere.

    One aspect of this chemistry is the production of oxidants in the so-called photochemical "smog." These oxidants are produced thru a rather complex chemistry involving free radicals (e.g. OH, $HO_2$, $RO_2$, etc.), and their interactions with organics and oxides of nitrogen. The oxidants produced include ozone, hydrogen peroxide, and organic oxidants [1, 2].

    Peroxyacyl nitrates is the name given to a rather unique class of organic oxidants. These species can be thought of as the mixed anhydrides of the organic peracids and nitric acid. Listed in Table 1 are the structures and names of some of the better known peroxyacyl nitrates, the most abundant and best studied being peroxyacetyl nitrate (PAN). The alkyl analogs are known as peroxyalkyl nitrates. A list of some of the more important peroxyalkyl nitrates and their structures is presented in Table 2. A number of these organoperoxy nitrate species are known to be produced during photochemical urban "smog" formation and have been shown to be potentially important toxic oxidants. As well, these species may play important roles in regional and global scale chemistries and the long-range

**Table 1.** Chemical formulas and names of some of the more important peroxyacyl nitrates (PANs)

| Peroxyacyl nitrate | Chemical formula | Abbreviation |
|---|---|---|
| Peroxyacetyl nitrate | $CH_3C=OO_2NO_2$ | PAN |
| Peroxyproprionyl nitrate | $CH_3CH_2C=OO_2NO_2$ | PPN |
| Peroxybutryl nitrate | $CH_3CH_2CH_2C=OO_2NO_2$ | PBN |
| Peroxybenzoyl nitrate | $C_6H_5C=OO_2NO_2$ | PBzN |
| Monochloroperoxyacetyl nitrate | $ClCH_2C=OO_2NO_2$ | chloro-PAN |

**Table 2.** Chemical formulas and names of some of the more important peroxyalkyl nitrates, and peroxynitric acid

| Peroxyalkyl nitrate | Chemical formula | Abbreviation |
|---|---|---|
| Peroxynitric acid | $HO_2NO_2$ | PNA |
| Methyl peroxy nitrate | $CH_3OONO_2$ | MPN |
| Ethyl peroxy nitrate | $CH_3CH_2OONO_2$ | EPN |
| Propyl peroxy nitrate | $CH_3CH_2OONO_2$ | PrPN |
| Phenyl peroxy nitrate | $C_6H_5OONO_2$ | PhPN |
| Monochloromethyl peroxy nitrate | $ClCH_2OONO_2$ | chloro-MPN |

transport of $NO_x$ [1, 2]. For these and other reasons, peroxyacyl nitrates have been the subject of numerous studies attempting to understand and characterize these species with regard to their chemical and physical characteristics. In this chapter we will attempt to overview and summarize our current understanding of this important class of compound's properties, the peroxyacyl nitrates. The complex chemistry of photochemical oxidant formation (i.e. ozone, etc.) has been described elsewhere [1, 2] and only will be discussed as it relates to the PANs. Thus, we will attempt to highlight how PAN chemistry is coupled to inorganic and organic peroxide formation, which has important concequences for tropospheric gas-phase and aqueous-phase regional and global scale chemistries [3].

**Discovery of Peroxyacetyl Nitrate (PAN)**

During the 1950s a new type of "photochemical smog" was being described by a number of researchers in the Los Angeles, California area [1, 2]. The term "smog" has been coined before to describe a combination of smoke and fog that occured in London and other coal burning regions. This London smog was associated with high levels of particulate matter and sulfur dioxide. This "reducing-type" of smog was contrasted with the "photochemical smog" which was observed in Los Angeles and its surrounding air shed. This type of smog was found to be oxidizing and caused eye-irritation. Photochemical smog was believed to be associated with motor vehicle emissions, in contrast to the coal derived pollution of London. Haagen-Smit [4, 5] and other researchers [1, 2, 6] quickly demonstrated that organic compounds and nitrogen oxides emitted from mobile and stationary sources could react in the sunlight to form ozone and other oxidizing substances, aldehydes and other eye irritants, which explained many of the characteristics of Los Angeles type smog.

Downwind of Los Angeles, many agricultural areas were affected by the photochemically derived pollutants, producing damage to the foliage and subsequent crop-loss [1, 2, 6–8]. Studies at the Agricultural Experimentation Center in Riverside, California clearly showed that much of this damage was due to the pollutant, ozone. However, a number of crops began to show an unusual bronzing and chlorosis of the leaves which could not be attributed to ozone, nitrogen dioxide, sulfur dioxide, or other known contaminants in air pollution [6, 9, 10]. This damage was attributed to an unknown oxidant, referred to as compound "X" [1, 2, 6]. This should not be confused with X-agent which has also been associated with photochemical oxidant formation [11].

The possibility that this was an organic oxidant was first proposed by Haagen-Smit [6, 12–14]. Since ozone was shown to be produced during the solar irradiation of automotive emissions (mixtures of hydrocarbons and nitrogen oxides), Haagen-Smit assumed that the unknown oxidant was a reaction product from the reaction of ozone with olefins [6, 15]. Plant pathology studies followed which showed that the irradiation of auto exhaust could indeed generate the observed plant damage found in southern California [6, 16]. However, the ozone/olefin reaction products were not observed to yield the compound X characteristic plant damage.

At the Franklin Institute in Philadelphia, a long-path gas cell had been built for infrared spectroscopic studies of air pollution [6]. Stephens, Hanst, and coworkers used this system to study the primary emissions in auto exhaust, and the secondary pollutants formed during solar irradiation [17–19]. These workers observed a rather unique set of infrared adsorption bands during the irradiation of hydrocarbon/$NO_x$ mixtures. This work and the history of these studies has been overviewed recently by Stephens [6]. These workers proposed five possible structures to explain the observed infrared spectra [6, 20]. The proposed structures are shown in Table 3. The actual structure of peroxyacetyl nitrate (PAN, see Table 1) was omitted in this paper. The researchers at that time favored the peroxyacetyl nitrite [6, 20]. However, by using gas chromatography to isolate PAN, and by examining its chemistry, the peroxyacetyl nitrate structure was finally deduced [6, 21].

Table 3. The originally proposed five possible structures for comound X. The actual chemical formula was $CH_3C=OO_2NO_2$, peroxyacetyl nitrate (PAN, see Table 1.) [6]

| Proposed Compound X Structures |
| --- |
| $CH_3C=ONO$ |
| $CH_3C=ONO_2$ |
| $CH_3C=OONO$ |
| $CH_3C=OONO_2$ |
| $CH_3C=OOONO$ (Favored) |

## Relationship to Ozone, Nitrogen Oxides, and Organics

It was recognized in the 1950s that the complex interactions of organics and nitrogen oxides in sunlight led to a rather unique chemistry [1, 2, 6, 22]. Researchers had shown that, in the absence of organics, the inorganic $NO_x$ system would lead to modest ozone formation based upon the so-called photostationary state [1, 2, 22]. The following reaction scheme describes the simple NO, $NO_2$, $O_3$ chemistry as it was understood at that time:

$$NO_2 + hv \rightarrow NO + O(^3P) \tag{1}$$

$$O(^3P) + O_2 \rightarrow O_3 \tag{2}$$

$$NO + O_3 \rightarrow NO_2 + O_2 \tag{3}$$

Upon balancing these chemical equations it became clear that the formation of ozone will be dependent on the light intensity and will be strongly moderated by the rapid reaction of ozone with NO, which leads back to nitrogen dioxide. With the dominant $NO_x$ emission from cars being NO, it was difficult to understand why the conversion of NO to $NO_2$ was occuring so rapidly in the Los Angeles Basin [1, 2, 22]. It was discovered that the photolysis of ozone to form $O(^1D)$ which reacts rapidly with water vapor to form 2 OH radicals could initiate chain reactions with the organic radicals in the atmosphere [1, 2, 22]. Thus, peroxy radicals (e.g. $HO_2$, $RO_2$, and $RCO_3$) can react with NO to form OH, alkoxy radicals, and $NO_2$, and subsequently ozone via reactions 1 and 2 [1, 2]. As will be discussed in detail in the following sections, peroxyacyl nitrates are a direct consequence of this peroxyradical chemistry and are in fact trapped radical species which are in thermal equilibrium with the peroxy radicals.

## Formation of Peroxyacyl Nitrates

The natural chemical reactions leading to the formation of PANs in the troposphere, as well as labortory synthetic methods are reviewed in the following section. These include both gas phase and liquid phase synthetic reactions.

### Natural Formation in the Troposphere

In the troposphere, PANs are formed during the photochemical oxidation of organic molecules that contain more than one carbon atom. The simplest peroxyacyl nitrate that has been observed in the troposphere is peroxyacetyl nitrate (PAN). The immediate precursor to PAN is the peroxyacetyl radical (PA). It can be formed by the oxidation of reactive organics directly [1, 2, 22], or by the reaction of acetaldehyde with OH radical via the following abstraction reaction involving the aldehydic hydrogen:

$$OH + CH_3CHO \rightarrow H_2O + CH_3CO \tag{4}$$

$$CH_3CO + O_2 \rightarrow CH_3CO_3 \tag{5}$$

Once formed the peroxyacetyl radical can react with $NO_2$ in the atmosphere to form PAN via reaction 6:

$$CH_3CO_3 + NO_2 \rightarrow CH_3COO_2NO_2 \qquad (6)$$

PAN is in equilibrium with the PA radical and $NO_2$ [1, 2, 23–26]. This equilibrium chemistry and its importance for the atmospheric lifetime of PAN and its analogs is discussed in subsequent sections in this chapter. One important reaction for the loss of PAN is the following reaction of the PA radical with NO:

$$NO + CH_3CO_3 + O_2 \rightarrow NO_2 + CH_3O_2 + CO_2 \qquad (7)$$

These reactions are pirmarily driven by the photochemical production of OH via photolysis of ozone, nitrous acid, and the chain reactions involving the hydroperoxyl radical ($HO_2$) [1, 2]. Recently, the nighttime reaction of nitrate radical ($NO_3$) with acetaldehyde has been proposed as another possible source of PAN via the following reaction [27, 28].

$$NO_3 + CH_3CHO \rightarrow CH_3CO + HNO_3 \qquad (8)$$

These reactions are the important natural sources of PANs in the atmosphere. The production of the other analogs of PAN (see Table 1) occur via similar reactions for the appropriate aldehyde precursors (i.e. proprionaldehyde for peroxyproprionyl nitrate, benzaldehyde for peroxybenzol nitrate, etc.). Numerous smog chamber studies [29–40] and chemical modeling efforts [41–83] have examined the PAN forming potential for a wide variety of hydrocarbons. These chemical reactions have been reviewed recently elsewhere and will not be dealt with in any further detail here [1, 2].

**Photochemical, Gas-Phase Synthesis**

Laboratory photochemical synthetic procedures for PAN syntheses have mimicked the natural reactions. The initial synthetic procedures used ethylnitrite photolysis in air to produce PAN [6]. The PAN was separated from the numerous side products using preparative gas chromatography [6, 84]. The reactions responsible for the production of PAN from ethylnitrite photolysis in air are:

$$C_2H_5ONO + h\nu \rightarrow C_2H_5O + NO \qquad (9)$$
$$C_2H_5O + O_2 \rightarrow CH_3CHO + HO_2 \qquad (10)$$

These reactions followed by the reaction of $HO_2$ with NO and the subsequent reaction of OH with acetaldehyde lead to the formation of the PA radical and formation of PAN. The method also leads to the formation of a variety of other products and thus requires the chromatographic separation of PAN from the by-products. This preparation procedure requires the freezing out of PAN to concentrate for gas chromatography. Explosive accidents have been reported in the literature using this methodology [85]. Care should be taken when handling PAN on metal surfaces, since PAN is a peroxy nitrate and like all nitrates has explosive potential.

Other gas-phase synthetic methods have been reported [86–90]. These reactions have usually been based upon a modification of the OH abstraction reaction using Cl, Br, or $NO_3$ radicals to remove the labile aldehydic proton from the corresponding aldehyde to form the appropriate PAN in the presence of oxygen and nitrogen dioxide. For examsple, the chlorine atom reaction follows the reaction pathway:

$$Cl_2 + h\nu \rightarrow 2Cl \qquad (11)$$

$$Cl + CH_3CHO + O_2 \rightarrow CH_3CO_3 + HCl \qquad (12)$$

These reactions, of course, are followed by the reaction of PA radical with $NO_2$ to form PAN. The use of the more selective bromine atom does appear to lead to a cleaner synthetic production of PAN via these reactions [86].

The reaction of nitrate radical has also been used to produce PANs [27, 28]. The reaction follows the same mechanistic pathway as OH or halide radical, with the nitrate radical abstracting the aldehydic proton and leading to nitric acid. The nitrate radical is usually generated by reaction of ozone and nitrogen dioxide in the dark, and is stored as nitrogen pentoxide, which is in equilibrium with $NO_3$ and $NO_2$.

In all cases, the gas phase synthetic methods require photochemical apparatus and usually need to perform chromatographic separations to produce PANs for laboratory or instrument calibration purposes.

**Liquid-Phase Synthesis**

As mentioned in the introduction, the PANs can be viewed as the mixed anhydride of peroxyacids and nitric acid. This fact has been used to make PANs in large quantities. Hendry and co-workers [24], first used this chemistry to produce liquid phase PAN by the strong acid nitration of peroxyacetic acid PAA in aqueous solution. They extracted the PAN from the aqueous solution by use of normal alkane solvents (i.e. hexane, octane). Other workers have also explored these reaction methodologies to produce PANs [91–94]. One of the problems with using high volatility solvents in these syntheses was the solvent contamination, and the potential safety hazards associated with volatile hydrocarbon solvents and an active oxidant (i.e. PAN).

It has been demonstrated that the use of heavy lipid solvents can overcome these difficulties, and can produce high purity PAN samples with minimal contamination from solvents [94]. The idea is a simple one. PAN has a vapor pressure of approximately 30 torr at room temperature [94]. If one chooses a lipid solvent that has a lower vapor pressure than PAN, then the resulting synthesis will result in a higher purity product without chromatographic separation. Using a simple vacuum distillation and n-tridecane as a solvent, for example, it has been demonstrated that very high purity PANs can be produced without any elaborate or expensive separation methodologies [94].

For calibration of instrumentation, this synthetic methology is the method of choice. By using diffusion tubes, samples of PAN in tridecane can be used to readily calibrate the various measurement devices without any interferences. As well, these samples can be stored as "PANsicles" (i.e. at temperatures of $-10\,°C$ or lower) for years with minimal loss of material [94]. Once synthesized, the samples can be thawed, vacuum distilled and used for laboratory or calibration use.

## Physical Properties

The physical properties of the PANs are reviewed in the following section. They include the current knowledge of the spectral, physical, and chemical solubility data that has been accrued for these compounds.

## Spectral Properties

The ultraviolet spectra for PAN was initially obtained by Stephens and co-workers on samples obtained from ethyl nitrite photolysis and gas chromatographic separation [21]. This early work has been repeated recently using high purity samples produced from liquid phase synthetic methods by aqueous extraction with heavy lipid solvents [95]. The uv spectral bands and band-strengths are given in Table 4 for PAN and PPN.

Table 4. Ultraviolet-visible absorption cross-sections for peroxyacetyl nitrate (PAN) and peroxyproprionyl nitrate (PPN) [95]

| $\lambda$ (nm) | PAN $\sigma\,(10^{20}\,cm^2)$ | | PPN $\sigma\,(10^{20}\,cm^2)$ | |
|---|---|---|---|---|
| 200 | 317 | $\pm 23$ | 269 | |
| 205 | 237 | $\pm 22$ | 211 | $\pm 4$ |
| 210 | 165 | $\pm 14$ | 155 | $\pm 6$ |
| 215 | 115 | $\pm 9$ | 101 | $\pm 6$ |
| 220 | 77 | $\pm 6$ | 69 | $\pm 5$ |
| 225 | 55 | $\pm 4$ | 47.9 | $\pm 3.4$ |
| 230 | 39.9 | $\pm 3.1$ | 34.7 | $\pm 1.4$ |
| 235 | 29.0 | $\pm 2.1$ | 24.2 | $\pm 1.8$ |
| 240 | 20.9 | $\pm 1.6$ | 17.3 | $\pm 1.3$ |
| 245 | 15.0 | $\pm 1.2$ | 12.4 | $\pm 0.8$ |
| 250 | 10.9 | $\pm 0.9$ | 8.9 | $\pm 0.6$ |
| 255 | 7.9 | $\pm 0.6$ | 6.5 | $\pm 0.4$ |
| 260 | 5.7 | $\pm 0.4$ | 4.6 | $\pm 0.4$ |
| 265 | 4.04 | $\pm 0.30$ | 3.24 | $\pm 0.33$ |
| 270 | 2.79 | $\pm 0.17$ | 2.29 | $\pm 0.22$ |
| 275 | 1.82 | $\pm 0.12$ | 1.51 | $\pm 0.15$ |
| 280 | 1.14 | $\pm 0.08$ | 0.99 | $\pm 0.15$ |
| 285 | 0.716 | $\pm 0.023$ | 0.60 | $\pm 0.11$ |
| 290 | 0.414 | $\pm 0.025$ | 0.33 | $\pm 0.04$ |
| 295 | 0.221 | $\pm 0.016$ | 0.16 | $\pm 0.04$ |
| 300 | 0.105 | $\pm 0.023$ | 0.097 | $\pm 0.009$ |

**Table 5.** Infrared band positions, integrated band strengths, and band assignments for peroxyacetyl nitrate (PAN) [94]

| $v$ (cm$^{-1}$) | $S$ (atm$^{-1}$ cm$^{-2}$) | Assignment |
| --- | --- | --- |
| 3022.3 | 26.7 ± 1.6 | CH$_3$ stretch |
| 1841.3 | 322 ± 9 | C=O stretch |
| 1740.6 | 808 ± 34 | NO$_2$ asym. stretch |
| 1430.1 | 27 ± 5 | CH$_3$ d-deform |
| 1372.2 | 47.9 ± 0.9 | CH$_3$ s-deform |
| 1302.1 | 405 ± 20 | NO$_2$ sym. stretch |
| 1163.3 | 477 ± 9 | C–O stretch |
| 1055.3 | 24.9 ± 1.4 | CH$_3$ rock |
| 990.4 | 34.8 ± 3.5 | CH$_3$ rock |
| 929.7 | 60.8 ± 1.8 | O–O stretch |
| 822.1 | 22.4 ± 1.4 | C–C stretch |
| 793.9 | 247 ± 6 | NO scissors |
| 717.6 | 17.3 ± 1.4 | NO$_2$ wag |
| 604.9 | 46 ± 4 | C–O bend |
| 574.2 | – | NO$_2$ stretch |
| 480.9 | | |

As can be seen from this table, there are no strong structural features to the PAN adsorbtion in the ultraviolet. PANs do not adsorb radiation above 2,900 Angstroms so that their tropospheric photochemistry is not important, in terms of photochemical loss mechanisms [1, 95].

The infrared spectra of PANs have been the diagnostic tool for these unique compounds since their discovery [6]. The infrared band positions and bandstrengths for PAN and PPN are given in Table 5 and 6 [94]. The characteristic ir band around 1,840 cm$^{-1}$ is used to identify the PAN structure. As will be discussed subsequently, this and other characteristic bands in the organo-nitrate regions (i.e. 800–1,000 cm$^{-1}$) have been used with long path infrared spectroscopy to measure PAN in the troposphere at the ppb level.

Laser Raman spectra have been determined by Bruckman and Willner [96]. The Raman bands and scattering strengths are presented in Table 7. Raman spectroscopy has not been used in any great detail for studies of PAN since most research has focussed upon gas phase studies. However, it should be noted that this is potentially a very strong research tool for aqueous PAN chemistry studies, wet aerosol surface interaction, or for other aqueous geochemical studies [97].

PAN has a rather simple hydrogen nuclear magnetic resonance (NMR) signal. The methyl protons are not split, and are all deshielded due to the electronegative nature of the molecule. The singlet is observed at 2.26 $\delta$ (relative to tetramethylsilane, TMS) in carbon tetrachloride solvent. PAN has been observed to be quite soluble in non-polar solvents, and after solvent stripping of gas phase synthesized PAN, the reactions of PAN with a number of organics in non-polar solvents have been reported using NMR to monitor the reaction progress [98–102]. For example, using NMR spectrscopy PAN has been found to be an effective epoxidizing agent for olefins [100], and was observed to acts as an oxidant in solution converting aldehydes to acids [98]. As well, NMR techniques have

**Table 6.** Infrared band positions, integrated band strengths, and band assignments for peroxyproprionyl nitrate (PPN) [94]

| $v$ (cm$^{-1}$) | $S$ (atm$^{-1}$ cm$^{-2}$) | Assignment |
|---|---|---|
| 2998.8 | | |
| 2959.0 | | CH$_3$ stretch |
| 2937.9 | | |
| 2900.8 | | |
| 1833.2 | 262 ± 39 | C=O stretch |
| 1738.6 | 537 ± 25 | NO$_2$ asym. stretch |
| 1469.0 | | |
| 1427.0 | | |
| 1349.0 | | |
| 1301.3 | | |
| 1152.2 | | |
| 1103.2 | | |
| 1048.1 | 131 ± 11 | |
| 992.2 | | |
| 967.2 | | |
| 925.1 | | O–O stretch |
| 852.5 | | |
| 799.2 | 210 ± 14 | NO scissors |
| 741.5 | | |
| 589.8 | | NO$_2$ stretch |
| 499.1 | | |

**Table 7.** Raman active band positions, relative intensities, and assignments for peroxyacetyl nitrates [96]. Spectra were taken of liquid samples at −40 °C

| $v$ (cm$^{-1}$) | Relative band strengths | Assignment |
|---|---|---|
| 3040 | Medium | CH$_3$ asym. stretch |
| 3000 | | |
| 2951 | Very strong | CH$_3$ sym. stretch |
| 1825 | Medium | C=O stretch |
| 1730 | Weak | NO$_2$ asym. stretch |
| 1430 | Weak | CH$_3$ d-deform |
| 1300 | Strong | NO$_2$ sym. stretch |
| 995 | Medium | CH$_3$ rock |
| 820 | Strong | C–C stretch |
| 795 | Medium | NO scissors |
| 715 | Weak | NO$_2$ wag |
| 605 | M | C–O bend |
| 575 | Weak | NO$_2$ stretch |
| 495 | Strong | |
| 375 | Medium | |
| 330 | Strong | |

been used to show that PAN can react with primary, secondary, and tertiary amines to form amides, as well as lead to free radical chemiluminescent reactions [99, 103]. Oxygen-17 NMR studies have also been reported for PAN [101], and have helped to confirm the peroxyacetyl nitrate structure.

## Solubility

PAN was initially thought to be rather soluble in water, since it was observed to undergo rapid base hydrolysis [2, 21, 103–105]. If passed through an aqueous bubbler at a pH of 12 or so, the following reaction was observed:

$$CH_3COO_2NO_2 + 2OH^- \rightarrow CH_3COO^- + NO_2^- + O_2 \text{ (gas)} + H_2O \quad (13)$$

The reaction is a nucleophilic attack by hydroxide ion at the carbonyl position of PAN, which leads to the observed acetate, nitrite, and evolved oxygen gas. It has been reported that the oxygen released is in the excited singlet delta state [104]. This observation will be discussed later with regard to the possible linear and cyclic structures of PAN. This reaction led to the further indications that PAN was indeed a peroxy nitrate.

However, further studies have found that PAN is not very soluble in water and, that under more normal atmospheric pHs (i.e. 4–5), does not undergo rapid hydrolysis. PAN and methyl nitrate have been observed to have aqueous solubilities of 3–5 Molar atm$^{-1}$ [3, 93, 106, 107]. They are more soluble than NO or $NO_2$, but are orders of magnitude less soluble that nitric acid [3]. Thus, the aqueous loss of PANs in the troposphere will not be important and, similar to ozone, troposheric wet deposition will be a very slow process for these compounds.

As mentioned previously, PANs have been found to be extremely soluble in non-polar organic solvents [94, 98–100]. This fact has been used to advantage in the aqueous synthetic chemical production of PANs. Since PANs are not water soluble, once produced in the aqueous layer they are readily trapped in the lipid solvent above the reaction media. The fact that PANs are quite soluble in non-polar organics, has important consequences for its toxic behavior in plants and animals. This property will be discussed further in the Sections "Structure" and "Toxicity".

The triple point, boiling point, and vapor pressure curves for Peroxyacetyl nitrate have been determined [90, 96]. The recommended triple point and boiling point are $-49.3 \pm 0.5$ °C and $105 \pm 2$ °C, respectively. The vapor pressure curve follows the equation:

$$\ln p = -4586/T + 18.78 \quad (14)$$

where p is pressure in torr and T is the temperature in degrees Kelvin.

## Chemical Properties and Stability

In the following section, the chemical properties of PANs will be discussed. The expected photolysis rates, thermal decomposition pathways, and chemical reactions are presented. The likely structural conformations of PAN that explain the chemical and physical data are overviewed and are compared to molecular modeling studies.

**Table 8.** Estimated photolysis rates (sec$^{-1}$) for peroxyacetyl nitrate (PAN) and peroxyproprionyl nitrates (PPN) calculated as a function of solar zenith angle. Solar rates assume a quantum efficiency of unity and are based on the adsorption cross-sections given in Table 4 to 300 nm and extrapolated to 330 nm [95]

| Solar zenith angle | PAN | PPN |
|---|---|---|
| 0° | $21 \times 10^{-8}$ | $23 \times 10^{-8}$ |
| 20° | $19 \times 10^{-8}$ | $21 \times 10^{-8}$ |
| 40° | $13 \times 10^{-8}$ | $15 \times 10^{-8}$ |
| 60° | $6 \times 10^{-8}$ | $7 \times 10^{-8}$ |

**Photolysis**

As can be seen from the ultraviolet-visible adsorption data for PAN (see Table 4), the photolysis of PANs is expected to be slow. Photolytic lifetimes for PAN have been calculated and are given in Table 8 for varying solar zentith angles [95]. It is clear that the photolysis of PANs in the troposphere will be a negligible loss process.

**Chemical Reactions**

The chemical reactions that have been reported are reviewed in the following discussion. They include the gas phase and liquid phase chemistries that can involve PAN reactions.

*Unimolecular Decompositions*

Peroxyacyl nitrates have been found to undergo two types of unimolecular decompositions. The first pathway was identified by Stephens [21]. It is a concerted thermal decomposition process that can be written as:

$$CH_3COO_2NO_2 \rightarrow CH_3ONO_2 + CO_2 \qquad (15)$$

Stephens proposed a cyclic intermediate to explain the observed reaction products.

The second pathway is the thermal decomposition reaction:

$$CH_3COO_2NO_2 \rightarrow CH_3COO_2 + NO_2 \qquad (16)$$

This is, of course, the back reaction of equation 6, since PAN is in equilibrium with the peroxyacetyl radical and nitrogen dioxide. The subsequent loss reactions are those of equation 7, and the following reaction with $HO_2$:

$$CH_3COO_2 + HO_2 \rightarrow CH_3COO_2H + O_2 \qquad (17)$$

This is analogus to the formation of hydrogen peroxide via the reaction:

$$HO_2 + HO_2 \rightarrow H_2O_2 + O_2 \qquad (18)$$

For sometime there was some disagreement with regard to the validity of the first pathway (reaction 15). Bruckmann and Wilner argued that the production of methyl nitrate which was observed in the thermal decomposition of PAN was not due to a cyclic concerted reaction, but rather via the following pathway. Starting with reaction 16, the peroxyacetyl radical was reacted with NO to form methyl peroxy radicals (see equation 7). It was proposed that the methyl nitrate was formed by the reaction of methoxy radicals, formed from methyl peroxy radical reaction with NO, and $NO_2$ as follows:

$$CH_3O + NO_2 \rightarrow CH_3ONO_2 \tag{19}$$

This reaction, however, is not quantitative and only occurs 90 percent of the time [26]. The other pathway leads to the formation of nitrous acid and formaldehyde via the reaction:

$$CH_3O + NO_2 \rightarrow HONO + CH_2O \tag{20}$$

This is a hydrogen abstraction reaction, whereas the reaction 19 is the gas phase addition process.

Using Fourier transform infrared spectroscopy, these pathways were evaluated recently [26]. In these studies, high purity PAN samples [94] were allowed to thermally decompose in cells that had been exposed to high concentrations of PAN to minimize wall reactions. NO was added to some of the studies to turn on the second pathway, and the proposed radical pathway for production of the observed methyl nitrate product. When NO was added, both nitrous acid and formaldehyde were observed in the correct proportions. However, in the absence of NO (and the free radical pathway), no nitrous acid or formaldehyde was observed. Thus, it has been confirmed that both thermal decompositional pathways are available for PANs.

The kinetic rates of these pathways have been measured by a number of workers, and have been reviewed recently [108]. The predicted lifetimes for PAN, if the peroxyacetyl radical is lost, range from 0.5 to 1 hour. However, the acutal lifetime of PAN will depend upon the relative concentrations of $NO_2$, NO, and the $HO_2$ radical, since the reverse reaction to form PAN is important. The lifetimes predicted for both unimolecular pathways are given in Figs. 1 and 2. These two unimolecular decomposition reactions are strongly temperature dependent. In each case, as the temperature decreases. PAN lifetimes are expected to approach months to years in the troposphere.

## Other Reported Reactions

A number of chemical reactions of PANs have been reported in gas and aqueous phases. Rate constants were determined for the gas phase reaction of PAN with CO, $SO_2$, isobutene, acetyldehyde, $NO_2$, water and ozone [25]. These reactions were all found to be sufficiently slow that the heterogeneous removal of PAN by the reactor walls could not be differentiated from the reactions. Based on this study, it is probably safe to say that PAN does not react significantly with any of these gases in the troposphere.

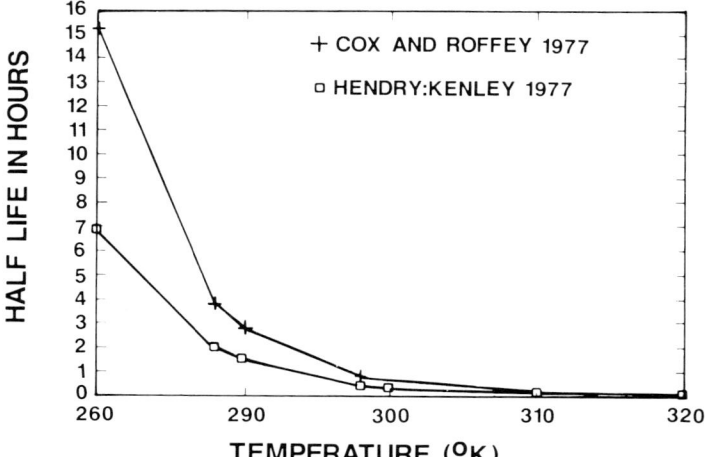

**Fig. 1.** Calculated half-lives for the thermal decomposition of PAN to form peroxyacetyl radical and nitrogen dioxide as a function of temperature. The half-lives are assuming the loss of the peroxyacetyl radical, and do not take into account the back reaction with $NO_2$ to reform PAN. These are then, lower limits for the PAN lifetime in the troposphere. If the ratio of $NO_2$ to NO and $HO_2$ is large the lifetime of PAN will approach infinity [23, 24]

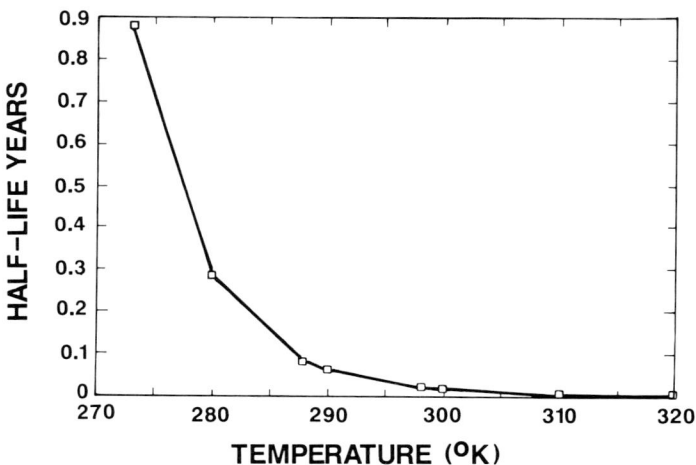

**Fig. 2.** Calculated half-lives as a function of temperature for the unimolecular decomposition of PAN to form methyl nitrate and carbon dioxide [26]

The reaction of OH with peroxyacetyl nitrate has been reported to be relatively slow [109–110]. The expected lifetime for PAN due to OH loss is given in Fig. 3. It is clear from this work that PAN loss by OH radical will also lead to a relatively long lived species in the troposphere.

PANs are known to be sensitive to surface reactions. Loss of PANs on a variety of natural surfaces have been reported [111]. This study and others have

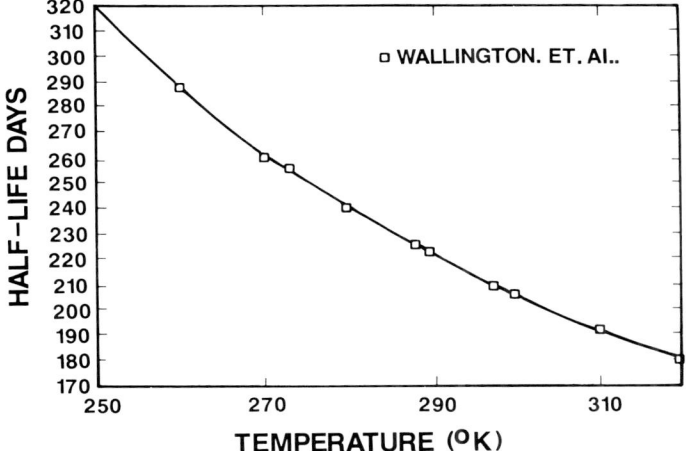

**Fig. 3.** PAN predicted tropospheric half-lives for reaction with OH as a function of temperature [109]

led to the use of teflon or glass columns for analytical measurements and sampling procedures (see Section "Sampling and Analysis Methods").

The only significant gas phase loss processes for PANs are stated earlier, the thermal decomposition of the PAN to the peroxyacyl radical and subsequent reaction of that radical with NO or $HO_2$. The gas phase reactions of PAN with ammonia, primary, secondary, and tertiary amines has been reported [99, 102]. The products observed in the case of ammonia were acetamide, oxygen, and nitrous acid. In the case of primary and secondary amines the substituted amides were observed. The tertiary amines reacted with PAN to form a very complex reaction mixture, probably due to the formation of diazo compounds from the reaction of nitrous acid and the amine. The tertiary amine reactions with PAN were observed to be chemiluminescent.

The reactions of PAN with amines were also observed to occur in non-polar solvents. These reactions were observed under high concentrations of amines, and are not likely to be important processes under the conditions expected in the troposphere. However, these results will be discussed further with regard to the possible structures of PAN.

As mentioned earlier, PAN is known to rapidly undergo hydrolysis in aqueous solutions under basic conditions to form acetate, nitrite and molecular oxygen [104]. In acidic solution the reaction is slow. Most reactions of PAN in aqueous solution are not likely to be important under atmospheric conditions due to the low water solubility of PAN, which was discussed earlier in this chapter. Thus, PAN will act as an oxidant in aqueous solution, but only under conditions where the PAN gas-phase concentration is very high compared to the levels expected in the troposphere.

The reactions of PAN in non-polar solvents at millimolar concentrations with secondary alcohols and olefins have been reported [98, 100]. The observed reaction products were the corresponding ketones and epoxidized olefins. Other studies have examined the reactivity of PAN with biological systems (see Section "Toxicity") and key biochemicals (e.g. sulfhydryl groups) [112].

**Structure**

As described in the introduction, the actual chemical formula of PAN was finally determined to be $CH_3C=OOONO_2$. This was based upon its spectral properties and, in particular, its chemical reactions [121]. A number of laboratory and theoretical studies have examined the structure of PAN and its properties [1, 21, 108, 113–116).

The available chemical and physical information of PAN suggests that it exists in two conformational forms: a linear conformation and a cyclic conformation. The cyclic structure was first suggested by Stephens [26] to explain the formation of methyl nitrate and carbon dioxide from the thermal decomposition of PAN.

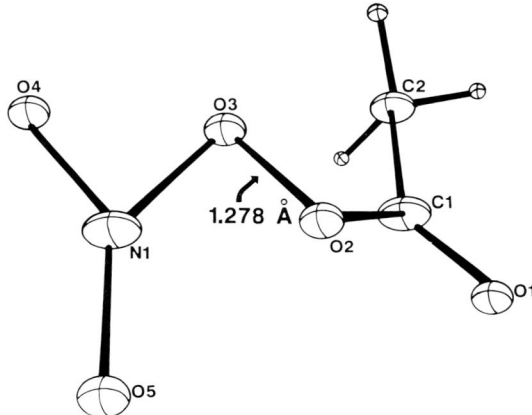

**Fig. 4.** Computer model calculation for the most stable conformation of the linear form of PAN [108]

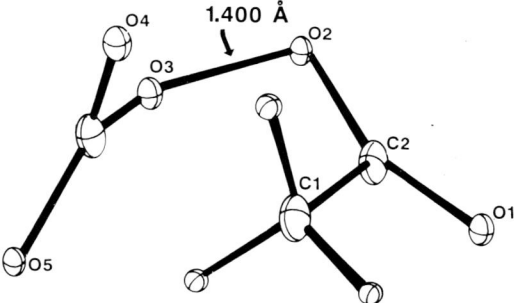

**Fig. 5.** Computer model calculation for the most stable conformation of the cyclic form of PAN [108]

The two conformational structures are given in Figs. 4 and 5. As mentioned earlier, the reaction of PAN in the gas phase and in non-polar solvents with ammonia has been observed to produce acetamide, molecular oxygen, and nitrous acid. This reaction is also consistent with two conformations, with the cyclic form acting as an important transition state for a number of its chemical reactions and explaining its solubility in non-polar solvents. As well, the cyclic form may explain the low water solubility of PAN. This cyclic form is apparently not a hydrogen bonded form since the studies of the thermal decomposition of PAN to methyl nitrate using deuterated PAN have shown no isotope effect on the reaction rate [26].

Using molecular modeling codes, it has been shown that the charge density on the methyl group is slightly positive due to the inductive removal of electron density by the peroxy nitrate group [108]. The oxygen is negatively charged and the interaction is likely a charge complexation.

The predicted cyclic structure has an extended bond-length for the peroxy bond of 1.40 Angstroms, when compared to the linear form which is shorter, 1.278 Angstroms (see Figs. 4 and 5). This stretched O-O bond length is consistant with the production of singlet delta molecular oxygen observed in the base decomposition of PAN in aqueous solution [104], since the bond length of the excited oxygen molecule is very comparable [1]. Thus, there is considerable physical and chemical evidence that PAN exists in two basic conformations, the linear one and the cyclic one. The cyclic conformation may explain many of the PAN molecules unusual chemical and physical properties.

## Sampling and Analysis Methods

In the following section, the measurement techniques that have been used for laboratory and field studies are reviewed. These include long-path infrared (IR) spectroscopy, gas chromatography with electron capture detection, chemiluminescent detection, chemical methods, mass spectrometry, and nuclear magnetic resonance techniques.

### Long-Path Infrared Spectroscopy

As discussed in section "Introduction", the discovery of PANs was accomplished using long-path infrared (LPIR) laboratory investigations of simulated smog [6, 18, 19, 117]. This method was extended to ambient atmospheric studies, where PANs were confirmed to be present in Los Angeles smog [6, 118]. Although PAN was detected using LPIR during strong smog episodes, the sensitivity of the method did not make it practical for routine ambient measurements. With the invention of the electron capture detector (ECD) in 1961 [119] and it's inherent sensitivity and selectivity for PAN measurement, gas chromatography with ECD detection became the dominant technique for ambient air pollution studies. Thus, the application of LPIR to measure PANs in the field was limited until the development of long-path Fourier transform infrared techniques in the early 1970s. LPIR techniques were, however, used to study PANs in the laboratory routinely during this period [21–25].

Stephens in 1961, described the LPIR instrumentation which was first used for air pollution research and PAN characterization [120]. Results were obtained using prism and grating IR spectrometers with multiple reflection long-path gas cells (White cell) [121]. With this instrumentation, pathlengths of 40–400 meters were obtainable, allowing PAN chemistry to be studied in the part per million (ppm) range. Hanst and co-workers continued to improve LPIR methods for air pollution research by further increasing the optical gas cell path lengths and by using Michelson interferometers and Fourier transform infrared techniques [122]. These developments resulted in higher spectral resolution (e.g. $0.025$ cm$^{-1}$), greater sensitivity, and increased time resolution for improved kinetic analysis of the photochemical reactions.

The current state-of-the-art long-path Fourier transform infrared (LP-FTIR) techniques and instrumentation are described by Tuazon et al. [123, 124] and Hanst et al. [125]. The long-path optical cell consists of an eight mirror gold-plated multiple reflection system capable of pathlengths up to 2 km. The cells can be used in either a closed cell for laboratory or smog chamber studies, or in an open configuration for ambient measurements. These systems can typically obtain ppb level detection sensitivities for PANs and other air pollutants. They also can be used for multicomponent analysis to determine several pollutants in the same sample including the PANs, aldehydes, $NO_x$, ozone, carbon monoxide, ammonia, nitrous oxide, methane, and numerous other important trace gas species.

The two IR bands that are commonly used to quantify PANs, are the $793.9$ cm$^{-1}$ band (NO scissors), characteristic of all peroxy nitrates, and the $1,165$ cm$^{-1}$ (C-O stretch) which is characteristic of peroxyacetyl nitrate (PAN). These bands were chosen due to their minimal interference from water and carbon dioxide in these regions of the infrared. Typical sensitivities for PAN detection are typically at the 1–5 ppb level for a one kilometer system [125, 126].

Matrix isolation techniques have also been used with conventional FTIR spectroscopy [127–129]. Atmospheric samples are cyrogenically trapped at 77 °K in a solid matrix of carbon dioxide. The frozen matrix is then analyzed by FTIR to determine the levels of trace gas species. This method has been shown to be very sensitive, with detection levels approaching 10 parts per trillion (ppt) for a number of trace gas species. Also, the high resolution of FTIR coupled with the collapsed rotational levels at low temperature leads to good selectivity. This technique may have advantages for upper atmospheric sampling from aircraft or balloon platforms. However, for routine PAN analyses it suffers from the fact that it is expensive and requires specialized sampling apparatus and cryogenic separation which inherently does not lead to good time resolution in analysis. Furthermore, it has not been evaluated whether PAN is lost or fractionated in the collection processes that employ cold traps or chemical traps to separate atmospheric water from the carbon dioxide frozen matrix.

FTIR techniques have some important advantages over other methods for PAN measurement. Since the infrared band strengths are well known, the FTIR spectrum is the primary standard for PAN calibration and purity evaluation [94, 95, 130–132]. In addition, FTIR measurement can allow real-time in situ detection of PAN which can be an advantage for those reseachers interested in toxicity studies. As well, the rapid time response and selectivity of analysis make FTIR an important tool for kinetic studies of PANs unimolecular decomposition and chemical reactions [94–96].

The major disadvantage of LP-FTIR usage in tropospheric measurements is the difficulties in measuring sub-ppb levels. The lowest detection limit reported in the literature for LP-FTIR analysis of PAN is 3 ppb using a 2 km pathlength [123]. For most field efforts, particularly remote atmospheric studies, FTIR has not been employed for PAN analysis due to the expense and the difficulties in setting up long-path instrumentation in these areas. However, it has been successfully used for numerous laboratory and smog chamber studies where the PAN concentrations are in the high ppb to ppm ranges [133–138].

**Gas Chromatography with Electron Capture Detection**

By far the most widely used method for the analysis of PANs, in both laboratory and field, is gas chromatography with electron capture detection (GC-ECD). It was fortuitous for the early PAN researchers that the electron capture detector was invented in 1961 by Lovelock [119], as the LPIR techniques discussed in the previous section were not sensitive enough at the time to measure PAN routinely at levels less than 50 ppb. Since PAN has a high electron capture cross-section, it was quickly demonstrated that GC-ECD could monitor this key pollutant at 1 ppb readily and routinely.

Due to the toxicity of PAN to plants at the tens of ppb level, the need for a sensitive and reliable method for PAN monitoring was strong. Stephens first demonstrated the potential of GC-ECD detection of PAN in the early 1960s [139]. He designed and built an automated analyzer which was used in early air monitoring, as well as plant pathology and toxicology studies. The basic features of the GC-ECD PANalyzers used today have not changed much since this first instrument.

The basic PANalyzer using GC-ECD detection has been described by Stephens and Price [130]. The ECD detector uses an ionizing source of radiation to produce a standing current of electrons. A typical radioactive source used in this detector is a nickel-63 foil, which emits beta particles. When nitrogen (or other gas chromatographic carrier gas) flows into the detector cell, it is ionized. By applying a pulsed voltage across two cell electrodes in the detector, the ion current can be monitored. When electron affinitive compounds are in the carrier gas, they can "capture" electrons and thus affect the current.

The GC-ECD equipment and accessaries are commercially available. The following components are necessary for routine, automated PAN GC-ECD analysis: 1) a gas chromatograph with ECD detector and non-polar column to separate PAN from the other active ECD components in the air, 2) high purity carrier gas, 3) an automated inert sampling valve with sample loop, and 4) an air pump for obtaining the sample. Typical columns are 40 inches in length using 1/8 the diameter teflon tubing and a non-polar column material, such as 10% Carbowax 400 on Supelcoport (60/80) mesh. As mentioned previously, the PANs are non-polar in behavior so many other non-polar column coatings and inert supports can and have been used for PAN GC-ECD analysis. Teflon is the material of choice for columns as it is relatively inert and minimum PAN decomposition will occur on the surface. It should be noted however that a number of rather porous teflon tubings and teflon wools on the market can lead to analytical difficulties due to oxygen adsorption. Since oxygen is the major souce of background from tailing of the oxygen peak, care should be used to make sure that the materials lead to minimal baseline levels where the PAN peaks are to be observed.

As mentioned in the earlier sections of this chapter, PANs are thermally labile, thus column temperatures are kept near ambient to insure reliable analysis. To allow for water to be passed thru the column in a reasonalbe time, the column is typically operated just above room temperature (i.e. 30 °C). Normally the EC detector is kept about 20 °C above the column temperature to minimize the adsorption of any molecules that may lead to reduced detector efficiency. Carrier flow rates are typically 20–50 ml/min. Under these conditions PAN measurements can be obtained approximately every 30 minutes or less at levels of detection as low as 70 ppt. A typical PAN chromatogram is shown in Fig. 6.

The methodology described above has been used for laboratory, urban, and/or rural measurements of PAN. For measurements of PAN in "clean air" (i.e. over the mid-ocean), cryogenic trapping and preconcentration can be used to measure levels of PAN as low as 5 ppt. The necessary cryogenic sampling processing and calibration techniques have been described elsewhere [140–142].

The GC-ECD PAN analysis method has several advantages over the other PAN analysis techniques. First, due to the large number of electronegative atoms in the molecule, the ECD sensitivity for PAN detection is quite good. Thus, although PAN may elute on the tail of the oxygen peak, it can readily be separated and identified using calibration standards [132]. The necessary instrumentation is relatively inexpensive, and if mini-gas chromatographs are used, the GC-ECD can be made small and light enough for use in field studies. It can be easily automated for continuous, unattended operation. Typically, very little column and analyzer performance degradation is seen after 6 months of continuous usage.

Historically, the only major drawback with the GC-ECD technique has been the accurate calibration of the instrument. Indeed, a great deal of effort has been

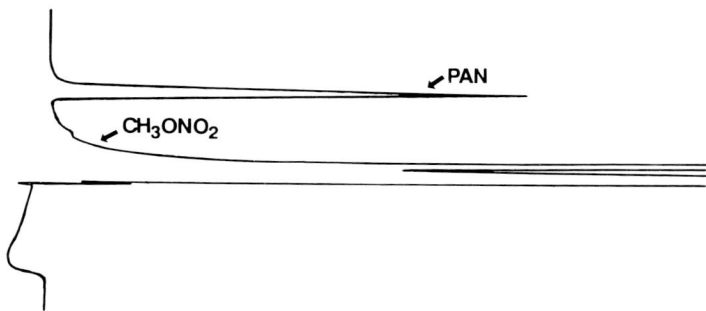

**Fig. 6.** Typical gas chromatographic analysis of PAN using ECD detection. PAN relative retention time is 4.5 minutes. Column conditions are described in the text in the gas chromatography with electron capture section

concerned with simple, cost-effective, and accurate methods for PAN synthesis of standards and their delivery to the instrument [87, 88, 94, 131, 132, 143–150]. With the advent of liquid-phase synthesis of PANs using heavy lipid solvents [94], the present calibration of the GC-ECD instrumentation has become easier, safer, and more reliable than with the gas phase synthetic methods described above.

By using a capillary diffusion tube filled with PAN in tridecane at 0 °C, a constant high purity source of PAN can be delivered directly to the GC-ECD PANalyzer for accurate calibration (typically 10%–15% error) [151]. The purity of the PAN standards are determined in the laboratory using direct spectroscopic IR analysis [94]. PAN standards are simultaneously measured by a $NO_x$ chemiluminescent analyzer and the GC/ECD analyzer to determine the calibration factors [140, 148, 152–155]. The $NO_x$ chemiluminescent instrumentation is described in detail in the following section. PAN has been shown to give an equivalent signal to $NO_2$ in these instruments, so that $NO_2$ gas standards can be used to cross calibrate the two methods.

PAN calibration of the GC/ECD analyzers can also be accomplished using wet chemical methods. After base decomposition, acetate and nitrite aqueous concentrations can be used to measure the gas phase PAN concentrations using ion chromatography or colorometric analysis [21, 103, 131, 132, 140, 147].

There are a number of reports in the literature that humidity can affect the calibration and analysis of PAN using GC-ECD [142–148]. These effects are likely due to water/column interactions. Since PAN has such a low water solubility, wetting the non-polar liquid support on the column would be expected to change the PAN/column adsorption properties. Peak area integration and drying techniques have been used to eliminate these problems [142–148]. Other explanations for these descrepancies, may be caused by unconditioned surfaces in the sampling, inlet lines, injection ports, columns, and detectors being less likely to decompose PAN when they are wet. Again this would be due to the low water solubility [142–148]. Conditioning the column with PAN calibration standards has been shown to eliminate these surface loss problems [21, 94, 95, 142–148].

To date many GC-ECD measurements of ambient PAN levels have been reported in the literature. Most of the reported studies have been in urban or rural atmospheres [54, 57, 149–187]. Fewer measurements are reported for PAN concentrations in remote regions of the world [140, 141, 188–198]. But as discussed in detail in Section "Importance of Peroxyacyl Nitrates in the Troposphere", PAN is now considered a major component of the tropospheric nitrogen oxide chemistry, and is receiving more attention on global scales. GC-ECD measurement techniques have also been used extensively in numerous laboratory studies examining the properties of PAN. For example, these PANalyzers have been used to examine the thermal decomposition [23], the absorption onto natural surfaces [111], the aqueous solubility and hydrolysis in acidic aqueous solutions [93], as well as the loss rates of PAN to teflon film surfaces [31, 32]. Formation of PANs in laboratory and smog chamber studies from the photochemical $NO_x$ driven oxidation of various hydrocarbons has been routinely studied using GC-ECDs [28, 29, 33, 35–37, 39, 72, 199–206]. The successful applications in these areas and in toxicity studies, all point to the usefullness of the GC-GCD technique for PAN measurement.

**Chemiluminescent Detection**

Three types of chemiluminescent detection methodologies have been reported for PAN measurement: 1) reaction with triethylamine, 2) conversion of PAN to NO and detection using the NO-$O_3$ reaction, and 3) the gas/solution surface reaction of PAN with luminol.

The reaction of PAN with triethylamine has been reported to yield a broad-band chemiluminescent peak at 650 nm, in both solution and gas phase reaction [102]. Ozone was also observed to produce an emission from this reaction, however, it occurs in a different spectral region (515 nm). A chemiluminescent spectrophotometric detector has been proposed for simultaneous PAN and ozone measurement using these reactions [102]. This instrument monitors the two oxidants by using two band pass filters which alternately measure the PAN emission at 650 nm and the ozone emission at 500 nm. Laboratory detection limits of 6 ppb for PAN were reported. This technique has not been used in field studies of PAN, and requires further study to determine other possible interferences from peroxides and organic peracids.

Winer et al. [153] first reported that commercial NO-$NO_2$ chemiluminescent analyzers respond to organic nitrates, namely PAN. This instrument relies upon the reaction of NO with $O_3$ to produce excited $NO_2$. The $NO_2$ emission is a braodband chemiluminescence, starting at approximately 600 nm and peaking at 1.27 microns in the infrared. A red sensitive photomultiplier is used to monitor the emission from the reaction of NO with $O_3$. To monitor $NO_2$, the gases are passed over a hot catalyst (gold, charcoal, etc.) which converts the $NO_2$ to NO. This can then be monitored by reaction with ozone [207]. The instrument then operates in two modes: 1) without a catalyst which measures NO only and 2) with a catalyst which measures NO and any compounds which are converted to NO by the catalytic decomposition. Not only PAN but a number of other nitrates including nitric acid have been found to be decomposed by the catalysts to form NO and are therefore measured by this instrument. Thus, the catalyst channel measures NO+$NO_x$, where $NO_x$ is $NO_2$, PAN, nitric acid, organic nitrates, and other nitrogen oxides. These instruments have, therefore, been commonly refered to as "$NO_x$ boxes".

The efficiency of PAN conversion for a number of $NO_x$ box converters has been quantified, and found to be typically greater than 98% [105, 155]. It has been suggested that PAN can be measured using $NO_x$ chemiluminescent detection by removing the PAN signal by reaction with base, and thus determining the PAN concentration by difference [105, 155]. This has some difficulties in that the other $NO_x$ species may also be removed by interaction with the base. Research instruments are capable of sub ppb detection of PAN using gold coated highly reflective reaction cells and improved optics for optimized light collection, and high output ozone sources to increase the reaction efficiencies with NO. However, commercially available instrumentation is limited to 1–2 ppb detection sensitivities, which limit the usefullness of this method for ambient measurements. As mentioned earlier, however, this method is useful for calibration of other techniques under controlled laboratory conditions using high purity PAN samples.

PAN has also been shown to react with luminol (5-amino-2,3-dihydro-1,4-phthalazine dione) in a gas/surface solution reaction to produce intense chemiluminescence with a maximum at 425 nm. Air is passed over a reaction cell consisting of a glass-fiber wick which is continuously coated with a solution of luminol. The chemiluminescent reaction on the wetted wick surface is monitored using a standard photodetector.

$NO_2$ also reacts with luminol to levels of 5 ppt [208]. Other oxidants have been reported to react with luminol (e.g. hydrogen peroxide, ozone), although these interferences can be lowered by changing the solvent system to take advantage of the different gas phase/solution solubilities of these oxidants. For example, organic solvents can be substituted for water to change the effective solvent extraction and subsequent solution reaction rates [208].

This method has an inherent advantage for PAN detection in that the 425 nm emission is easily monitored with most commercially available photomultiplier tubes. Thus, in contrast to the 1–2 ppb typical detection limit for $NO_2$ box systems, luminol detection has been demonstrated to be in the 5–10 ppt level. In addition, by coupling the luminol detector to a gas chromatograph, interferences from other oxidants can be avoided [209]. It is expected that improvements in detection sensitivity and instrument stability can be obtained by maximizing reaction cell optical collection efficiencies, by temperature control of the cell and photomultipler tube, and by use of photon counting and low-dark current photomultipliers. This instrumentation can also be designed to allow the direct measurement of other key organic oxidants, such as methyl hydroperoxide and peracetic acid. As described in Section "Relation to Peroxides, Peracids, and Organic Nitrates", these other oxidants are coupled to the decomposition reactions of PANs and the tropospheric chemistry of peroxy radicals [210].

**Chemical Methods**

Potassium iodide in aqueous solution will react with oxidants to produce iodine which can be determined spectrophotometrically or electrochemically. However, this method suffers from the lack of selectivity for PAN. Ozone, hydrogen peroxide, nitrogen dioxide, and other oxidants will all act as interferences. Under controlled laboratory conditions this method has been used for PAN analysis with moderate sensitivity.

As mentioned earlier in this chapter, PAN was found to undergo rapid alkaline hydrolysis to produce acetate, nitrite, and molecular oxygen [103–105]. The analysis of nitrite using the Saltzman colormetric method [211] or ion chromatography [132] has been used to determine PAN concentrations after base decomposition.

Wet chemical methods are not often used for the routine analysis of PAN in ambient air. This is principally due to the lack of selectivity, sensitivity of detection, and the greater time required for analysis. However, chemical methods do offer another alternative for the measurement of PAN in the laboratory and have been used to confirm results obtained by other methods.

### Mass Spectrometry

The electron impact and chemical ionization mass spectra of PAN have been reported [212–213]. Both these spectra were taken in the positive ion mode. Under these conditions the molecular ions were observed, but were of low intensity.

Mass spectrometric methods have not been used for ambient PAN analysis. Due to its high electron capture cross section, it is anticipated that using gas chromatography with mass spectrometric detection in the negative ion mode should lead to a sensitive and specific detection system for PAN measurement. Based upon the negative ion single ion monitoring sensitivities for similar electron capture sensitive compounds, detection limits would be expected to be in the 10–100 ppt region. This is a methodology which deserves further research to determine its capabilities for PAN studies in laboratory and field application.

### Nuclear Magnetic Resonance

As mentioned earlier, nuclear magnetic resonance (NMR) spectroscopy has been used to study PAN reactions in non-polar solvents with a number of organics in the millimolar concentration region [98–100]. The proton NMR spectrum of PAN is very simple. Since it contains only hydrogens on the methyl group the peak is a singlet [116]. In aqueous solution, PAN will hydrolyze to acetate and the NMR signal observed in $D_2O$ is identical to acetate [116]. In non-polar solvents such as carbon tetrachloride, the methyl group is observed at 137.5 cycles/s relative to Tetramethylsilane on a 60 megahertz spectrometer (i.e. 2.26 $\delta$) [98–100, 116]. This evidence was useful in elaborating the PAN structure as discussed earlier in this chapter [116].

Nuclear magnetic resonance techniques are not sensitive enough to be useful for routine ambient measurements. However, they should not be overlooked when studying the fundamental reactions of PAN in solutions.

## Importance of Peroxyacyl Nitrates in the Global Troposphere

PANs have been measured in urban, rural, and global environments [54, 57, 140, 141, 161–198, 214–230]. While originally thought to be only of importance in highly polluted atmospheres such as that found in Los Angeles, it has become inceasingly apparent that PANs may play important roles in regional and global scale chemistry [231].

The potential roles of PAN in the nitrogen oxide cycle, nighttime chemistry, non-methane hydrocarbon cycles, and their relation to perioxdes, peracids, and organic nitrates are reviewed briefly in the following section.

### Nitrogen Oxide Cycle

The importance of the nitrogen oxide cycle in tropospheric chemistry has been reviewed in detail elsewhere [1, 2, 231], and will be discussed only with regard to the increasing importance of PANs as a reservoir for $NO_x$.

Unlike the inorganic $NO_x$ species (e.g. NO, $NO_2$, $NO_3$, $HNO_3$, $N_2O_5$, etc.), the PANs can have an appreciable atmospheric lifetime. This is due to their physical and chemical properties which were reviewed in Sections "Physical Properties" and "Chemical Properties and Stability". PANs are photochemically inert, react slowly with OH radical, and are slow to wet deposit. Their principal loss mechanisms are thermal decomposition and reaction of the peroxyacyl radical with NO or $HO_2$. Under colder atmospheric conditions, PANs can be quite stable and can be transported long distances in the atmosphere, thus affecting regional and global scale levels.

As indicated before, PANs are in equilibrium with peroxyacyl radicals. Thus, as PAN concentrations in the troposphere change, they are a direct measure of the changing peroxyacyl radical levels. Typical diurnal patterns for PAN are indicate in Fig. 7 for a remote field site near Bandelier National Monument in New Mexico, USA. Note that the high levels are associated with a plume which has probably been transported for days from an urban source. Note also the strong diurnal variability of the PAN signal observed at the site. This is consistent with the photochemical production of peroxyacyl radicals in the troposphere and subsequent trapping of that radical by $NO_2$ to form PAN.

The levels of PAN observed at any site in troposphere will depend upon the thermal stability of PAN, the photochemical activity, and the local levels of NO, $HO_2$, and $NO_2$. It is interesting to examine concentration frequency plots of PAN levels at different sites as a function of the time of year to evaluate the

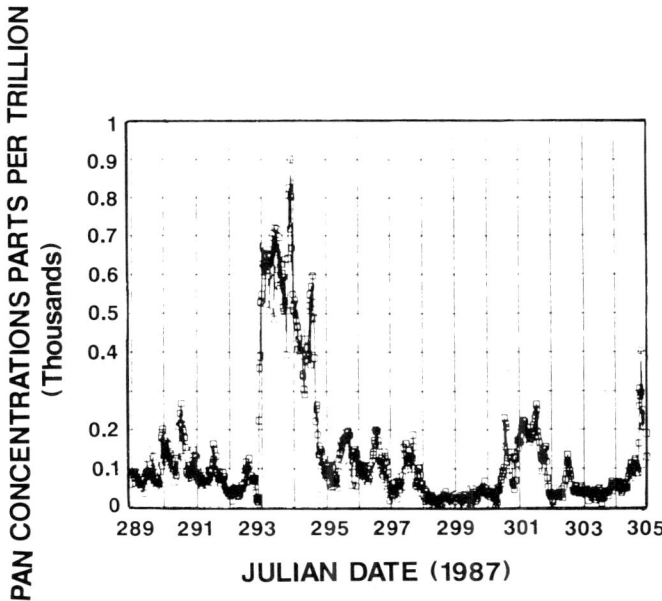

**Fig. 7.** PAN measurements from a remote site near Bandelier National Monument in New Mexico, USA. Note strong diurnal pattern for most days, and evidence for long range plume transport between Julian days 291 and 295 of 1987. Measurements were taken using a GC-ECD PANalyzer

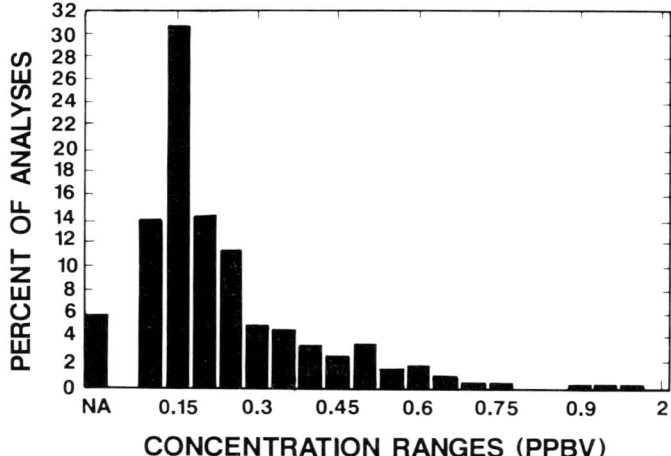

**Fig. 8.** Frequency distributions for PAN analyses for a site at Brookhaven National Laboratory on Long Island, New York, USA, for the period of March 14, 1985 to March 29, 1985. 472 analyses were taken, with a maximum of 1 ppb PAN measured. Only 6 percent of the time were PAN levels below the detection limit of 70 ppt

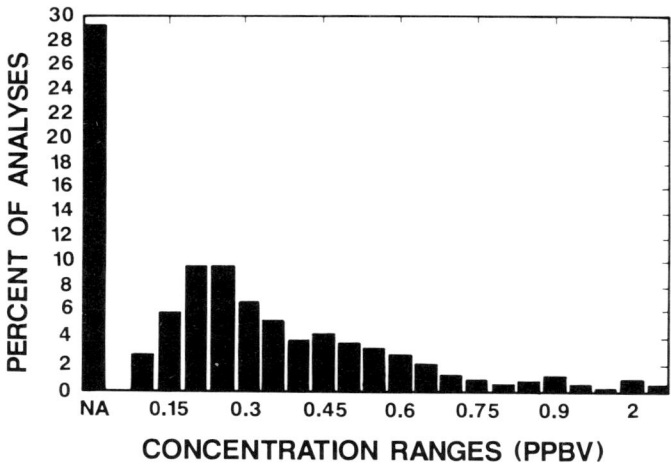

**Fig. 9.** Frequency distributions of PAN analyses at the Brookhaven National Laboratory site for the period of May 6, 1985 to May 15, 1985. 441 analyses were processed, with a maximum of 6 ppb PAN measured. During this period 29 percent of the time PAN levels were below the detection limit of 70 ppt

thermal stability factor. Given in Figs. 8 and 9 are frequency plots of PAN levels measured at Brookhaven National Laboratory on Long Island, New York, USA. These plots clearly indicate that during the colder months that the most of the time PAN levels are usually above detection limits of the GC/ECD instrument (70 ppt). During the colder month of March, only 6 percent of the time were PAN levels below the detection limit. During the warmer month of May, PAN concentrations were below the limit 29 percent of the time. This evidence is consistent with PAN levels being dependent upon temperature and indicate PAN is more stable during the colder months. However, the frequency of high PAN measurements (greater than 1 ppb) are less during the colder months and are found to increase during the spring and summer with the implication being that the photochemical production in the warmer (sunnier) months is greater and that local urban production and transport to the area can be important.

Analysis of data at other sites (Quakertown, Pennsylvania, USA; Whiteface Mountain, New York, USA; Rio de Janeiro, Brazil) [232] has confirmed these observations which indicate that the thermal stability of PAN, its photochemical production, and the peroxyacyl reactions with NO, $HO_2$ and $NO_2$ control its levels in the troposphere. Under cold conditions, particularly in the upper troposphere, PAN can be transported long distances in the atmosphere [231]. It is a greenhouse gas [233], like the other organic nitrates and may be playing an important role in the long range transport of $NO_x$ in the troposphere [231].

**Nighttime Chemistry**

Since PAN is in thermal equilibrium with the peroxyacetyl radical and $NO_2$ (see equations 6 and 16), PAN can be a source of peroxy radicals at night. The peroxyacetyl radical can react with NO to form $NO_2$ and methylperoxy radical (see equation 7). In turn, methylperoxy radical can react with NO to form $NO_2$, formaldehyde, and $HO_2$ via the following reaction:

$$CH_3O_2 + NO + O_2 \rightarrow NO_2 + CH_2O + HO_2 \tag{21}$$

Thus, the nighttime thermal decomposition of PAN in an atmosphere with NO at relatively high concentrations (e.g. 10 ppb), can lead to conversion of three NO molecules to $NO_2$, as well as form formaldehyde. The initial decomposition also leads to a molecule of $NO_2$. The $NO_2$s and formaldehyde produced via these dark reactions will in turn be photochemically active (whereas PAN is not), and can lead to enhanced photochemical smog production when the sun rises and the photochemistry is initiated.

Under cold conditions, this expected affect of PAN on nighttime chemistry in the troposphere will be slowed down and will not be as important. As well, under conditions of low NO (i.e. remote regions), the removal of the peroxyacetyl radical will be by reaction with $HO_2$ to form peroxyacetic acid. Thus, the production of $NO_2$ will not be chain related and only will depend upon the $HO_2$ concentration which is likely to be low in the dark in the absence of NO and $NO_3$ driven nighttime reactions.

The reaction of $NO_3$ with acetaldehyde and higher aldehydes has been proposed as a possible nighttime source of PAN [27, 28]. This reaction of course would have to compete with the reaction of $NO_3$ with reactive organics (olefins, isoprene, terpenes, etc.), $NO_2$ to form $N_2O_5$, and with water to form nitric acid. Although computer models have indicated that this may be an important mechanism for PAN formation, it is not clear from the available field data whether this is the case. Field studies examining this possibility will need to use conservative tracers to tag the air mass to differentiate PAN formation from $NO_3$, from transport of a polluted air mass.

It is clear that the nighttime chemistry of peroxyradicals, PANs, nitrate radicals, etc. is an area where we are just beginning to appreciate their potential importance. PANs are trapped peroxy radicals and may play an important role in this dark chemistry in the troposphere, as well as be an indicator of photochemical smog activity. This is an important area of research, which needs more attention in the future evaluation of the importance of PANs in tropospheric chemistry.

**Non Methane Hydrocarbon Cycles**

Not only are PAN and its analogs an indication of the $NO_x$ chemistry, they are directly tied to non-methane hydrocarbon cycles and are a measure of the hydrocarbon reactivity in the troposphere [1, 2]. As pointed out earlier, PANs can be produced from a wide variety of organic compounds. This can occur directly from the production of peroxyacyl radicals during the oxidation of the compound by ozone or OH radical. Otherwise PANs can be produced by the attack of OH or nitrate radical (nighttime) on aldehydes, which are secondary pollutants produced during the oxidation of primary emitted organic pollutants or naturally emitted reactive hydrocarbons.

For some time, global tropospheric chemistry models did not incorporate PANs or other higher analog organic nitrates into the codes. Recent data indicate that PAN and other hydrocarbons that have reduced reactivity can lead to increased levels of carbon monoxide, methyl hydroperoxide, and other species which can be important in global hydrocarbon cycles. The importance of these trace gas reactions must be taken into account with regard to global $NO_x$ cycles, ozone production, and trace greenhouse gas and aerosol production [234–236].

As we change fuel usage due to energy related constraints, we should also remember that aromatic fuels are likely to lead to peroxybenzoyl nitrate production [237]. This compound is a strong lachrymator and its toxic properties and chemistries are not well understood. Thus, changes in non-methane hydrocarbon emissions will definitely be reflected in the secondary production of PANs and other secondary pollutants in the atmosphere.

**Relation to Peroxides, Peracids, and Organic Nitrates**

The PANs are chemically tied to organic peroxides, peracids, and organic nitrates through their chemical decomposition and formation reactions [1–3, 26, 238–242]. As discussed previously, the reaction of the peroxyacetyl radical with NO or with $HO_2$ (see equations 7 and 17) will lead to the formation of smaller peroxy radicals or lead to the formation of peracetic acid. The reaction of methylperoxy radicals with $HO_2$ will lead to the formation of methylhydroperoxide in the troposphere. The thermal decomposition of PAN to yield methyl nitrate can also occur so that PAN chemistries can lead to organic nitrate production [26, 231].

Since PANs can decompose to form alkyl peroxy radicals, the presence of PAN is also an indicator that peroxyalkyl nitrates may be playing a more important role than previously indicated, particularly under cold weather conditions [1-3, 238, 241-242]. The ties of PANs to peroxyradical chemistry are an important reason for measurement of these compounds in the troposphere. Due to their equilibrium with peroxyacyl radicals, PANs can act as a direct indicator of the organic peroxy radical levels which play an important role in photochemical smog chemistry [1-3, 26, 238].

## Toxicity

As discussed in the introduction, PANs were first discovered because of their unusual type of visible plant damage (leaf bronzing), and their high toxicity to plants, particularly to a number of important agricultural crops. Since these early studies, toxicity and health effect studies have been carried out on plants [244-300], animals [301-320], and on man [321-330]. These studies and others have also explored the biochemical reactions of PANs as reactive oxidants and as potential mutagenic, carcinogenic, teratogenic toxic compounds [244-335]. The toxic effect of PANs will be briefly highlighted in the following sections.

### Contribution to Urban Oxidant Burden

There has been considerable concern that elevated levels of oxidants, particularly ozone can lead to health effects, agricultural losses, and ecosystem impacts. PAN concentrations are typically one-tenth the concentration of ozone in polluted atmospheres. For example, when ozone reaches 200 ppb levels, it is not uncommon to measure PAN levels in the 20 ppb range. PAN on a molecule per molecule basis has been observed to be a more active toxic agent than ozone. For example, plant damage is typically reported at levels of 10 ppb for PAN, while ozone must reach levels of 100 ppb for chlorosis to be observed [244-300].

Other peroxyacyl nitrates have not been studied to determine their relative toxic effects compared to PAN. There has been some concern that chlorinated organics (i.e. trichlorethylene, tetrachlorethylene, allyl chloride, etc.), will be oxidized to form chlorinated PANs [336-338]. These chloro-PANs have been found to be quite mutagenic [336]. These compounds, as well as their chemically coupled peracids, need to be examined in future studies. Due to their potential oxidative effects and their ability to interact with lipid membranes, PANs should not be ignored when evaluating the potential toxic effects of photochemical oxidants on man and his environment.

### Toxicity to Flora and Fauna

PANs have been known to cause visible damage (bronzing) to leaves of numerous types of plants, ranging from simple lichens to higher plants (e.g. conifers, orange trees, etc.) [244-300]. PANs have been found to be active mutagens in the Ames test [301–303, 307–309, 312, 313, 330], and are a major contributor to the gas-phase mutagenetic activity in the air we breathe.

Like most oxidants, PANs have also been found to have appreciable bactericidal activity [304, 310]. PAN has been known for sometime to be a potent lachrymator [21, 325, 326], and is responsible for an appreciable amount of the eye irritation potential along with associated aldehydes [326]. PAN has been proposed as being a significant factor in the epidemiology of skin cancer [327]. A number of health effect studies have indicated that PANs can act synergistically with other stress factors (i.e. chronic bacterial infection, carbon monoxide levels, heat stress, etc.) to amplify pulmonary and cardiovascular effects in man [323, 324, 328].

PANs have been demonstrated to be potentially important toxic agents in our air. Further studies are needed, particularly with other co-existing pollutants, to determine the potential increased hazards of PAN exposure to plants, animals, and man.

The related organic peroxides, organic peracids, and organic nitrates have not been examined in any detail with regard to their potential health effects and should not be neglected in future work. This is particularly true, since PANs will co-exist with peracids and other toxic agents which may affect the testing results due to their separate and combined biochemical interactions.

## Summary and Recommendations for Future Study

Our understanding of the chemistry and physical properties of PANs has improved substantially since its discovery in the early 1950s. It is now clear that PANs are not just a chemical curiosity associated with the Los Angeles smog problem, but are important trace gases in the troposphere. They have been clearly shown to play an important role in the $NO_x$ chemistry of the troposphere, as well as acting as a direct measure of the hydrocarbon and peroxyradical chemical activity in the air.

These important trace gas species must be considered when examining greenhouse effects and toxic impacts to ecosystems and man. Future efforts should not ignore these key organic oxidants.

## Acknowledgements

We wish to thank Dr. Edgar Stephens of the Statewide Air Pollution Research Center of the University of California, Riverside for helpful discussions. The authors wish to acknowledge the support of the Department of Energy's Office of Health and Environmental Research. In particular, we wish to thank Drs. Ballantine, Slade, and Patrinos for their continuing support and encouragement. Thanks also to Mr. Garth Tietjen for his graphic art contribution. This work was carried out at Los Alamos National Laboratory as part of our continuing research efforts into the transformation chemistry of organics in the troposphere, and their potential impacts on urban, regional, and global scales.

# References

1. Finlayson-Pitts, B.J., Pitts, J.N., Jr. (1986) Atmospheric Chemistry: Fundamentals and Experimental Techniques. John Wiley & Sons, New York, and references therein
2. Gusten, H. (1986) in The Handbook of Environmental Chemistry, Hutzinger, O. Ed., vol. 4A, p. 53
3. Gaffney, J.S., Streit, G.E., Spall, W.D., Hall, J.H.: Environ. Sci. Techn. 21:519
4. Haagen-Smit, A.J. (1952) Ind. Eng. Chem. 44:1342
5. Haagen-Smit, A.J., Fox, M.M. (1956) Ind. Eng. Chem. 48:1484
6. Stephens, E.R. (1987) EOS 68:89
7. Middleton, J.T., Haagen-Smit, A.J. (1961) J. Air Pollut. Contr. Assoc. 11:125
8. Middleton, J.T. (1961) Ann. Rev. Plant Physiol. 12:431
9. Middleton, J.T., Kendrick, J.B., Schwalm, H.W. (1950) Calif. Agric. 4:7
10. Middleton, J.T., Kendrick, J.B., Schwalm, H.W. (1950) Plant Dis. Rep. 34:245
11. Moriyama, H. (1971) Studies on X-agent; real cause of environmental pollution including photochemical oxidants. Igaku Shoin, Ltd. Tokyo
12. Haagen-Smit, A.J. (1952) Sci. Amer. 186:28
13. Haagen-Smit, A.J., Fox, M.M. (1955) Soc. Auto. Eng. 63:575
14. Haagen-Smit, A.J. (1950) Eng. and Sci. 14:1
15. Haagen-Smit, A.J., Bradley, C.E., Fox, M.M. (1952) Proc. Natl. Air Pollut. Symp. 2nd, pp. 54–56
16. Haagen-Smit, A.J., Darley, E.F., Zaitlin, M., Hull, H., Noble, W. (1952) Plant Physiol. 27:18
17. Stephens, E.R. (1958) Appl. Spec. 12:80
18. Stephens, E.R., Hanst, P.L., Doerr, R.C., Scott, W.E. (1956) Ind. Eng. Chem. 48:1498
19. Stephens, E.R., Hanst, P.L., Doerr, R.C., Scott, W.E. (1959) J. Air Pollut. Contr. Assoc. 6:159
20. Stephens, E.R., Scott, W.E., Hanst, P.L., Doerr, R.C. (1956) J. Air. Pollut. Contr. Assoc. 8:333
21. Stephens, E.R. (1969) Adv. Environ. Sci. Technol. 1:119
22. Leighton, P.A. (1961) Photochemistry of Air Pollution, Academic Press, New York
23. Cox, R.A., Roffey, M.J. (1977) Environ. Sci. Technol. 11:900
24. Hendry, D.G., Kenley, R.A. (1979) Nitrogenous Air Pollut.: Chem. Biol. Implic., Grosjean, Daniel (ED), Ann Arbor Sci., Ann Arbor, Mich., pp. 137
25. Pate, C.T., Atkinson, R., Pitts, J.N., Jr. (1976) J. Environ. Sci. Health, A11:19
26. Senum, G.I., Fajer, R., Gaffney, J.S. (1986) J. Phys. Chem. 90:152
27. Cantrell, C.A., Davidson, J.A., Busarow, K.L., Calvert, J.G. (1986) J. Geophys. Res. 91D:5347
28. Shepson, P.B., Edney, E.O., Kleindienst, T.E., Pittman, J.H., Namie, G.R. (1985) Environ. Sci. Technol. 19:849
29. Gery, M.W., Fox, D.L., Kamens, R.M., Stockburger, L. (1987) Environ. Sci. Technol. 21:339
30. Gery, M.W., Fox, D.L., Jeffries, H.E., Stockburger, L., Weathers, W.S. (1985) Int. J. Chem. Kinet. 17:931
31. McMurry, P.H., Grosjean, D. (1985) Environ. Sci. Technol. 19:1176–1182
32. Grosjean, D. (1985) Environ. Sci. Technol. 19:1059
33. Glavas, S., Schurath, U. (1985) Environ. Sci. Technol. 19:950
34. Leone, J.A., Flagan, R.C., Grosjean, D., Seinfeld, J. (1985) Int. J. Chem. Kinet. 17:177
35. Grosjean, D., Atmos. Environ. 18:1641
36. Spicer, C.W. (1983) Environ. Sci. Technol. 17:112–120
37. Winer, A.M., Darnall, K.R., Atkinson, R., Pitts, J.N., Jr. (1979) Environ. Sci. Technol. 13:822
38. Van Ham, J., Nieboer, H. (1977) Proc. 4th Int. Clean Air Congr. Kasuga, S., Ed., p. 504
39. Spicer, C.W., Miller, D.F. (1976) J. Air Pollut. Control Assoc. 26:45
40. Sturm, G.P., Jr., Dimitriades, B. (1973) U.S. Bur. Mines, Rep. Invest. (RI 7803), 14 pages
41. Dennis, R.L., Dodge, M.C., Seilkop, S.K. (1986) Proceedings of the 79th Annual Air Pollution Control Association Meeting, 2:(86/29.6) 16 pages

42. Rutledge, S.A., Hegg, D.A., Hobbs, P.V. (1986) J. Geophys. Res. 91:14385
43. Russell, A.G., Cass, G.R. (1986) Atmos. Environ. 20:2011
44. Russell, A.G., Cass, G.R., Seinfeld, J.H. (1986) Environ. Sci. Technol. 20:1167
45. Carney, T.A., Fishman, J. (1986) Tellus 38B:127
46. Kitada, T., Igarashc, K., Owada, M. (1986) J. Clim. Appl. Meteorol. 25:767
47. Nazaroff, W.W., Cass, G.R. (1986) Environ. Sci. Technol. 20:924
48. Carmichael, G.R., Peters, L.K., Kitada, T. (1986) Atmos. Environ. 20:173
49. Isaksen, I.S.A., Hov, O., Penkett, S.A., Semb, A. (1985) J. Atmos. Chem. 3:3
50. Akimoto, H., Yamazaki, H. (1985) Kokuritsu Kogai Kenkyusho Kenkyu Hokoku 73:79
51. Amble, E. (1985) Atmos. Environ. 19:905
52. Derwent, R.G. (1985) Pollut. Their Ecotoxicol. Signif. Nuernberg, H.W. ed., Wiley, Chichester, UK, p. 85
53. Sverdrup, G.M., Hov, O. (1984) Atmos. Environ. 18:2753
54. Bottenheim, J.W., Brice, K.A., Anlauf, K.G. (1984) Atmos. Environ. 18:2609
55. Kitada, T., Carmichael, G.R., Peters, L.K. (1984) J. Clim. Appl. Meteorol. 23:1153
56. Brodzinsky, R., Cantrell, B.K., Endlich, R.M., Bhumralkar, C.M. (1984) Atmos. Environ. 18:2361
57. Russell, A.G., Cass, G.R. (1984) Atmos. Environ. 18:1815
58. Anpo, F., Soga, I. (1984) Kokuritsu Kogai Kenkyusho Kenkyu Hokoku 64:191
59. Carmichael, G.R., Kitada, T., Peters, L.K. (1984) U.S. Environ. Prot. Agency, Res. Dev., Rep. EPA-600/9-84-006, p. 245
60. Brice, K.A. (1984) U.S. Environ. Prot. Agency, Res. Dev., Rep. EPA-600/9-84-006, p. 128
61. Fishman, J., Carney, T.A. (1984) J. Atmos. Chem. 1:351
62. Wright, R.S., Fox, D.L., Jeffries, H.E. (1982) Proc. 75th Ann. Meet. Air Pollut. Control Assoc. 3:(82-33.5)
63. Hov, H. (1984) J. Atmos. Chem. 1:187
64. Dunker, A.M., Kumar, S., Berzins, P.H. (1984) Atmos. Environ. 18:311
65. Russell, A.G., McRae, G.J., Cass, G.R. (1984) NATO Challenges Mod. Soc. 5:539
66. Builtjes, P.J.H., Van den Hout, K.D., Reynolds, S.D. (1984) NATO Challenges Mod. Soc. 5:507
67. Bhumralkar, C.M., Endlich, R.M., Nitz, K.C., Brodzinsky, R., Mayerhofer, P. (1984) NATO Challenges Mod. Soc. 5:35
68. Leone, J.A., Seinfeld, J.H. (1984) Int. J. Chem. Kinet. 16:159
69. Chisaka, F. (1982) Nippon Kikai Gakkai Ronbunshu, B Hen 48:2618
70. Guicherit, R., Van der Hout, D. (1982) Stud. Environ. Sci. 21:15
71. Derwent, R.G., Hov, O. (1982) Atmos. Environ. 16:655
72. Carter, W.P.L., Winer, A.M., Pitts, J.N., Jr. (1981) Environ. Sci. Technol. 15:831
73. Atkinson, R., Carter, W.P.L., Darnall, K.R., Winer, A.M., Pitts, J.N., Jr. (1980) Int. J. Chem. Kinet. 12:779
74. Derwent, R.G., Hov, O. (1980) Environ. Sci. Technol. 14:1360
75. Nieboer, H., Duyzer, J.H. (1978) Photochem. Smogform. Neth. Guicherit, R., Ed., Comm. Milieuprojecten TNO, Delft, Netherlands, p. 89
76. Falls, A.H., McRae, G.J., Seinfeld, J.H. (1979) Int. J. Chem. Kinet. 11:1137
77. Carter, W.P.L., Lloyd, A.C., Sprung, J.L., Pitts, J.N., Jr. (1979) Int. J. Chem. Kinet. 11:45
78. Hesstvedt, E., Hov, O., Isaksen, I.S.A. (1977) Geophys. Norv. 31:27
79. Graedel, T.E., Farrow, L.A. (1977) J. Geophys. Res. 82:4943
80. Gelinas, R.J., Skewes-Cox, P.D. (1977) J. Phys. Chem. 81:2468
81. Spicer, C.W. (1977) Adv. Environ. Sci. Technol. 7:163
82. Calvert, J.G., McQuigg, R.D. (1975) Int. J. Chem. Kinet. 7:113
83. Hanst, P.L. (1971) J. Air Pollut. Contr. Assoc. 21:269
84. Stephens, E.R., Burleson, F.R., Cardiff, E.A. (1965) J. Air Pollut. Contr. Assoc. 15:87
85. Stephens, E.R., Burleson, F.R., Holtzclaw, K.M. (1969) J. Air Pollut. Cont. Assoc. 19:261
86. Niki, H., Maker, P.D., Savage, C.M., Breitenbach, L.P. (1985) Int. J. Chem. Kinet. 17:525
87. Lonneman, W.A., Bufalini, J.J. (1982) Atmos. Environ. 16:2755
88. Gay, B.W., Jr., Noonan, R.C., Bufalini, J.J., Hanst, P.L. (1976) Environ. Sci. Technol. 10:82
89. Gay, B.W., Jr., Noonan, R.C., Bufalini, J.J., Hanst, P.L. (1978) NBS Spec. Publ. 526:227

90. Kacmarek, A.J., Solomon, I.J., Lustig, M. (1978) J. Inor. Nucl. Chem. 40:574
91. Nielsen, T., Hansen, A.M., Thomsen, E.L. (1982) Atmos. Environ. 16:2447
92. Nielsen, T., Hansen, A.M., Thomsen, E.L. (1982) Atmos. Environ. 16:2756
93. Holdren, M.W., Spicer, C.W., Hales, J.M. (1984) Atmos. Environ. 18:1171
94. Gaffney, J.S., Senum, G.I., Fajer, R. (1984) Atmos. Environ. 18:215
95. Senum, G.I., Lee, Y.-N., Gaffney, J.S. (1984) J. Phys. Chem. 88:1269
96. Bruckmann, P.W., Wilner, H. (1983) Environ. Sci. Technol. 17:352
97. Marley, N.A., Ott, M., Fearey, B.L., Benjamin, T.M., Rogers, P.S.Z., Gaffney, J.S. (1988) Rev. of Sci. Instrum. in press
98. Wendschuh, P.H. Pate, C.T., Pitts, J.N., Jr. (1973) Tetrahedron Lett. 31:2931
99. Wendschuh, P.H., Fuhr, H., Gaffney, J.S., Pitts, J.N., Jr. (1973) Chem. Soc. Chem. Comm. 3:74
100. Darnall, K.R., Pitts, J.N., Jr. (1970) J. Chem. Soc. D 20:1305
101. Solomon, I.J., Jarke, F.H., Kacmarek, A.J., Shamir, J. (1974) Proc. Annu. NSF Trace Contam. Conf., (1st), Fulkerson, W., ed., p. 440
102. Pitts, J.N., Jr., Fuhr, H., Gaffney, J.S., Peters, J.W. (1973) Environ. Sci. Technol. 7:550
103. Stephens, E.R. (1967) Atmkos. Environ. 1:19
104. Steer, R.P., Darnall, K.R., Pitts, J.N., Jr. (1969) Tetrahedron Lett. 43:3765
105. Grosjean, D., Harrison, J. (1985) Environ. Sci. Technol. 19:749
106. Lee, Y.N., Senum, G.I., Gaffney, J.S. (1983) CAGCP Symposium on Tropospheric Chemistry, Oxford, Eng. (BNL 33071R)
107. Lee, Y.N. in Gas Liquid Chemistry of Natural Waters, Newman, L., ed. (BNL-51757), vol. 1, paper no. 21
108. Gaffney, J.S., Hall, J.H., Rundberg, R.S., Erdal, B.R. (1986) ACS Environ. Chem. Preprint Ext. Abs., 26, no. 2, p. 165
109. Wallington, T.J., Atkinson, R., Winer, A.M. (1984) Geophys. Res. Lett. 11:861
110. Winer, A.M., Lloyd, A.C., Darnall, K.R., Atkinson, R., Pitts, J.N., Jr. (1977) Chem. Phys. Lett. 51:221
111. Garland, J.A., Penkett, S.A. (1976) Atmos. Environ. 10:1127
112. Leh, F., Mudd, J.B. (1974) Arch. Biochem. Biophys. 161:216
113. Ohkubo, K., Sato, H. (1979) Bull. Chem. Soc. Japan 52:1525
114. Harcourt, R.D., Kelly, D.P. (1974) Environ. Sci. Technol. 8:675
115. Domalski, E.S. (1971) Environ. Sci. Technol. 5:443
116. Nicksic, S.W., Harkins, J.H., Mueller, P.K. (1976) Atmos. Environ. 1:11
117. Hanst, P.L., Stephens, E.R., Scott, W.E. (1956) J. Air Poll. Contr. Assoc. 5:219
118. Scott, W.E., Stephens, E.R., Hanst, P.L., Doerr, R.C. (1957) Proc. Am. Petrol. Inst. 37:171
119. Lovelock, J.E. (1961) Anal. Chem. 33:162
120. Stephens, E.R. (1961) Infrared Physics 1:187
121. White, J.U. (1942) J. Opt. Soc. Am. 32:285
122. Hanst, P.L., Lefohn, A.S., Gay, B.W., Jr. (1973) Appl. Spectros. 27:188
123. Tuazon, E.C., Graham, R.A., Winer, A.M., Easton, R.R., Pitts, J.N., Jr., Hanst, P.L. (1978) Atmos. Environ. 12:865
124. Tuazon, E.C., Winer, A.M., Graham, R.A., Pitts, J.N., Jr. (1980) Adv. Env. Sci. Technol. 10:259
125. Hanst, P.L., Wong, N.W., Bragin, J. (1982) Atmos. Environ. 16:969
126. Stephens, E.R. (1964) Anal. Chem. 36:928
127. Adamson, P., Guenthare, H.H. (1980) Spectrochim. Acta, A 36A:473
128. Varetti, E.L., Pimentel, G.C. (1974) Spectrochim. Acta, 30A:1069
129. Griffith, D.W.T., Shuster, G. (1987) J. Atmos. Chem. 5:59
130. Stephens, E.R., Price, M.A. (1973) J. Chem. Ed. 50:351
131. Lonneman, W.A., Bufalini, J.J., Namie, G.R. (1982) Environ. Sci. Technol. 16:655
132. Grosjean, D., Fung, K., Collins, J., Harrison, J., Breitung, E. (1984) Anal. Chem. 56:569
133. Helas, G. (1982) NATO Adv. Study Inst. Ser., Ser. C 96:39
134. Hanst, P.L., Wong, N.W., Bragin, J. (1982) Atmos. Environ. 16:969
135. Akimoto, H., Bandow, H., Sakamaki, F., Inoue, G., Hoshino, M., Okuda, M. (1979) Kokuritsu Kogai Kenkyusho Kenkyu Hokoku 9:9 (Japanese)
136. Akimoto, H., Bandow, H., Sakamaki, F., Inoue, G., Hoshino, M., Okuda, M. (1980) Environ. Sci. Technol. 14:172

137. Tuazon, E.C., Winer, A.M., Graham, R.A., Pitts, J.N., Jr. (1978) Jt. Conf. Sens. Environ. Pollut., 4th Conf. Proc. p. 798
138. Fild, I., Lovell, D.J. (1970) J. Opt. Soc. Amer. 60:1315
139. Darley, E.F., Kettner, K.A., Stephens, E.R. (1963) Anal. Chem. 35:589
140. Singh, H.B., Salas, L.J. (1983) Atmos. Environ. 17:1507
141. Vierkorn-Rudolph, B., Rudolph, J., Diederich, S. (1985) Int. J. Environ. Anal. Chem. 20:131
142. Nakanishi, M., Matsuura, A., Watanabe, I. (1986) Taiki Osen Gakkaishi 21:29 (Japanese)
143. Stephens, E.R. (1983) Atmos. Environ. 17:1855
144. Fortunat, J.L., Landolt, W.F., Leuenberger, H. (1986) Environ. Sci. Technol. 20:1269
145. Meyrahn, H., Helas, G., Warneck, P. (1987) J. Atmos. Chem. 5:405
146. Kravetz, T.M., Martin, S.W., Mendenhall, G.D. (1980) Environ. Sci. Technol. 14:1262
147. Nielson, T., Hansen, A.M., Thomsen, E.L. (1982) Atmos. Environ. 16:2447
148. Joos, L.F., Landolt, W.F., Leuenberger, H. (1986) Environ. Sci. Technol. 20:1269
149. Glavas, S., Schurath, U. (1983) Chem. Chron. 12:89
150. Holdren, M.W., Spicer, C.W. (1984) Environ. Sci. Technol. 18:113
151. Prestbo, E.W., Gaffney, J.S. (1988) in preparation
152. Cox, R.A., Derwent, R.G., Holt, P.M., Kerr, J.A. (1976) J. Chem. Soc., Faraday Trans. 9:2061
153. Winer, A.M., Peters, J.W., Smith, J.P., Pitts, J.N., Jr. (1974) Environ. Sci. Technol 8:1118
154. Grosjean, D. (1983) Environ. Sci. Technol. 17:13
155. Grosjean, D., Harrison, J. (1985) Environ. Sci. Technol 19:862
156. Holdren, M.W., Rasmussen, R.A. (1976) Environ. Sci. Technol. 10:185
157. Lonneman, W.A. (1977) Environ. Sci. Technol. 11:194
158. Holdren, M.A., Rasmussen, R.A. (1977) Environ. Sci. Technol. 11:196
159. Lonneman, W.A. (1977) Environ. Sci. Technol. 11:196
160. Watanabe, I., Stephens, E.R. (1978) Environ. Sci. Technol. 12:222
161. Taylor, O.C. (1969) J. Air Pollut. Contr. Assoc. 19:347
162. Thompson, C.R., Hensel, E.G., Kats, G. (1973) J. Air Poll. Cont. Assoc. 23:881
163. Penkett, S.A., Sandalls, F.J., Lovelock, J.E. (1975) Atmos. Environ. 9:461
164. Nieboer, H., Van Ham,. J. (1976) Atmos. Environ. 10:115
165. Lonneman, W.A., Bulfalini, J.J., Seila, R.L. (1976) Environ. Sci. Technol. 10:374
166. Bos, R., Guicherit, R., Hoogeveen, A. (1977) Sci. Total Environ. 7:269
167. Nielson, T., Samuelsson, U., Grennfelt, P., Thomsen, E.L. (1981) Nature 293:553
168. Lewis, T.E., Brennan, E., Lonneman, W.A. (1983) J. Air Pollut. Control Assoc. 33:885
169. Spicer, C.W., Holdren, M.W., Keigley, G.W. (1983) Atmos. Environ. 17:1055
170. Temple, P.J., Taylor, O.C. (1983) Atmos. Environ. 17:1583
171. Grosjean, D. (1984) Atmos. Environ. 18:1489
172. Peake, E., MacLean, M.A., Sandhu, H.S. (1983) J. Air Pollut. Control Assoc. 33:881
173. Brice, K.A., Penkett, S.A., Atkins, D.H.F., Sandalls, F.J., Bamber, D.J., Tuck, A.F., Vaughan, G. (1984) Atmos. Environ. 18:2691
174. Hov, O. (1985) Atmos. Environ. 19:471
175. Anlauf, K.G., Bottenheim, J.W., Brice, K.A., Fellin, P., Wiebe, H.A., Schiff, H.I., Mackay, G.I., Braman, R.S., Gilbert, R. (1985) Atmos. Environ. 19:1859
176. Peake, E., MacLean, M.A., Sandhu, H.S. (1985) J. Air Pollut. Control Assoc. 35:250
177. Corkum, R., Giesbrecht, W.W., Bardsley, T., Cherniak, E.A. (1986) Atmos. Environ. 20:1241
178. Tsalkani, N., Perros, P., Toupance, G. (1987) J. Atmos. Chem. 5:291
179. Tanner, R.L., Miguel, A.H., de Andrade, J.B., Gaffney, J.S., Streit, G.E. (1988) Environ, Sci. Technol. in press
180. Peake, E., MacLean, M.A., Lester, P.F., Sandhu, H.S. (1988) Atmos. Environ. 22:973
181. Jin, S., Tian, Y. (1982) Huanjing Kexue 3:45
182. Westburg, H., Lamb, B. (1982) EPA Report EPA-600/3-85/013
183. Jin, S., Tian, B. (1985) Huanjing Huaxue 4:21
184. Anlauf, K.G., Bottenheim, J.W., Brice, K.A., Wiebe, H.A. (1986) Water, Air, Soil Pollut. 30:153

185. Tsani-Bazaka, E., Petrakis, M., Glavas, S. (1986) Proceedings of the 2nd International Conference on Environmental Contamination, 1986, CEP Consult., Edinburgh, UK, pp. 148–150
186. Ciccioli, P., Brancaleoni, E., Di Pablo, V., Liberti, M., Di Palo, C. (1986) Acqua Aria 7:675 (Italian)
187. Smith, R.G., Bryan, R.J., Feldstein, M., Levadie, B., Miller, F.A., Stephens, E.R. (1971) Health Lab Sci. 8:48
188. Lovelock, J.E., Penkett, S.A. (1974) Nature 249:434
189. Singh, H.B., Salas, L.J. (1983) Nature 302:326
190. Ferm, M., Samuelsson, U., Sjoedin, A., Grennfelt, P. (1984) Atmos. Environ. 18:1731
191. Singh, H.B., Salas, L.J., Ridley, B.A., Shetter, J.D., Donahue, N.M., Fehsenfeld, F.C., Fahey, D.W., Parrish, D.D., Williams, E.J., et al. (1985) Nature 318:347
192. Bottenheim, J.W., Gallant, A.G., Brice, K.A. (1976) Geophys. Res. Lett. 13:113
193. Singh, H.B., Salas, L.J., Viezee, W. (1986) Nature 321:588
194. Fahey, D.W., Hubler, G., Parrish, D.D., Williams, E.J., Norton, R.B., Ridley, B.A., Singh, H.B., Liu, S.C., Fehsenfeld, F.C. (1986) J. Geophys. Res. D: Atmos. 91:9781
195. Penkett, S.A., Brice, K.A. (1986) Nature 319:655
196. Rudolph, J., Vierkorn-Rudolph, B., Meixner, F.X. (1987) J. Geophys. Res. 93(D6):6653
197. Bottenheim, J.W., Gallant, A.G. (1987) Biogeochemical Cycles 1:369
198. Singh, H.B., Viezee, W. (1988) Atmos. Environ. 22:419
199. Stephens, E.R., Price, M.A. (1969) Atmos. Environ. 3:573
200. Spicer, C.W. (1977) Atmos. Environ. 11:1089
201. Pitts, J.N., Jr., Grosjean, D., Van Cauwenberghe, K., Schmid, J.P., Fritz, D.R. (1978) Environ. Sci. Technol.1 12:946
202. Altshuller, A.P. (1983) Atmos. Environ. 17:2383
203. Grosjean, D. (1985) Environ. Sci. Technol. 19:968
204. Kopczynski, S.L., Lonneman, W.A., Sutterfield, F.D., Darley, P.E. (1972) Environ. Sci. Technol. 6:342
205. Kopczynski, S.L., Kuntz, R.L., Bufalini, J.J. (1975) Environ. Sci. Technol. 9:648
206. Winer, A.M., Breuer, G.M., Carter, W.P.L., Darnall, K.R., Pitts, J.N., Jr. (1979) Atmos. Environ. 13:989
207. Fahey, D.W., Eubank, C.S., Hubler, G., Fehsenfeld, F.C. (1985) J. Atmos. Chem. 3:435
208. Wendel, G.J., Stedman, D.H., Cantrell, C.A. (1983) Anal. Chem. 55:937
209. Burkhardt, M.R., Maniga, N.I., Stedman, D.H., Paur, R.J. (1988) Anal. Chem. 60:816
210. Calvert, J.G., Lazrus, A., Kok, G.L., Heikes, B.G., Walega, J.G., Lind, J., Cantrell, C.A. (1985) Nature 317:27
211. Saltzman, B.E. (1964) Anal. Chem. 26:1949
212. Pate, C.T., Sprung, J.L., Pitts, J.N., Jr. (1976) Org. Mass. Spectrom. 11:552
213. Steer, R.P., Darnall, K.R., Pitts, J.N., Jr. (1969) Tetrahedron Lett. 43:3765
214. Guicherit, R., Schulting, F.L. (1985) Sci. Total Environ. 43:193
215. Galloway, J.N., Whelpdale, D.M., Wolff, G.T. (1984) Atmos. Environ. 18:2595
216. Peake, E., Sandhu, H.S. (1983) Can. J. Chem. 61:927
217. Spice, C.W. (1982) Sci. Total Enciron. 24:183
218. Grennfelt, P., Samuelsson, U., Nielsen, T., Thomsen, E.L. (1982) Phys. Chem. Behav. Pollut. EUR 7624:619
219. Bos, R. (1978) Netherlands TNO Report, (IG-TNO-G-866), 20 pp.
220. Loebel, J., Wipprecht, V., Schurath, U. (1980) Staub Reinhalt. Luft 40:243
221. Bottenheim, J.W., Strausz, O.P. (1980) Environ. Sci. Technol. 14:709
222. Bos, R., Goudena, E., Guicherit, R., Hoogeveen, A., De Vreede, J. (1978) Photochem. Smogform. Neth., Guicherit, R. Ed., Comm. Milieuprojecten TNO, Delft, Netherlands, pp. 20–59
223. Gaddo, P.P. (1980) Comm. Eur. Communities, (Rep.) EUR 6388:418
224. Bruckmann, P., Eynck, P. (1979) Schriftenr. Landesanst. Immissionsschutz Landes Nordrhein-Westfalen, Essen 49:19
225. Bruckmann, P., Muelder, W. (1979) Schriftenr. Landesanst. Immissionsschutz Landes Nordrhein-Westfalen, Essen 47:30

226. Bottenheim, J.W., Strausz, O.P. (1979) Atmos. Environ. 13:1085
227. Penkett, S.A. (1978) Environ. Pollut. Manage. 8:8
228. Penkett, S.A., Sandalls, F.J., Jones, B.M.R. (1976) VDI-Ber. 207:47
229. Stephens, E.R. (1975) Atmos Environ. 9:461
230. Penkett, S.A., Sandalls, F.J., Lovelock, J.E. (1975) Atmos. Environ. 9:139
231. Singh, H.B. (1987) Environ. Sci. Technol 21:320
232. Gaffney, J.S. (1988) unpublished results
233. Hummel, J.R., Reck, R.A. (1984) Atmos. Environ. 18:223
234. Kasting, J.F., Singh, H.B. (1986) J. Geophys. Res. D: Atmos. 91:13239
235. Aikin, A.C., Herman, J.R., Maier, E.J.R., McQuillan, C.J. (1983) Planet. Space Sci. 31:1075
236. Singh, H.B., Hanst, P.L. Geophys. Res. Lett. 8:941
237. Grosjean, D. (1985) Sci. Total Environ. 46:41
238. Gaffney, J.S., Senum, G.I. (1984) in Gas-Liquid Chemistry of Natural Waters. Brookhaven National Laboratory Rep. BNL 51757, v. 1, paper no. 5
239. Schurath, U., Wipprecht, V. (1980) Comm. Eur. Communities EUR6621:157
240. Hendry, D.G., Kenley, R.A., Davenport, J.E., Lan, B.Y. (1979) Report EPA/700/3-79/020 81 p
241. Niki, H., Maker, P.D., Savage, C.M., Breitenbach, L.P. (1978) Chem. Phys. Lett. 55:289
242. Hendry, D.G., Kenley, R.A. (1977) J. Am. Chem. Soc. 99:3198
243. Gaffney, J.S., Fajer, R., Senum, G.I., Lee, J.H. (1986) Int. J. Chem. Kinet. 18:399
244. Payer, H.D., Blank, L.W., Bosch, C., Gnatz, G., Schmolke, W., Schramel, P. (1986) Water, Air, Soil Pollut. 31:485
245. Ohashi, T. (1985) Tokyo-to Kogai Kenkyusho Nenpo 101–106 pp (Japanese)
246. Fujiwara, T. (1985) Taiki Osen Gakkaishi 20:149–157 (Japanese)
247. Temple, P.J., Taylor, O.C. (1985) J. Environ. Qual. 14:420
248. Prinz, B., Brandt, C.J. (1985) Pollut. Their Ecotoxicol. Signif. Nuernberg, H.W., ed., Wiley, Chichester, UK, pp. 67–84
249. Jonas, R., Heineman, K. (1985) Staub-Reinhalt. Luft 3:112 (German)
250. Lee, Y.H., Li, G.C. (1985) K'o Hsueh Nung Yeh (Taipei) 33:72 (Chinese)
251. Mayumi, H., Yamazoe, F. (1983) Nogyo Gijutsu Kenkyusho Hokoku B 1983:1 (Japanese)
252. Temple, P.J., Taylor, O.C. (1984) Atmos. Environ. 18:1491
253. Eversman, S., Sigal, L.L. (1984) Bryologist 87:112
254. Guderian, R., Tingey, D.T., Rabe, R. (1984) EPA-600/3-84-024 Report, 426 pp
255. Waldmann, G. (1984) Allg. Forst-Jagdztg. 155:80
256. Nouchi, I., Mayumi, H., Yamazoe, F. (1984) Atmos. Environ. 18:453
257. Taylor, O.C., Temple, P.J., Thill, A.J. (1983) Hort Science 18:861
258. Hershey, D.R., Paul, J.L. (1981) J. Plant Nutr. 3:935
259. Taylor, O.C. (1980) Prog. Ecol. 5/7:33
260. Metzler, J.T., Pell, E.J. (1980) Phytopathology 70:934
261. Nouchi, I. (1979) Taiki Osen Gakkaishi 14:489 (Japanese)
262. Posthumus, A.C., Cleij, O., Laurens, G.W.H., Schrijver, M.S., Tonneijck, A.E.G. (1980) Comm. Eur. Communities, (Rep.) EUR 6388:370
263. De Vos, N.E., Hill, R.R., Jr., Hepler, R.W., Pell, E.J., Craig, R. (1980) J. Am. Soc. Hortic. Sci. 105:157
264. Youngner, V.B., Nudge, F.J. (1980) Agron. J. 72:169
265. Sigal, L.L., Taylor, O.C. (1979) Bryologist 82:564
266. Pell, E.J., Gardner, W. (1979) Hort Science 14:61
267. Kohut, R.J., Davis, D.D. (1978) Phytopath. 68:567
268. Nouchi, I., Sawada, T., Ishiguro, T., Toyama, S., Iijima, T. (1977) Proc. 4th Int. Clean Air Congr., Kasuga, S., Ed., pp 95–98
269. Davis, D.D., Kohut, R.J. (1977) Int. Conf. Photochem. Oxid. Pollut. Control Proc. 2 (PB-264 233):647
270. Matsuoka, Y. (1977) Taiki Osen Kenkyu 11:195
271. Davis, D.D. (1977) Plant Dis. Rep. 61:640
272. Sawada, T., Nouchi, I., Oahashi, T., Oodaira, T. (1975) Annu. Rep. Tokyo Metrop. Res. Inst. Environ. Prot. pp 1–8

273. Posthumus, A.C. (1976) VDI-Ber. 270:153
274. Hanson, G.P., Addis, D.H., Thorne, L. (1976) Can. J. Genet. Cytol. 18:579
275. Kohut, R.J., Davis, D.D., Merrill, W. (1976) Plant Dis. Rep. 60:777
276. Starkey, T.E., Davis, D.D., Merrill, W. (1976) Plant Dis. Rep. 60:480
277. Pell, E.J. (1976) Phytopath. 66:731
278. Nouchi, I., Iijima, T., Oodaira, T. (1975) Taiki Osen Kenkyu 9:635 (Japanese)
279. Coulson, C.L., Heath, R.L. (1975) Atmos. Environ. 9:231
280. Thompson, C.R., Kats, G. (1975) Environ. Sci. Technol. 9:35
281. Chang, H.N. (1974) K'o Hsueh Nung Yeh (Taipei) 22:234 (Chinese)
282. Wood, F.A., Drummond, D.B. (1974) Phytopath. 64:897
283. Taylor, O.C. (1973) Advan. Chem. Ser. 122:9
284. Ben-Arie, R., Orkin, L., Kindinger, J.I. (1973) Plant Cell Physiol. 14:435
285. Thomson, W.W., Swanson, E.S. (1972) Proc., Electron Microsc. Soc. Amer. 30:360
286. Gordon, W.C., Ordin, L. (1972) Plant Physiol. 49:542
287. Ordin, L., Garber, M.J., Kindinger, J.I. (1972) Physiol. Plant. 26:17
288. Hope, H.J., Ordin, L. (1971) Plant Cell Physiol. 12:849
289. Hill, A.C. (1971) J. Air Pollut. Cont. Ass. 21:341
290. Ordin, L., Garber, M.J., Kindinger, J.I., Whitmore, S.A., Greve, L.C., Taylor, O.C. (1971) Environ. Sci. Technol. 5:621
291. Heggestad, H.E., Darley, E.F. (1969) Air Pollut., Proc. Eur. Congr., 1st, Stichting Centrum voor Landbouwpublikaties en Landbouwdocumentatie (Pudoc), Wageningen, Netherlands, pp 329–335
292. Thompson, C.R. (1969) Proc. Int. Citrus Symp., 1st. Chapman, H.D., ed., 2:705
293. Taylor, O.C. (1969) Proc. Int. Citrus Symp., 1st, Chapman, H.D., ed., 2:741
294. Dugger, W.M., Jr., Ting, I.P. (1970) Annu. Rev. Plant Physiol. 21:215
295. Hanson, G.P., Stewart, W.S. (1970) Science 168:1223
296. Ordin, L., Garber, M.J., Kindinger, J. (1970) Physiol. Plant 23:117
297. Ordin, L., Hall, M.A., Kindinger, J.I. (1969) Arch. Environ. Health 18:623
298. Dugger, W.M., Jr., Ting, I.P. (1968) Phytopathology 58:1102
299. Ordin, L., Hall, M.A., Katz, M. (1967) J. Air Pollut. Control Assoc. 17:811
300. Koukol, J., Dugger, W.M., Jr., Palmer, R.L. (1967) Plant Physiol. 42:1419
301. Shepson, P.B., Kleindienst, T.E., Edney, E.O., Nero, C.M., Cupitt, L.T., Claxton, L.D. (1986) Environ. Sci. Technol. 20:1008
302. Kleindienst, T.E., Edney, E.O., Namie, G.R., Claxton, L.D. (1986) Atmos. Environ. 20:971
303. Kleindienst, T.E., Shepson, P.B., Edney, E.O., Claxton, L.D., Cupitt, L.T. (1986) Environ. Sci. Technol. 20:493
304. Nover, H., Botzenhart, K. (1985) Zentralbl. Bakteriol., Mikrobiol. Hyg., Abt. 1, Orig. B 181:71
305. Shiraishi, F., Bandow, H. (1985) J. Toxicol. Environ. Health 15:531
306. Kleindienst, T.E., Shepson, P.B., Edney, E.O., Claxton, L.D. (1985) Mutat. Res. 157:123
307. Shepson, P.B., Kleindienst, T.E., Edney, E.O., Cupitt, L.T., Claxton, L.D. (1985) Environ. Sci. Technol. 19:1094
308. Kleindienst, T.E., Shepson, P.B., Edney, E.O., Cupitt, L.T., Claxton, L.D. (1985) Environ. Sci. Technol. 19:620
309. Shepson, P.B., Kleindienst, T.E., Edney, E.O., Namie, G.R., Pittman, J.H., Cupitte, L.T. (1985) Environ. Sci. Technol. 19:249
310. Nover, H., Botzenhart, K. (1983) Zentralbl. Bakteriol., Mikrobiol. Hyg., Abt. 1, Orig. B 177:298
311. Nogami, H., Azuma, E., Uozumi, M., Nakajima, T. (1982) Osaka-furitsu Koshu Eisei Kenkyusho Kenkyu Hokoku, Kogai Eisei 3:23 (Japanese)
312. Pitts, J.N., Jr., Grosjean, D., Van Cauwenberghe, K., Belser, W.L. (1978) Prepr. Am. Chem. Soc., Div. Environ. Chem. 18:341
313. Pitts, J.N., Jr., Van Cauwenberghe, K.A., Grosjean, E., Schmid, J.P., Fitz, D.R., Belser, W.L., Jr., Knudson, G.B., Hynds, P.M. (1979) Environ. Sci. Res. 15:353
314. Kruysse, A., Feron, V.J., Immel, H.R., Spit, B.J., Van Esch, G.J. (1977) Toxicology 8:231
315. Kruysse, A., Feron, V.J. (1976) VDI-Ber. 270:101

316. Campbell, K.I., Emik, L.O., Clarke, G.L., Plata, R.L. (1970) Arch. Environ. Health 20:22
317. Gross, R.E., Dugger, W.M., Jr. (1969) Environ. Res. 2:256
318. Peak, M.J., Belser, W.L. (1969) Atmos. Environ. 3:385
319. Campbell, K.I., Clarke, G.L., Emik, L.O., Plata, R.L. (1967) Arch. Environ. Health 15:739
320. Serat, W.F., Budinger, F.E., Jr., Mueller, P.K. (1967) Atmos. Environ. 1:21
321. Horvath, S.M., Bedi, J.F., Drechsler-Parks, D.M. (1986) J. Air Pollut. Control Assoc. 36:265
322. Drechsler-Parks, D.M., Bedi, J.F., Horvath, S.M. (1984) Am. Rev. Respir. Dis. 130:1033
323. Thomas, G.B., Fenters, J.D., Ehrlich, R., Gardner, D.E. (1981) J. Toxicol. Environ. Health 8:559
324. Thomas, G.B., Fenters, J.D., Ehrlich, R. (1979) Report (EPA/600/1-79/001), 44 pp
325. Okawada, N., Mizoguchi, I., Ishiguro, T. (1979) Nagoya J. Med. Sci. 41:9
326. Altshuller, A.P. (1978) J. Air Pollut. Control Assoc. 28:594
327. Lovelock, J.E. (1977) Ambio 6:131
328. Raven, P.B., Drinkwater, B.L., Horvath, S.M., Ruhling, R.O., Gliner, J.A., Sutton, J.C., Bolduan, N.W. (1974) Int. J. Biometeorol. 18:222
329. Raven, P.B., Drinkwater, B.L., Ruhling, R.O., Bolduan, N., Taguchi, S., Gliner, J., Horvath, S.M. (1974) J. Appl. Physiol. 36:288
330. Smith, L.E. (1965) Arch. Environ. Health 10:161
331. Fishbein, L. (1975) Environ. Qual. Saf. 4:200
332. Goldsmith, J.R. (1971) Calif. Med. 115:55
333. Hall, M.A., Brown, R.L., Ordin, L. (1971) Phytochem. 10:1233
334. Mudd, J.B. (1969) Air Pollut., Proc. Eur. Congr., 1st. Stichting Centrum voor Landbouwpublikaties en Landbouwdocumentatie (Pudoc), Wageningen, Netherlands, pp 161–166
335. Mudd, J.B., McManus, T.T. (1969) Arch. Biochem. Biophys. 132:237
336. Shepson, P.B., Kleindienst, T.E., Nero, C.M., Hodges, D.N., Cupitt, L.T., Claxton, L.D. (1987) Environ. Sci. Technol 21:568
337. Gay, B.W., Jr., Hanst, P.L., Bufalini, J.J., Noonan, R.C. Environ. Sci. Technol. 10:58
338. Gay, B.W., Jr., Hanst, P.L., Bufalini, J.J., Noonan, R.C. (1976) Prepr. ACS Natl. Meet., Div. Environ. Chem. 16:233

# Semivolatile Organic Compounds in the Atmosphere

*Ronald Harkov*

ENSR, Inc., 1 Executive Drive
Somerset, NJ 08873, USA

Introduction . . . . . . . . . . . . . . . . . . . . . . . . . . . 39
SVOCs Measurement Techniques . . . . . . . . . . . . . . . . . 43
Field Studies . . . . . . . . . . . . . . . . . . . . . . . . . . . 52
Environmental Fate . . . . . . . . . . . . . . . . . . . . . . . 57
Health Assessment . . . . . . . . . . . . . . . . . . . . . . . . 59
Future Developments . . . . . . . . . . . . . . . . . . . . . . 62
Conclusions . . . . . . . . . . . . . . . . . . . . . . . . . . . 64
Abbreviations . . . . . . . . . . . . . . . . . . . . . . . . . . 64
References . . . . . . . . . . . . . . . . . . . . . . . . . . . . 65

## Summary

Semivolatile organic compounds (SVOCs) in the atmosphere have been of some interest to scientists and regulators. The present chapter provides an overview of the sources, environmental fate, measurement and health effects of SVOCs in the air environment. Current trends in SVOC research have been highlighted and emphasis has been placed on currrent shortcomings on our present state-of-knowledge.

## Introduction

For many years, semivolatile organic compounds (SVOCs) in the air environment have been of interest to the scientific and regulatory communities. The polycyclic aromatic hydrocarbons (PAHs) were the initial group of SVOC characterized and studied by health and air pollution scientists [1]. In spite of the recent increase in the depth and breadth of our understanding of SVOCs in the atmosphere, most of the information that is presently available on this class of pollutants is concerned with the PAH. As a result of this situation, SVOCs will be defined with reference to certain PAH compounds. Perhaps in this way the present understanding of PAH in our environment can provide a foundation and reference point for the investigation of other groups of SVOC.

By nature of their definition, SVOCs are so-classified as a result of their vapor pressures. There are many and varied boundaries placed on what constitutes a

SVOC, but in this review it should be clearly understood that because the focus is on SVOCs present in *both* ambient and source environments, the range of vapor pressures to be included in this pollutant class will be much larger than those considered by most workers. SVOCs are defined as those organic substances with a vapor pressure 25 °C between napthalene (2E-4 atm) through approximately coronene (app. E-15 atm). As a comparison, Junge [2] defined SVOCs as having a vapor pressure between E-7 and E-10 atm, while Lewis [3] defined SVOCs as having a vapor pressure in the range of E-4 and E-9 atm. It is noted that Benzo(a)pyrene (BaP, E-12 atm) has been found in the vapor phase and napthalene (E-4 atm) has been found in particle phases of ambient air samples. The defined E-4 atm to E-15 atm range of SVOCs provides an approximation of the absolute boundaries to this class of substances. Thus those substances with a vapor pressure of E-4 to E-8 atm will be mostly in the vapor phase, while SVOC in the E-13 to E-15 atm range will generally be found in the particle phase.

It should be emphasized that very little is known about the actual phase (vapor/particle) distribution of SVOCs in the atmosphere. Most historic ambient air SVOC data has been collected utilizing the high-volume air system (83 m$^3$/hr) and glass-fiber filters (hivol/gff system). The hivol/gff system is highly artifactual because it will encourage; a) volatility losses of substances with a vapor pressure greater than benzo(a)anthracene (app. E-10 atm) [4, 5] and b) chemically induced reactions related to ozone, hydroxyl radical and $HNO_3$ as for example with BaP [6]. The most common approach adopted in the last 10 years to overcome these concerns has been to utilize the hivol/gff system followed by an adsorbent cartridge [7]. This approach can only develop pseudo-phase distributions, therefore reported phase distribution results from such studies are quite dubious. Sample collection and reactivity related issues have yet to be adequately dealt with by scientists concerned with SVOCs. It will be necessary to leave the above discussion for other portions of the this review.

*Why Study SVOCs?* As a pollutant class, SVOCs contain a significant number of substances which are of concern as potential causative agents in decrements in public health and as impairing other portions of the biosphere (i.e., impact of DDT on birds of prey). Four major sources of SVOC are (Table 1): Products of incomplete combustion (PICs); Petrochemical manufacturing, use and storage losses (Petro); Biogenically formed materials (Biogenics); Photochemical smog aerosols (smog aerosols).

Some SVOCs are mutagens, others are animal carcinogens, and still others are linked to a variety chronic diseases such as emphysema. As a major portion of the ambient aerosol, SVOCs are an important component of both "clean" and

**Table 1.** Examples of the four classes of SVOC

| | |
|---|---|
| *PICs* | benzo(a)pyrene, 2,3,7,8-tetrachlorodibenzodioxin, 1-nitropyrene |
| *Petro* | DDT, PCBs, Pentachlorophenols, Hexachlorobenzene |
| *Biogenics* | Terpenes, long-chain alkanes (LCAs) |
| *Smog aerosols* | Dicarboxylic acids, 2-nitrofluoranthene |

polluted environments. Major studies of the ambient aerosol in Europe, Japan and North America suggest that SVOCs as extractable organic matter (EOM) can constitute more than 50% of the inhalable particulate matter ($PM_{10}$) present in the urban atmosphere. For example, studies of $PM_{10}$ samples suggest that the urban aerosol in eastern North America is largely made up of $SO_4$, (as either $(NH_4)_2SO_4$, $(NH_4)HSO_4$ and/or $H_2SO_4$), organic carbon, elemental carbon and crustal-like materials (Fe, Si, Al, Zn, Ti, Mn, and Ca oxides) [8, 9].

Most of SVOCs that have been chemically identified in ambient aerosols are considered non-polar (i.e., soluble in aromatic solvents such as benzene or toluene). These substances include the PAH, long-chain alkanes (LCAs), nitro-PAH and some oxygenated-PAH, such as PAH-quinones. We know very little about the moderately polar (i.e. soluble in dichloromethane or chloroform after extraction with an aromatic solvent) and the most polar (i.e. soluble in methanol or acetone after the aromatic and chlorinated hydrocarbon extraction) SVOCs found in the atmosphere. As an example, in recent years there have been many studies which included both biological (Ames-mutagenicity test) assay assessments of SVOC samples. A recurring theme in most of these studies is that 20%–30% of the mutagenicity occurs in the most polar fractions, the chemical makeup of which is largely unknown [10–13]. Polar substances such as carboxylic acids have been measured in the ambient aerosol [8, 14, 15], but these substances are not biologically active.

Most of the generally utilized source emission testing methods for biologically active organic materials are not adequate for characterizing the range of SVOCs present in ducted emission sources. When methods such as the modified method five (MM5) procedure are utilized to collect SVOCs, the approach is not designed to control for chemical reactions on the filter, nor for the full range vapor pressures of substances likely to be qualitatively collected by the resin traps (XAD-2, Tenax-GC, or Florosil).

In summary, SVOCs are of interest to the scientific and regulatory communities as result of the following:
- Represents a major fraction of the ambient aerosol in both pristine and polluted environments
- There is no indication that past air pollution control strategies have reduced ambient concentrations of SVOCs
- Contains a large number of biologically active materials
- Inadequate knowledge of phase distributions and chemical reactivities in the environment
- Improper stationary source testing methods which may produce both chemical and physical artifacts

The challenge for the scientist and regulator is to close the reather large information gap related to the above issues.

SVOCs were first identified in the environment when benzo(a)pyrene was isolated from coal tars [16]. Since that time, SVOCs have been found in emissions from a large number of sources (Table 2). The classes of SVOCs that have been identified in source samples are shown in (Table 3) and indicate that a variety of substituted polycyclic and heterocyclic compounds have been identified in a large cross-section of air pollution sources. For PIC's, those sources subject to poor

**Table 2.** Some primary sources of SVOCs in the atmosphere

| | |
|---|---|
| *Combustion sources* | Power plants |
| | Space heaters |
| | Resource recovery facilities |
| | Hazardous waste incinerators |
| | Agricultural & forest residue removal |
| | Spark ignition motor vehicles |
| | Diesel motor vehicles |
| *Users* | Pesticide Applications |
| | Heat capacitors |
| | Roofing-tars |
| | Asphalt-roads |
| | Wood preservation |
| | Plasticizers |
| *Manufacturing* | Coke production |
| | Petroleum cracking |
| | Iron & Steel production |
| | Charcoal production |
| | Pesticide production |
| | Carbon black production |
| | Creosote production |
| | Coal tar production |

**Table 3.** Some SVOCs identified in combustion source samples

PAH
Nitro-PAH
Chlorinated-PAH
PAH-ketones
PAH-quinones
Methoxylated PAH
Dioxins
Dibenzofurans
Dicarboxylic acids
Thiophenes
Aza-arenes
Phthalates
Pentachlorophenols
Hexachlororbenzene
Chlorinated aromatic ethers
PCBs
Methoxyyphenols

combustion controls (wood and coal stoves) or ineffective inspection/maintenance programs (oil-space heaters, spark ignition and diesel motor vehicles) are significant sources of these SVOCs. Some industrial processes which are operated to reduce combustion efficiency, or for which SVOCs are the final products including coking, charcoal manufacturing, and the manufacture of coal-tar derivatives, can also be locally important sources of SVOCs.

Most pesticides are SVOCs and thus their manufacturing and use can be expected to be linked to localized and global atmospheric concentrations. Examples exist in the literature which have demonstrated that pesticide levels in the atmosphere are associated with the key agricultural seasons in a particular area [17–19] and that global background levels for any of these substances can be directly related to annual pesticide production estimates.

There are also a large number of sources of SVOCs, which may be of minor global importance yet are important localized contributors to ambient levels of SVOCs. The actual SVOC outputs and chemistry from these sources are ill-defined. These sources include waste treatment and storage facilities, such as lagoons and landfills, structural, industrial and commercial fires, explosions, and volcanoes.

A major source of SVOCs during the non-heating season is the formation of photochemical smog aerosols [20]. These compounds generally represent oxidized primary emission products and an abbreviated listing of some secondary SVOC classes present in the atmosphere is shown below.

Dicarboxylic acids
Nitro-PAH
Oxygenated-PAH
Long chain aldehydes
Long chain ketones
Aromatic acids
Hydroxynitro-PAH

Formation of SVOC polar aerosols during the non-heating season, as measured as EOM, has been to shown to increase up to 200% in the PM15 at urban NJ sites [21] and significant increases in suspended particulate matter in the Los Angeles air basin have been linked to increased levels of polar SVOC being present in the atmosphere [22]. It is presently unclear whether increased levels of SVOCs in the atmosphere during the photochemical smog season is associated with greater or lesser amounts of biologically active materials. Initial results from New Jersey and Los Angeles suggest that the biological activity of SVOCs are unchanged during photochemical smog episodes [23, 24]. Long range transport of SVOCs aerosols have revealed similar amounts of biological activity during the winter and summer [25]. Thus while increased photochemical activity may lead to changes in the level of SVOCs in the atmosphere, there is no indication that that overall resulting substances are a greater threat to the respiratory health of the exposed population.

## SVOCs Measurement Techniques

### Ambient

SVOCs have been collected primarily utilizing the hivol/gff air sampling system (Table 4). As was discussed earlier, such systems are apt to be highly artifactual for many SVOCs. The database from urban US sites for benzene-soluble organics (BSO) and BaP [26] could be the largest information source for this class

**Table 4.** Ambient sampling approaches for SVOC* approx. flow

| Approach | Rates (LPM) | Trap | Trap | Denuder | Target compound |
|---|---|---|---|---|---|
| Hivol | 1400 | Filters: Glass, Quartz, Teflon-coated glass | – | – | LCA's/ PAHs, EOM Mutagens |
| SSI-Hivol | 1100 | As above | – | – | LCA's/ PAHs, EOM Mutagens |
| Dichotomous sampler | 15 | Teflon membranes | – | – | nicotine |
| PS-1 | 150 | GFF | PUF | – | PCBs, PCDDs, Pesticides |
| Cal. DOH | 17 | GFF | $Al_2O_3$ | $Al_2O_3$ | PAH's-EOM |
| AGP | 18 | GFF | Tenax | – | PAH's, LCAs Carboxylic acids |
| Ontario Sci | 15 | Teflon | Tenax | Tenax | membranes Pesticides |
| U. Colorado | 10 | Quartz | – | – | LCA's & PAH's |
| SAPRC | 18,000 | GFF | – | – | Mutagens, EOM |
| MAVS | 14,000 | Teflon plates | – | – | Mutagens, EOM |

\* Explanation of abbreviations:
- SSI-hivol — size-selective hivol
- PS-1 — General Metal Works hivol equipped with gff followed by PUF
- Cal.DOH — Denuder, filter and solid absorbent system utilized by California Dept. Health Laboratory in Berkeley
- AGP — Atmospheric gas phase sampler developed at Univ. Antwerpe
- Ontario Sci. — Denuder, filter and solid adorbent system developed on a modified dichotomous sampler by Ontario Scientific
- U. Colorado — Thermal desorption based sample collection system developed at the Univ. Colorado
- SAPRC — Ultra-high volume, size-selective sampler developed at the Statewide Air Pollution Research Center, Riverside, Ca.
- MAVS — Ultra-high volume, size-segregated sampler developed by Battelle-Columbus Laboratories

of substance. Additional research has been directed towards a broader range of SVOCs at urban locations in North America and Europe. Many studies investigating the mutagenicity of the ambient aerosol have utilized the hivol/gff system [11, 23–25, 27]. It is reasonable to assume that these studies assessed more than the mutagenicity of the ambient aerosol, including gff-associated chemical reactivity artifacts. While there is inconsistent, information in the literature of gff chemical-induced mutagens being formed during sampling [28–31], there certainly appears to be greater volatility losses associated with the hivol/gff system than with other collection methods. Overall, chemical and physical changes in the makeup of SVOCs are expected to be greater during the warmer portions of the year, when corresponding photochemical activity is also at a peak.

Size-selective inlet/filter systems (SSI/filter) offer the potential advantage of collecting only those materials that are likely to be inhaled by an exposed population and provide for the concentration of only those substances with similar physical properties in the atmosphere. The latter issue might have an impact on the chemical makeup of the filter catch if trace substances associated with the larger particles could induce chemical changes in certain classes of the SVOC group. SSI-hivol/filter systems offer the opportunity of collecting reasonably large sample volumes (app. 1,600 m$^3$/day), while the dichotomous system collects only small sample volumes ($\sim$21 m$^3$/day). Thus the latter system requires much greater analytical sensitivities for the target compounds to be identified and quantified *and* cannot provide enough material for bioassay directed chemical assays. Advantages of the dichotomous system are; a) there is less opportunity for volatilization losses due to reduced volumetric flow rates and b) chemical artifacts are anticipated to be of lesser consequence on Teflon membrane filters. In addition, the dichotomous system provides a separation of coarse (2.5 $\mu$m to <10 $\mu$m aerodynamic diameter) and fine (<2.5 $\mu$m aerodynamic diameter) particle sizes, which may be of some importance for investigations of human exposure to specific SVOCs.

Two very large volumetric flow rate samplers (MAVS and SAPRC) have been designed to collect about 20,000 m$^3$/day and 26,000 m$^3$/day, respectively, for studies oriented towards bioassay-initiated assessments [32, 33]. Both samplers include SSI inlets with the MAVS providing specific particle size segregation. These samplers have been primarily utilized for mutagenicity assays and not for SVOC chemical identification [34].

During the last ten years two approaches for collecting SVOCs have been developed which will minimize volatization losses (Table 5). The first is the PS-1 system which is a modified hivol with a polyurethane foam (PUF) trap following the filter. This method was first developed by USEPA [7] to measure chlorinated pesticides and PCBs. It has since been applied to PAHs, PCBs, pesticides, polychlorinated dioxins (PCDDs) and polychlorinated dibenzofurans (PCDFs). Various workers have reported results from studies with the PS-1 system as having demonstrated vapor/particle distributions of specific SVOCs. However, it is clear that at the relativity high volumetric flow rate (app. 220 m$^3$/day) of the

**Table 5.** Commonly utilized SVOC collection media

| *Filters* | *Solid adsorbents* | *Liquid adsorbents* |
| --- | --- | --- |
| Glass fiber | $Al_2O_3$ | Ethylene glycol |
| Teflon-coated gff | Silica gel | Liquid nitrogen |
| Quartz fiber | XAD-2 | |
| | PUF | |
| | Tenax-GC | |
| | Florosil | |
| | Charcoal | |

PS-1 sampler, the gff prefilter is subject to SVOC blowoff. Thus the vapor/particle distributions reported from these investigations are pseudo-distributions, since they are artifactually related to the method of sample collection. Van Vaeck et al. [42] have developed an atmospheric gas phase sampler (AGP) which minimizes compound blowoff potential. The AGP sampler runs at about 18 lpm and is said to minimize blowoff by causing a different section of a $8 \times 10$ gff, to be exposed to a small volume of air; thus each portion of the filter receives approximately $2.6 \text{ m}^3$ of air every 2.4 hours of about $1 \text{ m}^3/\text{hr}$. It is probably reasonable to assume that at some point in the SVOC vapor pressure range (E-4 to E-15 atm), that a particular substance is not volatilized to a significant extent under the conditions of the AGP system. What has not been demonstrated is under what physical and chemical conditions and for what vapor pressures does the "set point" change. Thus benzo(a)anthracene (vapor pressure app. E-10 atm) may not be quantitatively retained from the particle-phase during the summer but may be quantitatively retained during the winter. The data to support this type of observation has yet to be presented by the Belgium research group.

A research group at the University of Colorado recently developed a semi-continuous sampling method for SVOCs [43]. Samples are collected at 10 lpm on a specially designed glass filter holder contained a small gff. Sampling times vary from 1 min to about 3–1/2 hours with the typical time being 10 min. The entire tube is then thermally desorbed into a gas chromatograph (GC) or a gas chromatograph/mass spectrometer (GC/MS) for analysis. While this approach is of some interest, in that blowoff losses are probably negligible, vapor/particle distributions for specific SVOCs cannot be estimated *and* the method is only applicable for those compounds in high enough concentrations or with significant vapor pressures to be detected in such small samples.

Recently, a number of different research groups have developed sampling approaches for SVOCs which can, in theory, differentiate between vapor/particle distributions of selected target groups [44–46]. The basic approach utilized by both research groups was the development of a medium flow (15–25 lpm) sampling device which contains three sections: vapor-phase denuder, particle filter, and blowoff solid adsorbent. These sections are to collect vapor, particle and blowoff losses of the SVOCs of interest. The Ontario Scientific method includes an annular denuder coated with silica and crushed Tenax-GC [45]. The denuder is followed by Teflon filter and Tenax-GC adsorbent traps. This approach utilizes a sampler with a volumetric flow rate of 15 lpm, and allows for the measurement of fine and coarse particles and blow-off losses. The approach has been utilized to study ortho-BHC, beta-BHC and gamma-BHC in the ambient air in Ontario [46, 47]. Overall the method would probably be of use for SVOCs with an approximate vapor pressure of E-8 to E-15 atm. SVOCs in the high vapor pressure range may not have enough adsorption sites on the denuder annuili to result in quantitative collection. Also, this method may not collect large enough samples of SVOCs such as polycyclic aromatic hydrocarbons which cannot be measured as sensitively as chlorinated pesticides on capillary GC with ECD.

The Appel et al. [44] group utilized an approach developed for use in the workplace environment (PAH in aluminium foundries) by Bjorseth and coworkers [48]. A denuder coated with $Al_2O_3$ is followed by a gff and a $Al_2O_3$ adsorbent bed. The collect system is inverted and samples are collected with a pump at 15 lpm. At present, the sampler has not been utilized with a size-selective inlet (SSI). Results from a limited sampling effort, targeted at a select group of PAH, indicated that less than 5% of the ambient levels of these substances were in the vapor phase. Thus Appel et al. [44] recommended utilizing only the filter and $Al_2O_3$ adsorbent trap for PAH sampling.

These approaches provide scientists with an opportunity to study vapor-particle distributions of selected SVOCs. However, the three different portions of the sampling system will require *three* separate chemical analyses. Thus such an approach can triple the cost of a study of a select group of SVOCs. Also, because different adsorbents are better utilized for specific groups of target SVOCs, broad range sampling is not likely to be accomplished with one type of denuder and trap. This situation also illustrates a potential cost escalation if more than one adsorbent per target compound class is utilized in a study. Finally, since each denuder must be made separately in the laboratory, uniform adsorption rates are not likely to found for each fabricated denuder. Thus the adsorption rate for each denuder must be assessed prior to its use in the field. Such QC measures are time consuming and costly. Denuder-based ambient air sampling systems currently can provide scientists with the best phase distribution information. It could be possible that in the future that the need for denuder sampling could be eliminated if vapor-particle distributions can be predicted based on particle loadings and ambient air concentrations and temperatures. At that point only filters and back-up traps would have to be collected for analysis.

**Source Sampling**

SVOCs have been primarily collected from combustion sources, with few biogenic and synthetic organic chemical manufacturing use or disposal loss emission studies present in the literature [12, 13, 49–57]. The collection of SVOCs from combustion sources requires focusing on phase-separation within the source emission streams, collection media-induced reactivity, target compound breakthroughs and sufficient sample sizes for analytical detection.

As with ambient air sampling approaches, source sampling requires careful consideration of the collection environment. Flue gases occur at elevated temperatures often significantly greater than 100 °C and have high concentration of reactive gases and/or droplets such as $NO_x$, $SO_x$, HCl, and $H_2SO_4$. Under these conditions SVOCs will be difficult to collect without significant sampling artifacts. An area that has not been adequately addressed is the formation of mutagens on the collection media. It would appear that short-term bioassays have the potential to be useful in regulatory enforcement activities *if* collection-media induced reactivity artifacts can be shown to be negligible.

Three basic approaches have been utilized to sample combustion source SVOC emissions that are ducted through a vent (Table 6). Particulate trains equipped with non-heated gff have been utilized to collect PAH from fossil fueled power plants. Thus samples collected for regulatory compliance tests for particles can be utilized to estimate SVOC emissions if they are properly handled after col-

**Table 6.** Some examples of SVOC source collection techniques in the literature

| Method | Source type | Target compound |
|---|---|---|
| I-*Stationary sources* | | |
| Method 5 | Fossil Fuel | PAH |
| PAH-Train | Fossil Fuel, Industrial, Incinerators | PAH |
| MM-5 | Incinerators | PCDDs, PAH, PCBs |
| Dilution Tunnels | Incinerators | PAH |
| Cold traps | Coal-fired Power plants | PAH |
| II-*Mobile* | | |
| Dilution Tunnels | Diesel, Spark-ignition | PAH, nitroPAH |
| Cold traps | Diesel, Spark-ignition | PAH, nitroPAH |
| III-*Fugitive* | | |
| Hivol/gff | Agricultural Residue | PAH, oxygenated-PAH, substituted phenols |
| PS-1 | Hazardous Waste Sites | PCBs, PCDDs |
| Hivol/gff | Tunnel studies | PAH, nitroPAH, Azaarenes |

lection. These types of SVOC emission results can only be considered qualitative at best, as a result of ineffective collection of vapor phase SVOCs. The next approach is similar to that utilized in ambient air studies; a filter is followed by an adsorbent trap to collect vapor phase and/or blowoff losses of SVOCs present in the effluent stream. Adsorbents which have been utilized to collect SVOCs include: florosil, XAD-2, and Tenax-GC. These approaches are also called proportional sampling since only a fraction of the effluent stream is collected by the sampling device. Most samples have been collected with gff although quartz fiber filters and teflon-coated gff have also been utilized. A final approach includes dilution sampling. In this case, flue gases are mixed with either clean air or pure nitrogen to a certain ratio (often 10:1 clean to flue) and the particles are collected on a filter (gff, quartz fiber filter, Teflon-coated gff). The goal in dilution sampling is two-fold: a) to reduce the complexity of the sample collection and b) to simulate flue gas condensation so that sampling artifacts are reduced or eliminated. At the present time there is some uncertainty as to whether the dilution systems are more or less artifact prone than utilizing filter plus adsorbent type sampling trains [58, 59]. Different results from mutagenicity studies do not necessarily imply that one source testing method is less artifactual than another [58]. At the present time, dilution sampling techniques for stationary combustion sources are being utilized solely in the research arena, since commercial systems are currently unavailable. There are some limitations to dilution sampling techniques in that the equipment tends to be physically large and is thus not amenable to many real world testing environments. There are also problems collecting flue gas samples isokinetically, thus limiting dilution system applications for regulatory enforcement.

The main SVOC sampling technique currently utilized in stationary combustion source testing is the US-EPA modified method five (MM5). The preferred adsorbent trap method is XAD-2, although other absorbents have been utilized. This approach has been utilized to sample a wide variety of SVOCs in stationary combustion sources; PCDD/PCDF, PCBs, PAHs, chlorinated pesticides, and chlorophenols. Also, MM-5 samples can be collected isokinetically making the train useful for regulatory enforcement activities. While some validation efforts have been carried out for the MM-5 [60], there is some question about artifact formation when stack sampling focuses on reactive compounds such as 1-nitropyrene or BaP.

Mobile source sampling for SVOCs has been carried out utilizing primarily two approaches; dilution sampling and whole-effluent sampling utilizing condensation knock-out traps. Most work in the US and Sweden has been accomplished utilizing dilution tunnels, while the West Germans and Japanese have utilized whole-air collection [55, 61, 62]. Because the source can be brought to the sampling device, and due to the fact that enforcement activity for SVOCs in motor vehicle tailpipe emissions are not an issue, dilution sampling is an attractive collection approach. Combining such sampling with modern dynamometer approachs provides a very useful method for studying SVOC emissions under different driving conditions. This approach has been extensively utilized in the investigation of SVOC mutagens present in diesel exhausts [31].

Indirect source sampling approaches for SVOCs basically utilize ambient air collection methods under very specific conditions. These studies include motor vehicles in tunnel air, indoor bus depots; field applications of pesticides, landfill PCB emissions; agricultural residue burning operations; indoor smoking environments; and kerosene heating environments. In most instances, these studies can provide a qualitative indication of SVOC emission rates, but they cannot be utilized to develop mass emission rates. The tunnel studies appear to be an exception to this situation and these efforts provide a very eloquent approach for quantifying "real-world" motor vehicle SVOC emissions [13, 56, 122].

Overall the most critical aspects of SVOC sampling of combustion sources primarily revolve around trap issues; temperature effects on pollutant phase distributions and chemically induced artifacts formed on the collection media. In gross terms, the former issue has been successfully addressed utilizing adsorbent traps and/or dilution tunnels. Actual SVOC reactivities in flue gases have yet to be adequately addressed. The latter area has been inadequately investigated for a broad range of SVOCs. This issue is of great concern when biologically driven chemical assays are carried out on SVOC source samples.

**Analytical Techniques**

The analytical approach for source or ambient air SVOC samples is essentially identical. The steps involved in SVOC analysis generally follow the process of storage, extraction, separation and detection. Each step is associated with a precision, accuracy and recovery. When an environmental sample contains sub-ppb levels of a SVOC, expected precision, accuracy and recoveries are low [65].

While good quality control results are achievable for trace SVOCs, when large numbers of samples are processed in a specific laboratory, quality control results are often poor. Depending on the substance, system precision and accuracies of $\pm 50\%$ and recoveries of 50%–90% should be common in trace SVOC analysis when significant numbers of samples are processed on a nearly continuous basis.

With the preceding discussion of quality control in mind, storage of samples can become an important issue. Unless otherwise verified, SVOC source/ambient air samples should be stored in the dark and at low temperature (0 degree C or less). Under such conditions, PAH can be stored for as long as 4 weeks without measurable losses [personal communication (PC) by A. Greenberg]. Particle associated PAH have been stored in the dark at room temperature, at NBS for two years with no detectable concentration changes (PC Willie May). Other more reactive SVOCs, may require almost immediate (24–72 hr) extraction and analysis. Long term storage (>1 month) of SVOC samples is generally not advisable, since biodegradation of certain compounds may occur. Overall, storage related losses of SVOCs are not a significant problems as long as reasonable care is taken such as placement of the sample in the dark and in a refrigerator or freezer. For those compounds which are particularly reactive (such as epoxides) sample storage should be eliminated or kept to a minimum.

Extraction procedure for SVOCs can be accomplished in a number of different ways. The most common method is Soxhlet extraction utilizing a non-polar (Benzene, Toluene) or a moderately polar (Dichloromethane) solvent. Other extraction techniques include sonication, mechanical shaking and thermal desorption [43, 64]. The use of binary or tertiary solvent mixtures affords one the opportunity to develop broad scale investigation of SVOCs far beyond those studies when only non-polar solvents were utilize to extract these materials [22]. By utilizing various solvents in a well defined scheme a degree of SVOC separation can be achieved during the extraction phase of an analytical study. In past SVOC studies, typical multisolvent SVOC extraction schemes involved sequential extraction utilizing solvents of increasing polarity [27] and acid/base extraction [65, 66]. Thermal desorption is an intriguing approach for removing SVOCs from a collected sample matrix [43]. Besides the advantage of the rapid extraction time needed for thermal desorption, this technique can reduce the possibility of any artifact formation during elevated temperatures associated with Soxhlet extraction [67]. However, thermal desorption techniques are generally applicable for high volatility compounds, and may not be appropriate for studies that are concerned with the identification of unknown SVOCs. The limits of application of the thermal desorption techniques lie in the narrow window of compounds that can be extracted from the sample matrixes and the need to establish specific extraction efficiencies for a significant number of compounds.

The separation and detection techniques utilized in a specific analytical scheme are intimately associated (Table 7). Many recent studies have been carried out utilizing a multi-step separation procedure prior to analyte detection. Various examples include: a) the use of TLC prior to HPLC combined with UV/Fluorescence for PAH, nitro-PAH, and PAH-quinones [68, 69], b) Application of color chromatography prior to capillary column GC-ECD for 1-nitropyrene [70], c) The use of HPLC prior to HRGC/MS for borad spectrum

**Table 7.** Examples of separation/detection approaches utilized to study SVOCs in the air environment

| Separation method | Detection method | Target |
|---|---|---|
| Paper chromatography | Colorimetry | Nitroso-compounds |
| Column chromatograph | LRMS, Cap. Col. GC | Lipids |
| Capillary column GC | ECD | 1-nitropyrene |
| TLC | Fluorometry | PAH, BaP |
| Liquid chromatography | UV/Fluorescence | PAH |
| HPTLC | Colorimetry, UV | PAH, nitroPAH |
| HPLC | UV/Fluor, MS, GC/MS | PAH, nitroPAH |
| Packed-column GC | FID, ECD, LRMS | Long-chain alkanes, Carboxylic acids |
| HPLC-Cap. Col.-GC | HRMS | PCDD, PCDF |

SVOC analysis [12] and multistep separation associated with PCDD/PCDF, including column chromatography, HPLC and HRGC/HRMS [71]. Recent applications of HPLC/MS to river and drinking water samples [72] can perhaps be considered to beginning of a new phase in identifying unknown SVOCs in environmental samples.

Present analytical schemes are well developed for many groups of SVOC including:

Long chain-alkanes
Polycyclic aromatic hydrocarbons
Nitro-PAH
Carboxylic acids
Chlorinated pesticides
PCDD/PCDFs
PCBs
Aza-arenes
Phthalates

With the exception of the carboxylic acids, most of the substances which have reasonably well defined analytical procedures are relatively non-polar and are in the upper half of the SVOC volatility range (E-4 to E-10 atm). Presently, we have a poor appreciation of the chemistry of moderately polar and polar compounds and for those substances with relatively low vapor pressures. Some substances that are known to possess these properties such as hydroxy-nitro-PAH and substituted phenols have been implicated as mutagens in a number of measurement studies [12, 73]. It is often typical for many source and ambient air studies of SVOCs that include biologically driven chemical assays that some 15%–30% of the mutagenic activity occurs in the most polar fractions, the chemistry of which is largely unknown. Only those analytical procedures which can be used to investigate low volatility and more polar portions of the SVOC class will be useful for identifying the many unknown mutagens present in such complex mixtures. Certainly analytical techniques based on HPLC/MS or supercritical GC [74] hold great promise for furthering our knowledge of SVOCs in the atmosphere. However, these approaches to SVOC sample analysis need many improvements, before they are useful for the identification of unknown and for greater analytical sensitivity. Covey et al. [75] provide an excellent review of the state of the art of HPLC/MS and its current applications.

## Field Studies

With some notable exceptions, SVOCs have not been generally studied under programs solely designed to measure these substances (Table 8). Field study approaches for SVOCs will be discussed as they pertain to either ambient air or source emissions. There are very few instances when multimedia studies (air, water, soil & biota) of SVOCs have taken place or when both ambient air and source testing occurred within the same airshed [76]. To truly understand the role of SVOCs in the biosphere, more integrated studies of these compounds are necessary.

Table 8. Some important SVOC field studies

| Authors | Name | Target compounds |
|---|---|---|
| I. *Ambient* | | |
| Hidy et al. [8] | ACHEX | Polar Organics |
| Wolff et al. [82] | GM-Detroit SAS | PCA of Mutagens |
| Greenberg et al. [63, 68, 69] | ATEOS & related projects | PAHs, substituted PAH & mutagens |
| Faoro [26] | NASN | BSO, BaP |
| Van Vaeck et al. [5, 42, 79] | Belgium aerosol studies | PAHs, LCA, Carboxylic acids |
| EPA [18, 19] | Community Pest. studies | Chlorinated Pesticides |
| Lewtas [76] | Intergrated Cancer Project | Woodsmoke constituents |
| Pitts et al. [24] | Geographic dist. of airborne mutagens | Mutagens |
| Daisey et al. [27] | NYSA | Mutagens |
| Harkov & Greenberg [78] | NJ CEIP BaP study | BaP |
| II. *Source* | | |
| Pierson et al. [13] | Ford Tunnel study | PAHs and mutagens |
| Rappe [71] | MSW incineration | PCDD/PCDFs |
| Mast et al. [12] | Rice-straw smoke | SVOC smoke constituents |
| Many authors [31, 50, 55, 70, 86, 94, 96, 127] | Diesel exhaust | Mutagens, PAHs and subsituted-PAHs |
| Grimmer [86] | PAH sources | PAHs and substituted-PAHs |
| Hangebrauck et al. [49] | PAH sources | PAHs |
| Visalli [90] | ASME-dioxin study | 2,3,7,8-TCDD |
| Env Canada [91] | NITEP | PCDDs, PCDFs, PCBs, PAHs, and Chlorophenols |
| III. *Multimedia* | | |
| USEPA [89] | National Dioxin Study | 2,3,7,8-TCDD |
| Lioy et al. [99] | THEES | BaP |

## Ambient Air

Most of these studies fall into one of three categories:
- SVOC measurements to obtain baseline ambient concentrations
- SVOC measurements to relate ambient concentrations to some physical characteristic of the environment
- SVOC measurements to relate to the biological acitivity (mutagenicity) of an atmosphere

Some studies such as the ATEOS project and associated efforts [64, 69, 77] have been an attempt to gather information on SVOCs as they relate to all three study goals. Both the NASN and Community Pesticide studies were efforts which were largely carried out to solely establish baseline ambient concentrations of the target substances. In the case of NASN, BSO were collected from 1960 through 1972 mostly in urban areas with coke oven sources. While many NASN sites were oriented towards urban areas with coke oven operations, the overall information developed from this monitoring is not be useful for anything more than establishing long-term ambient air concentration trends [26]. The Community Pesticide Study established short-term concentration levels for a select group of pesticides in ambient air [18, 19]. These studies decumented ambient levels of the target compounds in cities and rural towns during 1970 through 1972. Although urban/rural comparisons were made and conclusions were drawn relating ambient air concentrations to pesticide use, this type of study is mostly useful for establishing baseline ambient concentrations.

A number of measurement studies were carried out to develop an understanding of the relationship of ambient aerosols (including SVOCs) to some physical phenomena in the environment. In the case of ACHEX, ambient aerosols were sampled during periods of visibility reduction in the Los Angeles air basin [8]. In addition to attempting to relate aerosol concentration and size to visibility status, thze chemical nature of SVOCs as smog episodes intensified and subsided was also studied [22]. The results from the ACHEX study provided evidence that polar organics were a major component of ambient aerosols during *both* high smog and low smog levels in the Los Angeles air basin, and that the levels of polar SVOCs were directly related to the level of photochemical activity in the air environment. In the NJ-CEIP study [78], BaP was measured at 27 NJ sites during 1982. Ambient levels were shown to differ between urban/rural sites, and winter/summer seasons and the measured concentrations seemed to be best accounted for by emission inventories rather than by atmospheric chemistry. The Belgium aerosol studies [42, 79, 80] provided atmospheric concentration data for PAHs, Aza-arenes, long chain alkanes and carboxylic acids at urban and rural sites in Belgium. In addition, these workers were able to characterize the particle size distributions of the target compounds and their respective pseudo-phase distributions by utilizing the AGP sampler. These field and laboratory studies have established the influence of atmospheric temperatures on the phase-distribution of many SVOCs and that the seasonal distribution of the target compounds were correlated with the particle size characteristics of the ambient aerosol.

A number of studies have been carried out to relate levels to the biological activity (mutagenicity) in the atmosphere. These studies have shown that the distribution of SVOC mutagens in the atmosphere can vary by season [11, 27] by geographic location [11, 24], diurnally [81] and by pollution episodes [11, 23]. In addition, during the Detroit aerosol study of 1981, mutagenic activity in the atmosphere was related to meteorology and atmospheric chemistry [82]. Results from this study indicated that three major components in the environment could be identified that were responsible for the mutagenic activity of the ambient aerosol: $SO_2$ sources, motor vehicles and atmospheric water. While not a receptor modeling study, this research provides a first attempt at resolving pollution source contributions to the measured ambient SVOC mutagenicity levels in a specific locale.

Two recent studies provide a holistic investigation of ambient levels of SVOCs. The NJ ATEOS study and associated investigations measured SVOCs at four locations in NJ [63, 69, 77]. This study provided comparisons between geographic, seasonal, and pollution episode levels of selected SVOCs [83]. Morandi et al. [9] and Harkov and Shiboski [84, 85] provided some understanding of the source resolution of the measured EOM. Finally, mutagenicity assays [11] were carried out and were related to various physical and chemical changes in the air environment. Most recently, investigations at the Newark monitoring site have been focused on nitro-PAHs, PAH-Quinones and the distribution of SVOC mutagens within specific EOM subfractions [69]. By utilizing HRGC/HRMS and FTIR, the establishment of relationships between specific classes of SVOCs and the mutagenicity of individual subfractions has been a major thrust of these efforts. In the Integrated Community Air Project (ICAP), USEPA is attempting to combine ambient and source testing results to determine which are the most important airborne mutagens and a quantification of their major sources in a specific community [76]. The first phase of the study was to investigate two communities which habe major woodstove emission sources of SVOCs during the winter season. While this study may provide interesting insight into source-receptor mutagenicity relationships, it is unlikely to be useful in the more typically complex urban/industrial airshed.

**Source Studies**

Source measurements of SVOCs have been traditionally carried out for research purposes, since regulatory limits on the emission of specific substances have generally been non-existent. However, based on the regulatory direction related to hazardous waste disposal in commercial treatment, storage and disposal facilities, especially for PCBs, the RCRA has spurred the increasing need for regulatory-based SVOC measurements in combustion source emissions. Since the development of the RCRA regulations, increasing emphasis on diesel-powered motor vehicles and municipal solid waste incineration in North America has lead to an additional level of attention on the SVOCs associated with combustion source flue gas discharges. With this information in mind, SVOCs in source emissions have been evaluated for the following purposes:

- Qualitative identification of SVOCs relating the biological activity of a sample to a specific class of substances.
- Performance evaluation of specific combustion sources and various operating conditions

**Table 9.** Sources included in the USEPA national dioxin study

---
Secondary copper smelters
Municipal solid waste incinerators
Hazardous waste incinerators
Wood-fired boilers
Sewage sludge incinerators
Drum Incinerators
Hospital incinerators
Kraft paper mill boilers
Co-fired boilers
Coal-fired boilers
Industrial carbon regenerators
Municipal carbon regenerators

---

- Development of SVOC source identification for motor vehicles
- Use of ambient air sampling approachs for unique, fugitive source emission studies

Two classic source assessment efforts with regard to PAH emissions have been carried out. The early effort was by Hangebrauck et al. [49] and there been a continuing amount work in this area by Grimmer and co-workers [86]. In these cases, a significant amount of information has been developed to provide assistance in the development of emission inventories for PAHs [78, 87, 88]. Each of these research groups recognized the importance of collecting both particle and vapor-phase PAHs in source effluents that occur at elevated temperatures. However, chemically associated collection artifacts were not seriously addressed by either group of researchers. Most recently the US-EPA has carried out a similar effort for PCDD/PCDFs utilizing the MM-5 sampling train [89]. This database can provide the beginnings of an emissions inventory for this class of SVOC (Table 9).

Recently, a number of combustion source studies have been directed towards an evaluation of the impact of various combustion variables on the emissions of PCDD/PCDFs [90–93]. These studies compared combustion temperatures, oxygen, carbon monoxide, hydrochloric acid, flue gas and moisture levels and PVC composition in the refuse [90, 91] or investigated the role of metal (copper) catalyzed [92, 93] effects on PCDD/PCDFs emissions from municipal solid waste and hospital waste incinerator. Overall these studies have shown that the formation and/or distribution of PCDD/PCDFs in incinerator waste streams are a complex and poorly understood phenomenon. For example, Visalli [90] concluded that the data that he reviewed indicated that PCDD/PCDFs emissions increased when oxygen levels decreased below 5%, conversely, Hagenmaier et al. [90] observed that these substances increased substantially at oxygen surplus conditions.

During the past oil crisis of the mid-1970's, predictions were made that diesel-powered motor vehicles could make up as much as 25% of light-duty motor vehicle population in the US. As a result of this anticipated penetration of light-duty diesels into the US motor vehicle population, concern was focused on the potential impacts of the increased diesel-particulate emissions on human health. As a result of this issue, a significant level of effort was placed on characterizing light-duty diesel emissions and their biological activity [94]. These studies can be characterized as providing important linkages between the biological activity of the collected source samples and their chemistry. Schuetzle and Lewtas [95] provide an informative overview of the application of biologically-driven chemical assays. In addition, much of the work on diesel-particulate focused on the utilization of state-of-the-art analytical chemistry techniques, including complex sample fractionation schemes, to assist in the identification of unknown diesel SVOC mutagens. From these efforts, nitro and dinitro derivatives of PAHs were shown to account for a significant fraction of the biological activity associated with diesel exhaust particulate [96]. However, for the more polar and less volatile fractions of diesel exhaust particles, accounting for approximately 25% of mutagenicity of a sample, specific identification of the possible substances responsible for this activity was limited due to the currently available analytical techniques [96].

A number of researchers have attempted to measure in-use motor vehicle emissions through the use of tunnel-oriented studies [13, 56, 122]. These efforts provide a very eloquent methodology for resolving emission rates of various SVOCs by accounting for the makeup of the motor vehicle population with regard to the concentration of specific substances in the tunnel atmosphere. SVOC motor vehicle emissions that have been measured by these techniques include PAHs, nitro-PAHs, PAH-quinones, and Aza-arenes. Surprisingly, in-use motor vehicle SVOC emission rates of specific substances and overall sample mutagenicity, correlate very well with results gathered from dynamometer-based studies [13].

Fugitive SVOC combustion emissions from agricultural residue removal operations have also been characterized [12, 97, 98]. These studies are similar to the motor vehicle-tunnel investigations in that ambient air collection techniques were utilized to collect samples within the polluted area, and average background concentration corrections were made utilizing samplers in non-impacted areas. While the fugitive combustion emission estimates are less quantitative than those developed from the tunnel studies, they provide a very useful approach for qualitative inferences on the nature of SVOC emissions from these types of source categories.

As discussed earlier, SVOCs are somewhat unique among organic pollutants in that they tend to be cycled through various compartments in the biosphere. As such, SVOCs are very much a multimedia environmental issue, yet there are few examples where a single project was initiated which encompassed assessment of various media levels of a SVOC. The recently completed national dioxin study, focused on multimedia assessment including air, soil, water, and fish [89]. However, there do not appear to be any efforts to provide linkages between sources, environmental levels and human or ecological exposures. A more modest effort has been recently initiated in NJ, where a total human exposure study related to BaP has been carried out around a grey iron foundry [99]. In this study, multiple sources of human exposure (indoor/outdoor air, food, water and soil) are being simultaneously assessed for this SVOC.

## Environmental Fate

SVOCs are anticipated to have complex environmental distributions simply by virtue of their diverse vapor pressures, fugacities and octanol-water partition coefficients [100]. The spatial distribution and concentration of a particular SVOC in the atmosphere is a function of local source strengths, atmospheric stability and the chemical and photochemical activity of a particular atmosphere. The interaction of SVOCs with gaseous and droplet reactive inorganic pollutants such as nitric acid, ozone, hydrochloric acid and sulfuric acid and with sunlight, humidity and temperature is of great interest with regard to the fate of specific SVOC *and* the biological activity of a specific environmental sample.

The chemical reactivity and physical distribution of the PAH in the atmosphere are the best characterized of all SVOCs [64]. The major recent studies of reactivity and physical losses of PAHs have been carried out at the Statewide Air Pollution Research Center, Riverside, Ca (SAPRC); Univ. North Carolina (UNC) and Univ. Antwerp (UA), with other research groups providing important contributions to these investigations [5, 6, 31, 101–121]. A recent article by Pitts [101] provides an excellent overview of SAPRC studies of the nitration of gaseous PAH to form mutagenic nitroarenes. The remaining discussion will focus on the physical distribution and the chemical and photochemical reactivity of SVOCs.

### Phase Distribution

As stated earlier in this review, little is known about the actual phase distribution of SVOCs in the atmosphere. What is currently known about the physical distribution of SVOCs in the atmosphere can best be described as a pseudo-phase distribution. Pseudo-phase distributions (PPD) for SVOCs (mostly PAH) have been described by many investigators including Yamasaki et al. [4]; Van Vaek et al. [79]; Thrane & Mikalson [41] and Bidleman et al. [38]. The PPD of PAH and other SVOCs reported by these workers generally follow the categories found in Table 10. For those substances that are soley particle-associated during collection with the hivol air system, these materials are in fact particle-associated in the atmosphere. However, the remaining of phase distributions of the SVOCs are highly uncertain. As an example, Johnson et al. [46, 47] found significant particle-associated alpha-HCH (up to 35%) and HCB (up to 100%) utilizing a denuder-based sampling approach; a result which contrasts with those reported by Bidleman et al. [138] utilizing a high vol/PUF system, as shown in Table 10. Additional SVOC-specific measures with denuder-type samplers are not available. However, the Van Vaeck et al. [79] AGP sampler is a low volume system which should cause smaller filter blow-off artifacts than the hivol/gff system. Results for compounds such as anthracene and phenanthrene indicate much higher particle distributions (16%–79%) than the with hivol/PUF system [4, 41].

Although phase distributions of SVOCs in the atmosphere are primarily temperature related, particle adsorbing characteristics are also important. Recent information provided by Yokley et al. [102]; Behymer and Hites [104]; Grosjean et al. [102] indicate that higher levels of carbonaceous material in the aerosol in-

**Table 10.** Pseudo-phase distributions of selected PAH and other SVOC*

| Vapor (>95%) | Vapor/Particle | Particle (>95%) |
|---|---|---|
| Phenanthrene | Pyrene | Benzo(ghi)perylene |
| Anthracene | Benzo(a)fluorene | BaP |
| alpha-HCH | Benz(b)fluorene | BeP |
| HCB | Chrysene | Corenene |
| Naphthalene | Benzo(a)anthracene | Benzo(b)fluoranthene |
| Dibenzothiophene | p,p-DDT | Benzo(k)fluoranthene |
| | Aroclor-1254 | Nonadecanoic acid |
| | Palmitic acid | eicosanic acid |
| | pentadecanoic acid | octacosane |
| | eicosane | heptacosane |
| | nonadecane | |

* Based on [4, 38, 41, 79]

creases the amount of PAHs that are particle adsorbed. Since many SVOCs of interest are emitted into the atmosphere from combustion sources this observation has limited "real world" importance, although it is of definite interest for laboratory studies.

**Reactivities**

The reactivities of selected SVOCs have been presented or reviewed by many authors [101–120]. Laboratory oriented studies have focused primarily on reactive gases (ozone, $SO_2$, $NO_x$), with lesser attention associated with nitric acid, nitrogen pentoxide and the hydroxyl and nitrate radicals. While other reactive inorganics associated with SVOC source emissions such as ammonia, chlorine, hydrochloric acid and sulfuric acid may be of importance in altering the nature of the SVOCs present in an atmosphere, little evidence or effort has been directed towards evaluating their significance.

The SAPRC group [101, 107, 109, 112–115] have developed information on the vapor phase reactivities of a number of SVOCs including napthalene, biphenyl, phenanthrene, anthracene, furan, thiophene and pyrole to nitrate and hydroxyl radicals and nitrogen pentoxide and ozone. Additional work has been directed toward the study of less volatile SVOCs such as reactions of PAHs with nitric acid and ozone and PCBs, PCDD/PCDF's. These laboratory studies have generally been carried out at elevated reactant concentrations (by more than a factor of 10), when compared with known atmospheric levels. While the investigations of nitration reactions of gaseous PAH, by the SAPRC group, appear to have developed a very convincing theory on the origins of certain nitroarenes such as 2-nitrofluoranthene [101], it is unclear how laboratory experimentation with unrealistic concentrations of reactants can be translated to a broad spectrum of SVOCs. The hivol sampling work by Grosjean et al. [102] with realistic reactive gases (ozone and $NO_x$) and 1-nitropyrene and BaP provides a case in point.

The issue of the reactivity of SVOCs in the atmosphere *and* during collection and analysis is of concern primarily because such changes can effect our understanding of the relative importance of these pollutants in the biosphere. Ambient air studies have shown that the mutagenic activity of SVOCs can vary diurnally, seasonally, geographically and as a function aerosol aging and photochemical activity. The work by the UNC group [10, 118, 119] has provided a very interesting system to evaluate the potential changes in PAH chemistry resulting from emissions from wood combustion. Woodsmoke aging at elevated temperature, humidity and sunlight caused a rapid reduction in PAH levels. Thus under environmental conditions typical of summertime in the eastern US, PAH half-lives were predicted to be on the order of one hour. These changes accompany a 50%–70% reduction in the individual sample mutagenicity, as measured by the Ames-test. In contrast, to these results we note the BaP data of Harkov & Greenberg [78], and a review of PAH reactivity by Greenberg and Darack [6]. In New Jersey, summertime levels of BaP are roughly 1/10 wintertime levels. Harkov & Greenberg [78] have shown through an air emissions inventory approach that virtually all the BaP levels can be accounted for as a function of annual changes in fossil and biomass fuel combustion during the year. The work of Kamens et al. [10, 118, 119] provides a very convincing alternative explanation for this phenomenon. First, higher PAH concentrations in the atmosphere decay at the same rate as lower levels, assuming environmental conditions are similar. Next, since heating season environmental conditions are often reflective of lower humidities, solar angles and temperatures, the reactivity of PAH are further reduced. *In toto,* these observations can also contribute to an annual 10-fold difference in summer and winter BaP levels. It is reasonable to consider that during the course of the year the interplay of the shifting emissions inventories *and* atmospheric reactivity has an influence on the concentrations of BaP in the New Jersey atmosphere. As much as these observations are useful for understanding the fate of PAH in the atmosphere, it is equally of interest for the entire class of SVOCs. For some SVOCs, such as PAHs, reactivity will alter the biological activity of specific compounds, as in the conversion of fluoranthene to 2-nitrofluoranthene. Overall, aerosol aging appears to be associated with an increase in polar and less mutagenic substances at the expense of *both* non-polar and indirect mutagenic materials. Thus in a gross sense, the reactivity of SVOCs, *en masse,* results in an overall loss of biological activity in the ambient aerosol.

**Health Assessment**

The primary reason SVOCs have been studied, is their potential impacts on human health, and the most pressing health effect concern for these materials has been as cancer-causing agents. Indeed, probably more is known about the carcinogenic potential of one SVOC, BaP, than of any other organic pollutant [31]. On September 24, 1986, USEPA provided guidance on carcinogen risk assessments [123]. USEPA proposed that four steps be involved in carcinogen risk assessment:
- Hazard evaluation
- Exposure assessment

- Dose extrapolation
- Risk approximation

These steps will provide a logical overview of the health assessment of SVOCs. Step three will be excluded since it relates mostly to toxicological modeling.

**Hazard Evaluation**

Since SVOCs are a diverse group of chemicals, only selected information will be provided in this review (Table 11). Many SVOCs have been shown to be carcinogens, mutagens, immunosuppressors, and teratogens [124–130]. A number of chlorinated-SVOCs such as PCBs, DDT, and TCDD have been shown to have a wide range health effects, in studies conducted with laboratory animals [131, 132]. Polycyclic aromatic hydrocarbons have been primarily studied with regard to their role in cancer, although they have been linked to other health effects such as emphysema [31]. Not all SVOCs have been linked to decrements in human health and some classes of these materials, such as quinones, contain compounds which can be carcinogenic and anticarcinogens [128].

SVOCs, as complex mixtures, have been shown to associated with excess human cancer mortality in a number of high exposure situations:

Coal tar workers
Coal gasification
Cigarette smokers
Chimney sweeps

While individual components of these complex mixtures of SVOCs have been shown to be carcinogenic, the overall carcinogenic potential of the mixture is of great interest. Alpert [135] provides an interesting comparison of a number of SVOC complex mixtures and their toxicity in carcinogen and short-term assays. Not surprisingly, it is difficult to predict the carcinogenic potency of a particular complex mixture by virtue of its performance in a single toxicity assay.

Table 11. Comparison of different risk values

| Compound | Risk level for 1 ng/m$^3$ |
|---|---|
| A) Some quantitative risk values for the SVOC from USEPA CAG [145] | |
| Aldrin | $3.3 \times 10^{-6}$ |
| BaP | $3.3 \times 10^{-6}$ |
| Hexachlorobenzene | $4.8 \times 10^{-7}$ |
| PCBs | $1.2 \times 10^{-6}$ |
| TCDD | $4.5 \times 10^{-2}$ |
| DDT | $9.7 \times 10^{-8}$ |
| B) Some quantitative risk values for competitive sports (Wilson & Crouch [146]) | |
| Boating | $5 \times 10^{-5}$ |
| Hunting | $3 \times 10^{-5}$ |
| Swimming | $3 \times 10^{-5}$ |
| Bicycle Racing | $9 \times 10^{-5}$ |
| Ski Racing | $2 \times 10^{-5}$ |

**Exposure Assessment**

The information that is available to develop an exposure assessment for SVOCs must consider multiple exposure pathways for these substances. These pathways include air, water, food and perhaps soils or dust. An ongoing study in New Jersey [99] has attempted to document total exposure to BaP in the vicinity of a grey iron foundry. Generally, the primary sources of ambient field data for SVOCs are large, comprehensive measurement studies such as those for PAH [68] in air, organochlorines in food [136–138] and PCBs in soil [139]. The shortcomings of these databases is that they do not allow for accurate localized estimation of exposures, since environmental concentrations of the SVOCs of interest are likely to have spatial and temporal variation. Modeling of exposures to SVOC have been carried out by a number of workers [100]. These multimedia models provide a generalized estimate of exposure, but do not provide very accurate results.

Harkov [140] provided air exposure estimates for some selected PAH, based on monitoring results from studies carried out in New Jersey. In this effort, consideration of the following issues were utilized to develop reasonable estimates of exposure:
- Time spent indoors and outdoors
- Deposition percentage of the target substance
- Absorption percentage of the target substance
- Indoor-outdoor PAH ratios

Recent data by Bond et al. [141] provides data on estimating lung deposition and half-lifes for selected SVOCs (1-nitropyrene, BaP, 2-aminoanthracene, and phenanthridone). While Hattis et al. [142] provided useful information on particle deposition in human lungs. Combining these diverse sources of information it is possible to develop a fairly reasonable estimate of doses for airborne PAH. This type of estimate of dose is necessary to calculate excess risk for environmental exposures in a target population.

**Risk Approximation**

Carcinogenic risk assessments are normally carried out utilizing some numerical models that relate animal or human occupational studies to typical environmental exposure levels. While there are a number of models available to make a carcinogenic risk assessment [143], the multistage, nonthreshold model is the most commonly utilized. This model is one of the most conservative, that is, it can lead to a higher level risk for each estimated exposure than other models, such as the Weibull or log-probit [144]. Since the actual shape of the dose-response curve at low levels of exposure (such as those commonly occurring in the ambient environment) are unknown, there is a general level of uncertainty in all carcinogen risk models. The use of different models can yield results which vary by three or more orders of magnitude. In spite of these shortcomings, these models are increasingly being utilized to assist in environmental decision making. Harkov [140] in analyzing cancer risks associated from SVOCs, VOCs and trace metals from data collected in New Jersey concluded that SVOCs pose the greatest public health risk of all the pollutants measured in this study. This conclusion was based

**Table 12.** Hazard evaluation of selected SVOC

|  | Source |
|---|---|
| *Carcinogens* | |
| Adrin | CAG [135] |
| BaP | |
| Kepone | |
| DDT | |
| Dieldrin | |
| PCBs | |
| TCDD | |
| 1-nitropyrene | Hirose et al. [133] |
| Benzo(a)anthracene | IARC [124] |
| *Mutagens* | |
| 9-nitroanthracene | Claxton [127] |
| Chrysene | |
| Perylene | |
| *Immunosuppressors* | |
| DDT | Caren [131] |
| PCBs | |
| TCDD | |
| BaP | NAS [31] |
| *Negative Carcinogens* | |
| Aldicarb | Shelby & Stasiewicz [134] |
| Diazinon | |
| Malathion | |
| 1-nitronaphthalene | |

on following carcinogenic risk assessment procedures as described above *and* due to the mutagenicity of the EOM, which indicated that all samples were mutagenic at all sites during every season. Table 12 provides some lifetime risk levels for selected SVOCs at the 1 ng/m$^3$ level in the atmosphere and provides comparison with some common physical activities.

Overall, the entire process of carcinogenic risk assessment is fraught with uncertainty. While there are opportunities to reduce the overall uncertainty in a risk assessment, for the two most important sources of ambiguity; high to low dose and animal to human dose extrapolation, there are very limited options for resolving these issues. In spite of this situation, SVOCs are major health concern in our environment, and their continued study is a strong indication of their importance.

## Future Developments

A number of different areas hold great promise for the investigation of SVOCs in the atmosphere. In the area of SVOC collection techniques two approaches are noteworthy: stationary source dilution sampling and the use of diffusion denuders for ambient sampling. Application of dilution sampling for stationary source sampling opens up the possibility of collecting SVOC samples which are more representative of the chemistry of the actual dispersion environment than with conventional techniques. If broad-scale application of biological assays for combustion assessments such as the Ames-test become common, then increasing use of dilution sampling will occur. Standard filter and adsorbent source sampling systems such as the MM5, will remain important for those relatively unreactive SVOCs such as PCBs. However, when detailed chemical analysis is to be driven by biological assays, dilution sampling is a preferred method. For stationary source sampling, the evolution of these dilution methods must proceed to those which are more amenable to the small working areas typical of most stacks.

Diffusion samplers for ambient air have been utilized to a very limited extent. Concerns with diffusion samplers are that they are not commercially available and suffer from the possibility of nonuniformity between sampler and/or tube designs. This latter problem can definitely be overcome through careful application of the desired adsorbent. The diffusion sampler is a long way from being utilized routinely so that commercial production will take place. At this point diffusion samplers are primarily of research interest to develop better phase-distribution information for a wide range of SVOCs in the atmosphere.

Advanced analytical chemistry techniques for SVOCs include LC/MS-MS and FTIR/MS. While these instrumental methods have not generally been applied to air samples, they hold great promise for identifying the less volatile, higher polar and/or moderately polar SVOC found in source and ambient samples. A principal problem in utilizing these techniques is collection of an adequate sample size to reach analytical detection limits. Since the compounds of interest (mutagens) may occur at extremely low concentrations, it may be very difficult to provide quantification with these methods. This is the case since the analyst needs to proceed from an unknown substance of unknown concentration to a known substance at known concentration. Without standards these techniques will be of limited value since they are not as useful for compound identification as HRGC/HRMS for the more volatile materials. However with enormous gap in our present understanding of the identity of SVOC mutagens in the atmosphere, increased emphasis on new analytical chemistry techniques will be needed to enhance the current state-of-knowledge of these important substances.

There has been a recent evolution in approaches for quantifying human exposure to environmental pollutants. The TEAM study for VOCs opened up new possibilities for the application of personal dosimetry for environmental exposure assessments [147, 148]. A more modest study in New Jersey, THEES, is attempting to build on the experience of the TEAM, to carry out personal dosimetry for a SVOC, BaP [99]. In this study, exposures are estimated utilizing personal monitors, indoor-outdoor air sampling, and by analysis of food, water and soils for BaP. Additionally, some effort is being made to investigate BaP-DNA adducts in blood and BaP metabolites in urine. The interest in personal dosimetry is a reflection of the need to develop better exposure data for standard setting and health risk assessments. In the past, efforts to link pollutants with decrease in human health have often been difficult due to the lack of relevant exposure data. These studies illustrate that by the development of accurate exposure information it may be possible to develop a more precise understanding of the impact of SVOCs in the environment on human health.

Reactivity studies of SVOCs such as those carried by SAPRC need to be better linked to realistic environmental concentrations of the reactants *and* to better source data. That is, without adequate source testing data it will be difficult to attribute the presence of specific compounds in the atmosphere as due to chemical and/or photochemical reactions. Artifactual source testing methods will produce unrealistic source data for SVOCs. The ultimate disposition of all the various SVOCs in the environment is due to a large extent on their chemical and photochemical half-lives in the atmosphere. Physical removal mechanisms, along with those of chemical origin need to better understood for SVOCs. Finally, the continual use of biologically driven chemical assays will assist in the development of a better understanding of the importance of various SVOCs in the atmosphere.

## Conclusions

Semivolatile organic compounds are important to environmental scientists due to the large number of these materials in the biosphere and that they are biologically active. Various approaches to studying these materials will assist in developing a more cohesive appreciation of the sources, environmental levels, human exposures and other factors related to these pollutants. Many important efforts are required to continue in the study of SVOCs in the US, Canada, Europe and Japan. As the SVOCs are components of the combustion of fossil and biomass fuels, it is likely that they will continue to be of importance in a wide range of human endeavors. Although environmental regulations can limit human exposures to many SVOCs, the reality is that people throughout the world will be exposed to these ubiquitous substances. With this in mind, it becomes important for environmental scientists to utilize the full benefit of the tools available for understanding the role these pollutants play in the biosphere. To this end, it is hoped that the present review will assist these scientists in their study of SVOCs.

## Abbreviations

| | |
|---|---|
| ACHEX | aerosol characterization experiment |
| AGP | atmospheric gas phase sampler |
| ASME | American society for mechanical engineers |
| ATEOS | atmospheric toxic elements and orgtanic substances |
| atm | atmosphere |
| BaP | benzo(a)pyrene |
| BaP-DNA | benzo(a)pyrene-DNA |
| BeP | benzo(e)pyrene |
| BSO | benzene soluble organics |
| C | degrees centigrade |
| CAG | carcinogen assessment group |
| DDT | |
| ECD | electron capture detector |
| EOM | extractable organic matter |
| FID | flame ionization detector |
| FTIR | fourier transform infrared spectrometry |
| GC | gas chromatography |
| GC/MS | gas chromatography/mass spectrometry |

| | |
|---|---|
| gff | glass fiber filters |
| HBC | hexachlorobenzene |
| hivol | high volume air sampler |
| HPLC | high pressure liquid chromatography |
| HPLC/MS | high pressure liquid chromatograph/mass spectrometry |
| hr | hour |
| HRGC | high resolution gas chromatography |
| HRMS | high resoltuion mass chromatometry |
| IARC | International agency for research on cancer |
| ICAP | integrated cancer assessment project |
| LCAs | long chain alkanes |
| LC-MS-MS | liquid chromatography-dual mass spectrometry |
| lpm | liter per minute |
| LRMS | low resolution mass spectrometry |
| MAVs | massive air volume air sampler |
| min | minute |
| MM5 | USEPA modified method five |
| NASN | national atmospheric surviellance network |
| NJ | New Jersey |
| NJ-CEIP | NJ Coastal energy impact study |
| NYSA-NY | summer aerosol study |
| PAH | polycyclic aromatic hydrocarbons |
| PCA | principle components analysis |
| PCBs | polychlorinated biphenyls |
| PCDD | polychlorinated dibenzodioxins |
| PCDF | polychlorinated dibenzofurans |
| PICs | product of incomplete combustion |
| PM10 | inhalable particulate matter, $<10\ \mu m$, ad |
| PM15 | inhalable particulate matter, $<15\ \mu m$, ad |
| ppb | part per billion |
| PPD | pseudophase distribution |
| PS-1 | PUF system one sampler |
| PUF | polyurethance foam |
| QC | quality control |
| RCRA | resource conservation recovery act |
| SAPRC | California Statewide air pollution research center |
| SAS | summer aerosol study |
| SSI | size selective inlet |
| SVOC | semivolatile organic compound |
| TEAM | total exposure assessment methodology study |
| TCDD | tetrachlorinated dibenzodioxins |
| THEES | total human environmental exposure study |
| TLC | thin layer chromatography |
| UA | University of Antwerpe |
| UNC | University of North Carolina |
| USEPA | United States Environmental Protection Agency |
| UV | ultraviolet |

# References

1. Cook, W.J. et al. (1933) J. Chem. Soc.:395
2. Junge, C.E. (1977) In: Suffet, I.H., Fate of pollutants in the air and water environments. J. Wiley, NY, NY pg 7–25
3. Lewis, R.G. (1986) In: Measurement and monitoring of toxic air pollutants. Air Poll. Cont. Assoc. VIP-7 907 pp

4. Yamasaki, H. et al. (1984) Env. Sci. Tech. 16:189
5. Van Cauwenberghe, K. et al. (1980) In: Recent advances in mass spectrometry, Vol. VIII Heyden, London, Great Britain
6. Greenberg, A., Darack, F. (1985) In: Molecular structure and energetics. VCH Pub, Deerfield, Fla
7. Lewis, R.G. et al. (1980) In: Sampling and analysis of toxic organics in the atmosphere. ASTM STP 721 pp 36–47
8. Hidy, G.M. et al. (1981) The aerosol characterization experiment (ACHEX). Wiley-Interscience, NY, NY
9. Morandi, M.T. et al. (1987) Atmos. Env. 21:1821
10. Kamens, R. et al. (1985) Env. Sci. Tech. 29:63
11. Louis, J. et al. (1987) in: Lioy, P.J., Daisey, J.M.: Toxic air pollution. Lewis Pub, Chelsea, Mi. 294 pp
12. Mast, T.J. et al. (1984) Env. Sci. Tech. 18:338
13. Pierson, W.R. et al. (1983) Env. Sci. Tech. 17:31
14. Kawamura, K., Kaplan, I.R. (1987) Env. Sci. Tech. 21:105
15. Yokouchi, Y., Ambe, Y. (1986) Atmos. Env. 20:1727
16. Goulden, F., Tipler, M.M. (1949) Brit. J. Can. 3:157
17. Arthur, R.D. et al. (1976) Bull. Env. Cont. Tox. 15:129
18. Kutz, F.W. et al. (1976) In: Lee, R.E., Air pollution from pesticides and agricultural practices. CRC Press, Clev., Oh 265 pp
19. Stanley, C.W. et al. (1971) Env. Sci. Tech. 5:430
20. Grosjean, D. (1977) In: Ozone and other photochemical oxidants. NAS Press, Wash. DC
21. Daisey, J.M. et al. (1982) Atmos. Env. 16:2162
22. Appel, B.R. et al. (1976) Env. Sci. Tech. 10:359
23. Daisey, J.M. et al. (1984) Proc. 77th Annual APCA meeting, San Fran., Ca. #84–80.1
24. Pitts, J.N. et al. (1980) Geographic and temporal distribution of atmospheric mutagens in California. CARB Report No. 47-138-30, 92 pp
25. Alfheim, I., Moller, M. (1979) STOTEN 13:275
26. Faoro, R.B. (1975) JAPCA 25:638
27. Daisey, J.M. et al. (1980) Env. Sci. Tech. 14:1487
28. Clark, C.R. et al. Atmos. Env. 15:397
29. Fitz, D.R. et al. (1984) Atmos. Env. 18:205
30. Crosjean, D. et al. (1983) Env. Sci. Tech. 17:673
31. NAS (1983) Polycyclic aromatic hydrocarbons. Nat. Acad. Press, Wash, DC
32. Henry, W.M., Mitchell, R.I. (1978) EPA 600/4-78-009
33. Fitz, D.R. et al. (1983) JAPCA 33:877
34. Kolber, A.R. et al. (1983) EPA-600/S2-83-045
35. Hunt, G.T. (1986) In: Meas. Toxic Air Poll. APCA Pub VIP-7, 907 pp
36. Lewis, R.G. et al. (1985) Env. Sci. Tech. 19:986
37. Fairless, B.J. et al. (1987) Env. Sci. Tech. 21:550
38. Bidleman, T.F. et al. (1986) Env. Sci. Tech. 20:1038
39. Foremen, W.T., Bidleman, T.F. (1987) Env. Sci. Tech. 21:869
40. You, F., Bidleman, T.F. (1984) Env. Sci. Tech. 16:189
41. Thrane, K., Mikalsen, A. (1977) Atmos. Env. 15:909
42. Van Vaeck, L. et al. (1979) Env. Sci. Tech. 13:1494
43. Greaves, R.C. et al. (1985) Anal. Chem. 75:2807
44. Appel, B.R. et al. (1981) In: Hidy, G.M., The aerosol characterization experiment. Wiley-Interscience, NY, NY
45. Johnson, N.D. et al. (1985) In: Proc. 78th annual APCA meeting, Detroit, Mi, #85–81.1
46. Johnson, N.D. et al. (1986) In: Proc. 79th annual APCA metting, Minn, Mn, #86–32.1
47. Johnson, N.D. et al. (1987) In: Meas. Tox. Air Poll. APCA Pub VIP-8, 775 pp
48. Bjorseth, O. et al. (1980) Electr. Acta 25:117
49. Hangebrauck, R.P. et al. (1967) Sources of polynuclear hydrocarbons. USDHEW-PHS, 44 pp
50. Schuetzle, D. Env. Health Persp. 47:65
51. Black, F., High, L. (1979) SAE #790422, 16 pp

52. EPRI (1978) Polycyclic aromatic hydrocarbons and the electric power industry. EPRI EA-787-SY 25 pp
53. Merrill, R.G., Harris, D.B. (1987) In: Proc. 78th annual APCA meeting, NY, NY #87-64.7
54. Warman, K. (1985) In: Handbook of polycyclic aromatic hydrocarbons, Vol. 2. Marcel Dekker NY, NY 416 pp
55. Stenberg, U.R. (1985) In: Handbook of polycyclic aromatic hydrocarbons, Vol. 2. Marcel Dekker NY, NY 416 pp
56. Handa, T. et al. (1984) Env. Sci. Tech. 18:895
57. EPA (1987) Locating and estimating emissions from sources of polycyclic aromatic hydrocarbons. EPA-450/4-84-007 p
58. Merrill, R.G. et al. (1983) In: Short-term bioassays of complex environmental mixtures-III
59. Cook, M. et al. (1984) Waste oil heater: organic, inorganic and bioassay analyzes of combustion samples. USEPA report
60. Margeson, J.H. et al. (1987) JAPCA 37:1067
61. Ahlberg, M. et al. (1983) Env. Health Persp. 47:85
62. Alsberg, T. et al. (1985) Env. Sci. Tech. 19:43
63. Harkov, R. et al. (1987) Env. Mon. Assess. 9:83
64. Harkov, R. (1986) J. Env. Sci. Health 21:409
65. Nishioka, M.G. et al. (1985) Env. Int. 11:137
66. Alfheim, I. et al. (1984) CRC Crit. Rev. Env. Cont. 14:91
67. Lafleur, A.L., Pangaro, N. (1981) Anal. Letters 14:1613
68. Greenberg, A. et al. (1985). Atmos. Env. 19:1325
69. Greenberg, A. et al. (1987) In: Proc. 80th annual APCA meeting, NY, NY, #87-96.1
70. LaCourse, D.L., Jensen, T.E. Anal. Chem. 58:1894
71. Rappe, C. (1984) Env. Sci. Tech. 18:78a
72. Crathorne, B. et al. (1984) Env. Sci. Tech. 15:797
73. Nishioka, M.G. et al. (1985) In: 10th Int. Symp. Poly. Arom. Hydro., Oct. 1985
74. Lee, M.L., Markides, K.E. (1987) Science 235:1342
75. Covey, T.R. et al. (1986) Anal. Chem. 58:1451a
76. Lewtas, J. (1987) In: Meas. Tox. Related Air Poll., APCA Pub. VIP-8 775 pp
77. Lioy, P.J., Daisey, J.M. (1986) Env. Sci. Tech. 20:8
78. Harkov, R., Greenberg, A. (1985) JAPCA 35:238
79. Van Vaeck, L., Van Cauwenberghe, K. (1984) Atmos. Env. 18:323
80. Cauwenberghe, K., Van Vaeck, L. (1983) Mut. Res. 116:1
81. Pitts, J.N. et al. (1982) JAPCA 35:638
82. Wolff, G.T. et al. (1986) Atmos. Env. 20:2231
83. Daisey, J.M. (1987) In: Toxic Air Pollution. Lewis Pub, Chelsea, Mi. 294 pp
84. Harkov, R. et al. (1987) In: Meas. Tox. Air Poll., APCA Publ. VIP-8. 775 pp
85. Harkov, R. et al. (1986) J. Env. Sci. Health 21:177
86. Grimmer, G. (1983) Polycyclic aromatic hydrocarbons. CRC Press, Boca Raton, Fla. 261 pp
87. Daisey, J.M. et al. (1986) JAPCA 36:17
88. Ramdahl, T. et al. (1982) In: The 7th Int. Symp. on polycyclic aromatic hydrocarbons. Battelle Press
89. EPA (1987) National dioxin study. EPA/530-SW-87-025
90. Visalli, J. (1987) JAPCA 37: 1451
91. Env. Can. (1985) The national incinerator testing and evaluation program: two stage combustion. Report EPS 3/UP/1
92. Hagenmaier, H. et al. (1987a) Env. Sci. Tech. 21:1080
93. Hagenmaier, H. et al. (1987b) Env. Sci. Tech. 21:1085
94. Cuddihy, R.G. et al. (1983) Env. Sci. Tech. 18:14a
95. Schuetzle, D., Lewtas, J. (1986) Anal. Chem. 58:1060a
96. Salmeen, I. et al. (1984) Env. Sci. Tech. 18:375
97. Standley, L.J., Simoniet, B.R.T. (1987) Env. Sci. Tech. 21:163
98. Simoniet, B.R.T. (1984) STOTEN 36:61
99. Lioy, P.J. et al. (1988) Arch. Env. Health. In press

100. Cohen, Y. (1986) Pollutants in a multimedia environment. Plenum Press, NY 338 pp
101. Pitts, J.N. (1987) Atmos. Env. 21:2531
102. Grosjean, D. et al., Env. Sci. Tech. 17:673
103. Yokely, R.A. et al. (1986) Env. Sci. Tech. 20:86
104. Behymer, T.D., Hites, R.A. (1985) Env. Sci. Tech. 19:1004
105. Ohe, T. (1984) STOTEN 39:161
106. Katz, M. et al. (1979) In: Polynuclear aromatic hydrocarbon. Ann Arbor Press, Ann Arbor MI
107. Atkinson, R. et al. (1984) Env. Sci. Tech. 18:110
108. Bierman, H.W. et al. (1985) Env. Sci. Tech. 19:244
109. Pitts, J.N. et al. (1980) Science 210:1347
110. Nielsen, T. (1984) Env. Sci. Tech. 18:157
111. Glotfelty, D.E. (1978) JAPCA 28:917
112. Atkinson, R. (1987) Env. Sci. Tech. 21:305
113. Arey, J. et al. (1986) Atmos. Env. 20:2339
114. Arey, J. et al. (1987) Atmos. Env. 21:1437
115. Atkinson, R. et al. (1985) Env. Sci. Tech. 19:87
116. Gibson, T.L. (1985) In: Proc. 78th annual APCA meeting. Detroit, Mi. #85–36.2
117. Gibson, T.L., Wolff, G.T. (1985) In: Proc. 78th annual APCA meeting. Detroit, Mi. #85–22.2
118. Kamens, R.M. et al. (1988) Env. Sci. Tech. 22:103
119. Kamens, R.M. et al. (1984) Env. Sci. Tech. 18:523
120. Van Vaeck, L., Cauwenberghe, K. (1984) Atmos. Env. 12:323
121. Korfmacher, U.A. et al. (1980) Env. Sci. Tech. 14:1094
122. Hering, S.V. et al. (1984) STOTEN 36:39
123. EPA (1986) Guidelines for carcinogen risk assessment. Red. Reg. 51:33992
124. IARC (1982) Chemicals, industrial processes and industries associated with cancer in humans. Supp. #4, Lyon, Fra. 292 pp
125. Haseman, T.K. et al. (1984) J. Tox. Env. Health 14:621
126. Santodonato, J. et al. (1978) J. Env. Path. Tox. 5:1
127. Claxton, L.P. (1983) Env. Mut. 5:609
128. Smith, M.T. (1985) J. Tox. Env. Health 16:665
129. Lai, D.Y. (1984) J. Env. Sci. Health 2:135
130. Kaden, D.A. et al. (1979) Can. Res. 39:4152
131. Caren, L.D. (1981) Bioscience 31:582
132. Kurzel, R.B., Certrulo, C.L. (1981) Env. Sci. Tech. 15:626
133. Hirose, M. et al. (1984) Can. Res. 44:1158
134. Shelby, M.D., Stasiewicz, S. (1984) Env. Mut. 5:871
135. Alpert, R.E. (1983) Env. Health Persp. 47:339
136. Gartell, M.J. et al. (1985a) J. Assoc. off. Anal. Chem. 68:1184
137. Gartell, M.J. et al. (1985b) J. Assoc. off. Anal. Chem. 68:1163
138. Davies, K. (1988) Chemosphere 17:263
139. Creaser, C.S., Fernandes, A.R. (1986) Chemosphere 15:499
140. Harkov, R. (1987) In: Toxic air pollution. Lewis Pub, Chelsea, Mi. 294 pp
141. Bond, J.A. et al. (1986) In: Aerosols: research, risk assessment and control strategies. Lewis Pub., Chelsea, Mi. 1221 pp
142. Hattis, D. et al. (1987) JAPCA 37:1060
143. Selken, R.L. (1987) Env. Sci. Tech. 21:1033
144. Fishbein, L. (1980) J. Env. Path. 7:1275
145. EPA (1985) Potency values of 55 chemicals as evaluated by the carcinogen assessment group (CAG) as carcinogens. EPA -600/8-85/004f.
146. Wilson, R., Crouch, E.A.C. (1987) Science 236:267
147. Hartwell, T.D. et al. (1987) Atmos. Env. 21:1995
148. Wallace, L.A. et al. (1985) Atmos. Env. 19:1651
149. Ott, W.R. (1985) Env. Sci. Tech. 19:880
150. Muller, J. (1984) STOTEN 36:339

# Arctic Haze

*Glenn E. Shaw*
Geophysical Institute, University of Alaska Fairbanks
Fairbanks, AK 99775-0800, USA

*M.A.K. Khalil*
Oregon Graduate Center, Institute of Atmospheric Sciences
19600 NW Von Neumann Drive, Beaverton, OR 97006-1999, USA

Introduction . . . . . . . . . . . . . . . . . . . . . . . . . . . . . . 70
Brief History . . . . . . . . . . . . . . . . . . . . . . . . . . . . . 71
    Mitchell's Early Observations of Arctic Haze . . . . . . . . . . . 71
    Early Observations at Barrow, 1972–75 . . . . . . . . . . . . . . 71
    Early Chemical Studies at Barrow . . . . . . . . . . . . . . . . . 72
    Summary of the State of Knowledge by Late 1970s . . . . . . . . . 72
Meteorological Considerations . . . . . . . . . . . . . . . . . . . . . 73
    The Arctic Air Mass and the Arctic Front . . . . . . . . . . . . . 73
    Currents of Polluted Air into the Arctic . . . . . . . . . . . . . 77
Phenomenology of the Arctic Haze . . . . . . . . . . . . . . . . . . . 80
    Seasonal Variability . . . . . . . . . . . . . . . . . . . . . . . 80
    Vertical Morphology . . . . . . . . . . . . . . . . . . . . . . . 84
    Size Distribution of Arctic Haze . . . . . . . . . . . . . . . . . 85
    Volatility-Hygroscopicity of Haze Particulates . . . . . . . . . . 87
Aerosol Chemistry . . . . . . . . . . . . . . . . . . . . . . . . . . . 88
    Chemical Tracers . . . . . . . . . . . . . . . . . . . . . . . . . 88
    Bromine . . . . . . . . . . . . . . . . . . . . . . . . . . . . . 89
    Sulfate . . . . . . . . . . . . . . . . . . . . . . . . . . . . . 89
    Size-Dependent Chemistry of Arctic Haze . . . . . . . . . . . . . 91
    Light-Absorbing Carbon . . . . . . . . . . . . . . . . . . . . . . 91
Trace Gases and Arctic Haze . . . . . . . . . . . . . . . . . . . . . . 92
Optics and Radiation . . . . . . . . . . . . . . . . . . . . . . . . . 99
    Radiation Budget . . . . . . . . . . . . . . . . . . . . . . . . . 99
    Satellite Sensing . . . . . . . . . . . . . . . . . . . . . . . . 101
Deposition . . . . . . . . . . . . . . . . . . . . . . . . . . . . . . 102
    Physics of Particle Removal in the Arctic . . . . . . . . . . . . 102
    Particle Removal – Evolution Sequences for Arctic Air Pollution . 103
    Experimental Data on Deposition . . . . . . . . . . . . . . . . . 104
    Arctic-Wide Deposition . . . . . . . . . . . . . . . . . . . . . . 106
    Additional Remarks on Wet Scavenging . . . . . . . . . . . . . . . 106
Trends . . . . . . . . . . . . . . . . . . . . . . . . . . . . . . . . 107
Conclusion . . . . . . . . . . . . . . . . . . . . . . . . . . . . . . 108
Acknowledgements . . . . . . . . . . . . . . . . . . . . . . . . . . . 108
References . . . . . . . . . . . . . . . . . . . . . . . . . . . . . . 109

## Summary

Arctic haze is the pseudonym for large-scale industrial air pollution found all through the arctic air mass. It is perhaps in areal extent the most extensive air pollution system so far identified, affecting as it does approximately 9 percent of the earth's surface (an area larger than the African continent) when at its maximum size and strength in January to April. It is the result of strong meridional air currents that surge northward over central to western Eurasia carrying polluted air to the arctic basin. The arctic air mass is dark, cold, stable and stratified, all of which are characteristics that slow the removal of material from the atmosphere. Arctic haze has built up since the industrial revolution, but the pace has quickened since the Second World War as industry evolves northward. Pollution material in arctic haze is of a submicron size and contains a fraction of black carbon: it interacts strongly with solar radiation. In springtime, the atmosphere is heated by an additional 5–20 wm$^{-2}$.

The chemistry of arctic haze is fascinating since it represents an affected air mass in a sort of equilibirium state. Sulfate, mostly as $H_2SO_4$, graphitic carbon, and a wide range of heavy metals in higher concentration than in strongly enriched crustal material are the predominant constituents of the particulate arctic air pollution. In addition, the polluted air reaching the arctic brings with it dozens of man-made gaseous pollutants; these can serve as indicators of the original arctic pollution.

## Introduction

Arctic haze is the designation given to a complex mixture of trace particles and gases in the polar atmosphere. It affects mainly the uninhabited regions of the arctic basin, which constitute roughly 9% of the earth's surface. Following nearly two decades of research, one can now state unequivocally that the greatest fraction of arctic haze is from human activity taking place at the mid-latitude regions.

The following are believed to constitute the steps in the life cycle of arctic haze Polluted air builds up over industrialized regions and is periodically swept poleward in strong, anticyclonic meteorological systems; the usual situation is to have a high pressure system to the east and a low pressure system to the west. Upon entering the polar zones, the overturning by turbulence quiets down and strength of transport declines due to the weakening of pressure gradients.

The pollution, or rather the most removal-resistant remnant of the pollution including submicron aerosols and a variety of gaseous components with slow reactive activities, then spreads out and meanders around the arctic basin, sometimes forming into bands which show up against the sky when viewed "edge on" from aircraft. The laminar flow, the polar winter darkness and the near total lack of cloud droplets, ice crystals and precipitation, take away the usual mechanisms by which air is cleansed; the result is that the pollution remains airborne for an unusually long time. A corollary is that the haze reaches to and affects much greater geographical areas than in normal instances of mid-latitude transport where the air is well scrubbed. The absence of solar heating and the strongly negative radiation balance contribute to turning the arctic atmosphere into a stagnant pond of cold, dense air.

Finally, the removal-resistant debris leaks out of the Arctic and the air warms, becomes more turbulent, and picks up water vapor; clouds form and precipitate, carrying the material to the surface. Very little is known about this final removal stage of the arctic haze.

Arctic haze is possibly the most extensive (by areal extent) air pollution phenomenon so far yet identified as a "system" and is of concern not only because of its obvious transboundary characteristic, but because the sooty haze over the reflecting arctic ice pack might constitute a significant heating mechanism and have a global climatic influence.

## Brief History

### Mitchell's Early Observations of Arctic Haze

The term "arctic haze" was coined by Murry Mitchell, Jr. in the 1950s to describe bands of dusky material that he and his colleagues observed against the clear blue skies of the Alaskan Arctic. The haze bands of uncertain origin were verbally reported to Mitchell by weather observers on the Ptarmigan weather reconnaissance missions that flew regularly into the High Arctic from bases in Alaska; their purpose was to keep track of the large-scale synoptic patterns in an area of the world that is both strategic and unpopulated. Mitchell published an account of the haze bands [1] believing that it was an unusual phenomenon because no obvious source of pollution or blowing sand was anywhere in the area. From the way the haze scattered light, Mitchell was able to deduce that the material was submicron particulates. The published account did not receive any particular response and Mitchell's paper was forgotten, though in later years, some of the Ptarmigan data were reanalyzed [2].

### Early Observations at Barrow, 1972–75

In spring 1972, one of the authors (G. E. S.) rediscovered "arctic haze" while carrying out measurements of the optical transparency of the atmosphere from Barrow on Alaska's northernmost point. The purpose was to measure the spectral optical thickness of the artic atmosphere, in the belief that this would be of some interest in establishing a baseline level of aerosol, since the Alaskan Arctic is very remote from known sources of pollution and crustal aerosol that typically derives from deserts. Instead of the expected low values of turbidity, the measurements suggested that the arctic atmosphere was polluted. The optical depth was 0.10 to 0.15, two or three times the annual mean for the atmosphere at Tucson, Arizona, where the author had carried out a previous extensive series of measurements.

Blowing dust, subvisible cirrus, suspended ice crystals, and pollution from nearby sources were quickly discounted as causes and the conclusion was reached that the arctic atmosphere was charged with submicron aerosols, at least during the spring months: in midday the solar radiation was diminished by approximately 30% due to the scattering and absorption by the arctic haze (solar elevation angle 28°).

By carrying sun photometers aloft in a small aircraft, Holmgren and colleagues [3] and Shaw [4] determined that the haze dropped off in intensity with altitude roughly as $e^{-h/H}$, with the scale height $H \sim 1$ km. The lowest hundred meters or so were relatively depleted (by about a factor of two) in comparison to the region just above the turbulent boundary layer. Individual layers or bands of haze modulated the smooth exponential decrease; some of the layers were associated with elevated humidity. The haze was relatively homogeneous over distances of a few hundred kilometers, but no obvious source was apparent.

Analysis of actinometer records from McCall Glacier in the Brooks Range confirmed that the turbidity was higher than expected for a clean location and that it reached its maximum in spring [5].

**Early Chemical Studies at Barrow**

G. Shaw, K. Rahn and R. Borys carried out a series of airborne aerosol collection experiments near Barrow in 1976 to determine the composition of the aerosols in arctic haze.

The aim of the Alaskan chemical studies was relatively modest: to *broadly* classify the nature of the aerosol source (it must be remembered that in 1976 the origin of the arctic haze was unknown). Analyses of the first ten or so airborne filter collections taken between April 10–30, 1976, provided a clear-cut result: the material was crustal, indicating that the haze had most probably originated somewhere in distant dust storms. Microscopic analysis of the individual particles, which were congregated around a micron in volume distribution, confirmed this.

Kenneth Rahn pressed the issue of meteorology and discovered that around the time of the aerosol collection at Barrow, a series of strong pulses of airborne dust had been widely documented over the Gobi and Takla Makan deserts in Eurasia in association with passage of strong springtime frontal systems [6]. One to two weeks earlier than the collection period, heavy falls of yellow-colored sand termed "Kosa" were noticed in Nagasaki. Furthermore, the computer-generated back air trajectories from the isobaric meteorological charts confirmed that a rapid river of air had traveled from the deserts of Asia to the region of collection in Alaska at exactly the right time.

It turned out that the series of airborne experiments conducted in 1975 were made during a somewhat unusual meteorological situation, one which favored the transport of crustal material from the Asian deserts. The crustal dust masked what turned out to be the more representative nature of arctic haze, namely a chemical signature with a strong pollution component.

Over the next several years, surface-based sampling at Barrow uncovered the existence of a strong seasonal variation with maximum haze concentration in spring and indicated that the haze was enriched in a number of elements that are commonly found in industrial-produced air pollution [7].

**Summary of the State of Knowledge by Late 1970s**

By the late 1970s after a few years of chemical, meteorological and atmospheric spectroscopic investigations at Barrow and a few other places, the information about arctic haze could be summarized as follows:
- Arctic haze is most intense around the last winter-early spring months, virtually disappears by mid-May, and is absent all through the summer.

- The haze is spatially uniform over scales larger than hundreds of kilometers, and is primarily confined to the lower 2 to 3 km of the atmosphere, frequently occurring in layers 50–100 m thick that are visible against the sky.
- The haze particles are submicron size, and are apparently water-active since the haze optical extinction increases in humid layers.
- Haze is not correlated with or chemically similar to several sources of air pollutants on the north slope of Alaska.
- No sources could be absolutely implicated for the haze, but there was some evidence by the late 70s that industrial sources, in central Eurasia, eastern Europe, and eastern North America were major contributors.
- There was no connection between the appearance of leads or cracks in the pack ice and haze.
- Though the Ptarmigan flights had observed haze to be widely distributed, it was seldom found over Greenland.

The above conclusions were becoming clear by the time of the first symposium on arctic haze, held during April 1977 in Norway. Overviews of the early state of knowledge by about 1980 are in a special issue of Atmospheric Environment (1981). The first symposium initiated a large increase in research devoted to air pollution in the Arctic; of particular importance was the northward evolution of sampling stations in the Canadian and Scandinavian Arctic and on the Greenland Ice Sheet. The participants of the 1977 Norwegian symposium agreed among themselves to establish an informal network in the Arctic in which the data would be freely shared; this proved to be an enormously successful enterprise and within a few years the "big picture" of arctic haze phenomenology started to become apparent.

## Meteorological Considerations

### The Arctic Air Mass and the Arctic Front

The arctic air mass, an understanding of which is crucial for a discussion of arctic haze, can be envisioned as a dome of cold, dense air extending up to the troposphere, which in winter is only at about 8–10 km high [8]. The dome is slightly asymmetrical with lobes reaching down over the continents where the heat balance is strongly negative in winter, but in summer the air mass contracts strongly and is confined to the northernmost areas of the pack ice. One finds a notable absence of clouds in this air mass system in winter; the only substantial clouds are associated with intrusions of low-pressure systems which invade from oceanic sectors and rapidly die out over the ice. Paths of these arctic cyclones are shown in Fig. 1. Other than these rare intrusions of cyclonic system clouds, there are only thin cirrus decks over the Arctic in winter; precipitation is low, typically less than ten cm of water equivalent during the polar night. The arctic air mass, then, is a cold, dense, quiescent, nearly cloud-free system with very little or no solar radiation for months at a time in winter (see Fig. 2). In summer, the system is much more turbulent, brightly illuminated, charged with clouds, especially

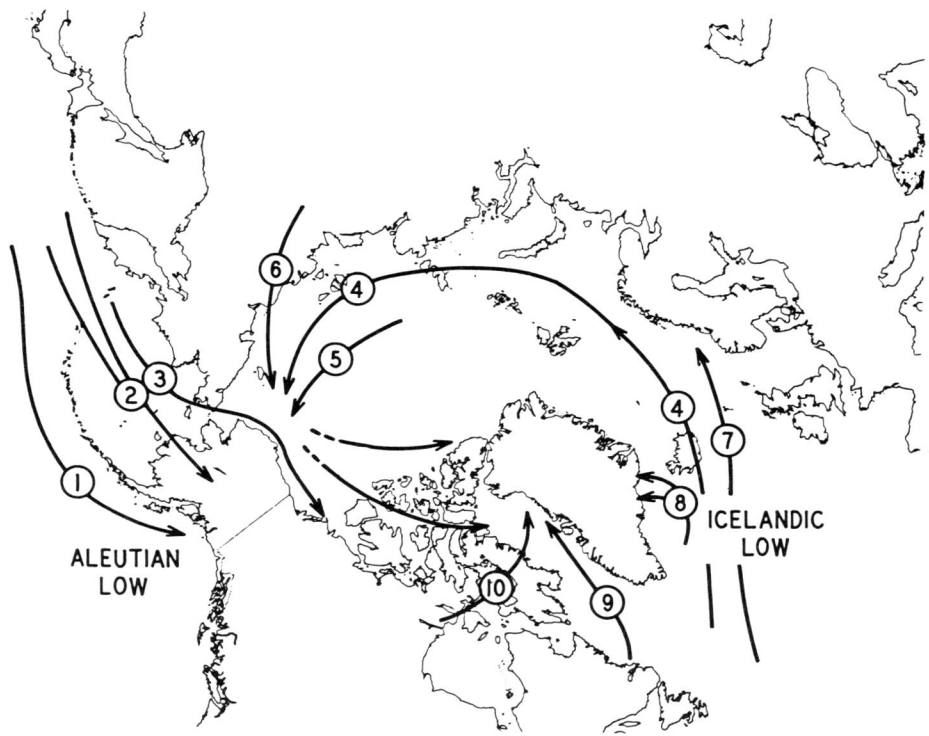

**Fig. 1.** Cyclone tracks in the Arctic (from reference 88)

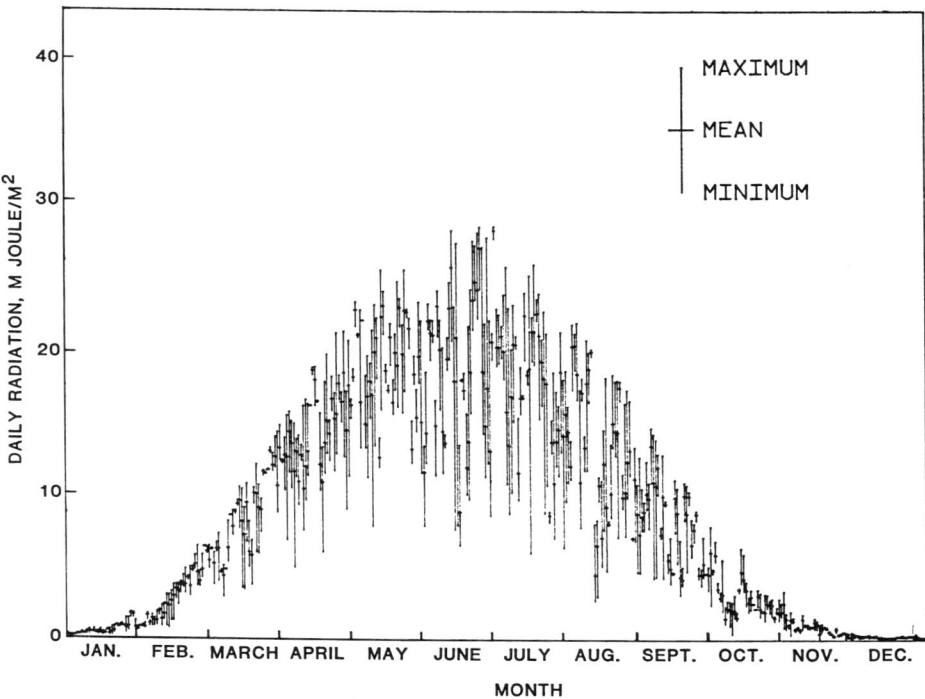

**Fig. 2.** Solar radiation at Fairbanks (Courtesy of G. Wendler, Geophysical Institute, University of Alaska Fairbanks)

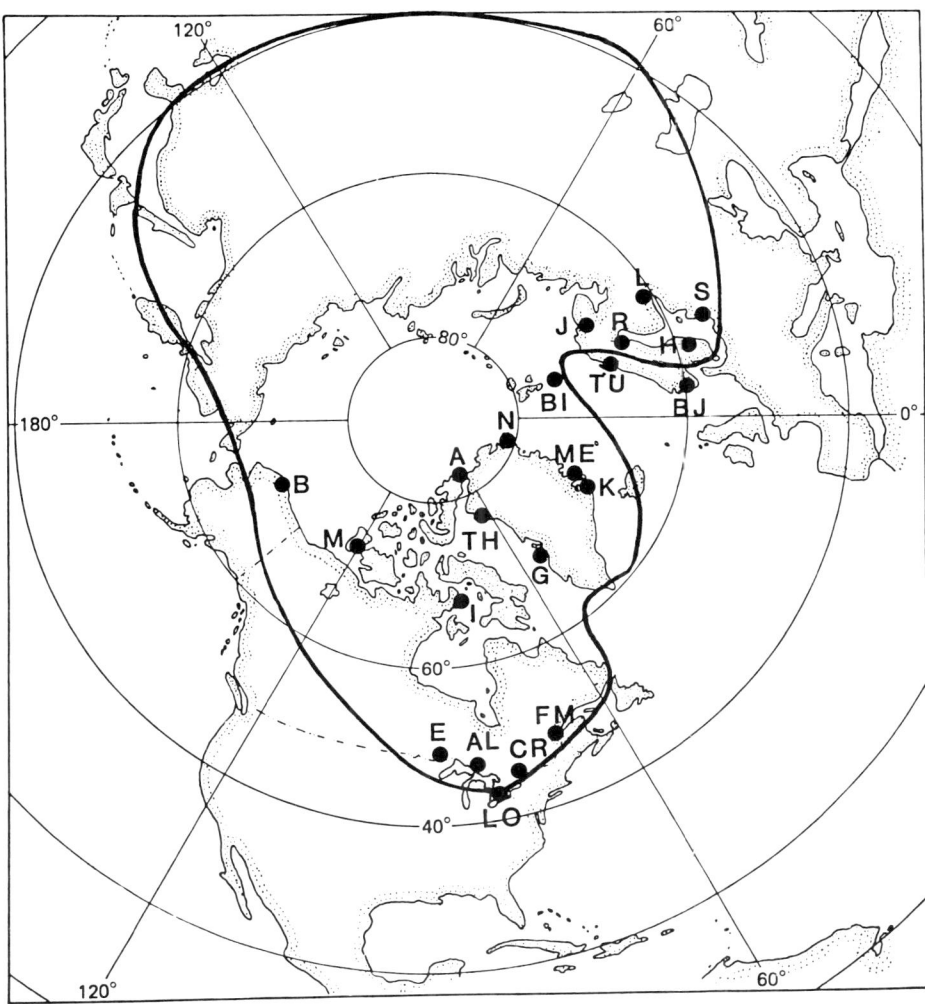

**Fig. 3.** The mean position of the arctic front in January (from Barrie and Hoff, reference 38) The location of air monitoring stations: A-Alert; AL-Algoma; B-Barrow; BI-Bear Island; BJ-Birkenes; CR-Chalk River; E-ELA-Kenora; FM-Forêt Montmorency; LO-Long Point; H-Hoburg; I-Igloolik; J-Jergul; K-Kap Tobin; L-Lesogorski; LO-Long Point; M-Mould Bay; ME-Mestersvig; N-Nord; R-Rickleâ; S-Suwalik; TH-Thule; TU-Tustervaten

with persistent stratus, and it is debatable whether one can actually detect a unique "arctic" air mass at that time of year.

Perhaps the most convenient definition of the arctic air mass is that contained within the boundaries of the arctic front, the mean position of which is shown in Fig. 3, along with some of the air monitoring stations in the arctic network. Notice the strong asymmetry of the front, which has lobes in winter extending southward of the major industrial regions in Eurasia and North America.

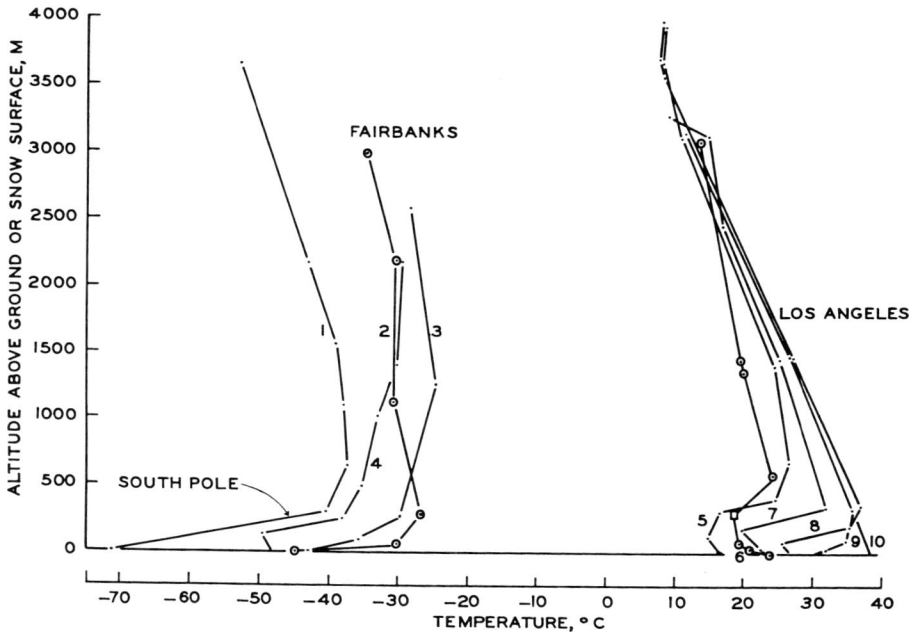

Fig. 4. Comparison of temperature inversions over the South Pole (2,792 m), Fairbanks, Alaska (135 m) and Los Angeles, California (38 m). The stations are plotted at a common altitude to allow direct comparison of temperature profiles; times GMT. Data from radiosonde records. Curves: South Pole (1) 2315, 21 Apr 1958: Fairbanks (2) 1200, 27 Jan 1962; (3) 000, 22 Dec 1961; (4) 1200, 26 Dec 1961: Los Angeles (Santa Monica) (5) 1200, 13 Aug 1963; (6) 0000, 23 Sept 1963; (7) 0000, 25 Sept 1963; (8) 0000, 26 Sept 1963; (9) 1200, 26 Sept 1963; (10) 0000, 27 Sept 1963 (from reference 88)

Another notable feature of the arctic air mass system is the strong temperature inversion (an "inversion" is a region where temperature increases with altitude) that builds up because of the strongly negative heat balance in winter. A comparison of temperature profiles over Los Angeles, Fairbanks, and the South Pole is in Fig. 4; note particularly not only the extremely low temperatures at all heights in the Alaskan and antarctic soundings, but the strengths of the low-lying inversions. In such inversions the air stagnates and turbulence is highly suppressed, although some turbulence is excited from the propagation of gravity waves that develop in regions of strong wind shear, particularly just above the low-lying inversion.

In summary, we see that one can define a type of unique air mass system that forms over the arctic basin and extends down over the continental land masses of Eurasia and North America during winter, but which mainly disappears as an entity, or at least contracts strongly poleward, following large-scale melting of the surface snow in about May, and preceding the disappearance of the sunlight and freeze-up and snowfalls in autumn. The air mass can be characterized by its great dynamic stability, its low temperatures, its surface-based temperature inversion, its low frequency of cloudiness and by the small amount of precipitation that falls through it. In addition, the arctic front, which roughly defines the southern boundary of the arctic air mass, extends poleward of major industrialized source regions in eastern North America and central Eurasia, so that one can imagine the air mass becoming polluted from its own sources and because of the general lack of mechanisms which remove particles and gases: all the removal mechanisms are strongly suppressed in the arctic air mass during winter.

**Currents of Polluted Air into the Arctic**

One can get an idea about the sources of the arctic haze from simply considering areas of known pollution or blowing dust with patterns of the general circulation of the atmosphere. Such an approach is somewhat oversimplified, and in particular it ignores what turn out to be some very interesting transport pathways around anticyclonic systems which, though they are transient, are responsible for carrying a large fraction of the pollution from low to high latitudes. These transient synoptic features tend to be smoothed out in climatological maps of the general circulation. Nevertheless, many interesting sources and transport pathways of the arctic haze do become clear from analyses of the major circulation patterns; for example, note the northeastern flow of air extending from central Europe to northcentral Eurasia in January, which is entirely absent in summer (Fig. 5). There is also evidently a possibility of pollutants from northeastern North America being forced northeastward by southwesterlies between the Icelandic Low and subtropical Atlantic high. One can also sense a possibility of transport from the Orient and eastern Asia to the Gulf of Alaska; this particular current is strongly affected by the Aleutian Low and is heavily scavenged by precipitation. It has been identified, however, as an effective pathway for transporting crustal material from the high-latitude deserts of Asia to the Pacific Rim and into the Arctic; this pathway seems to operate at high altitudes, approximately at the 500 mb level, and requires strong frontal systems and dust carried to high altitudes over the source areas.

The map of the annual emissions of $SO_2$ (Fig. 6) prepared by Len Barrie [8] is very helpful, when considered in conjunction with the surface pressure map (Fig. 5), in elucidating the major sources and currents of pollution-derived material affecting the arctic basin. First, notice that the Eurasian $SO_2$ emissions in areas liable to influence the Arctic are about a factor of 2 to 4 times larger than for North America. Also, notice that the major Eurasian $SO_2$ emissions are placed 5°–10° of latitude poleward in comparison to those in North America; this contrast is made all the more intense by the fact that the arctic air mass has a strong lobe down over the Eurasian land mass. Thus, on the basis of 1) the relatively strong source region in the central and western Eurasia sector, 2) the occurrence of a deep lobe of the arctic air mass over much of this source, 3) the occurrence of a poleward flowing circulation over this source area, and 4) the absence of precipitation, clouds and turbulence along the pathway, one can provisionally conclude that Eurasia is of greater importance than North America as a pollution source region for the Arctic.

In an analysis of the meteorological conditions during and preceding episodes of arctic haze in Alaska, Raatz and Shaw [9] confirmed that the Eurasian source

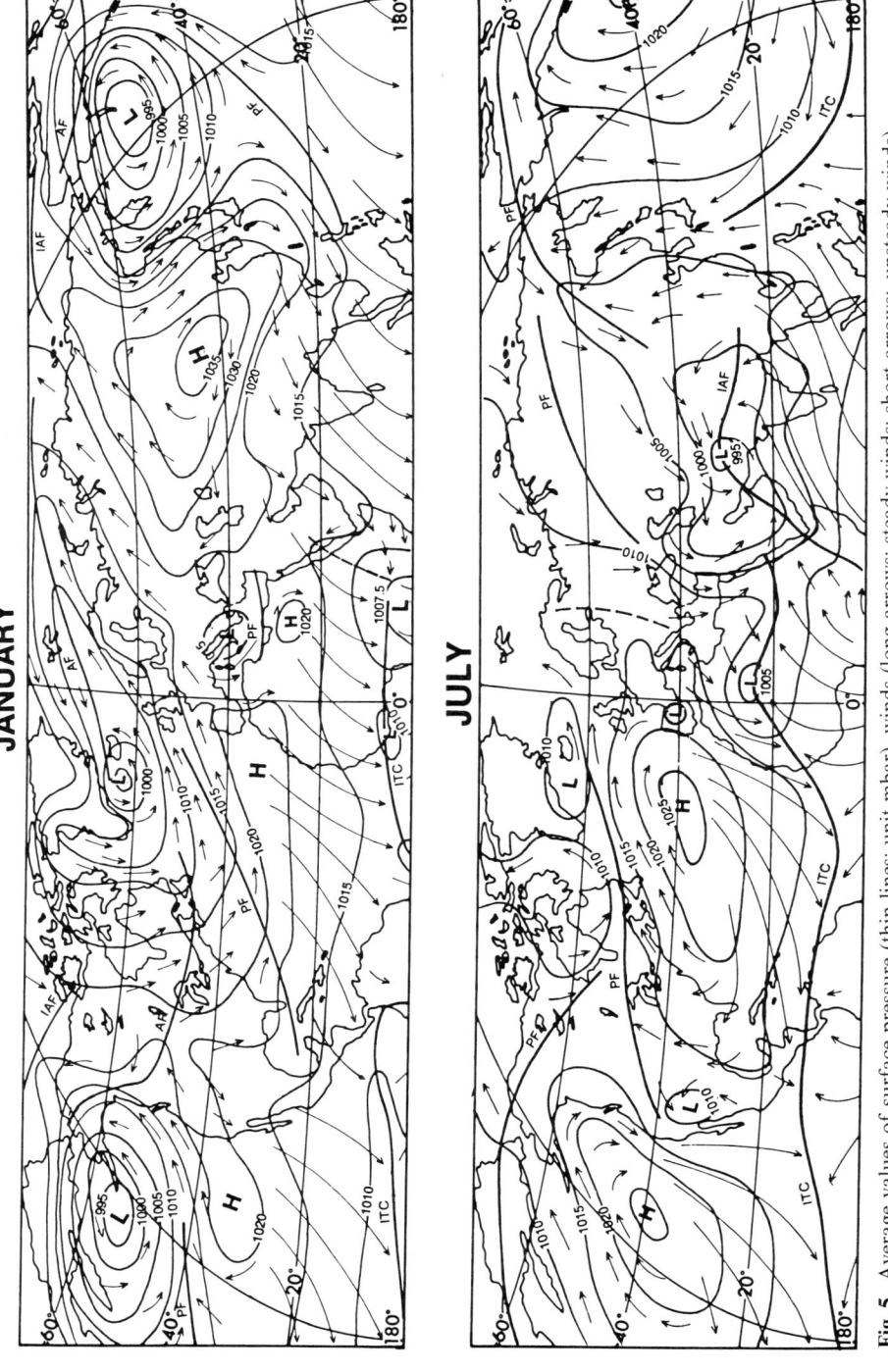

**Fig. 5.** Average values of surface pressure (thin lines; unit mbar), winds (long arrows: steady winds; short arrows: unsteady winds), frontal zones and convergence zones in the northern hemisphere during January and July (from reference 89)

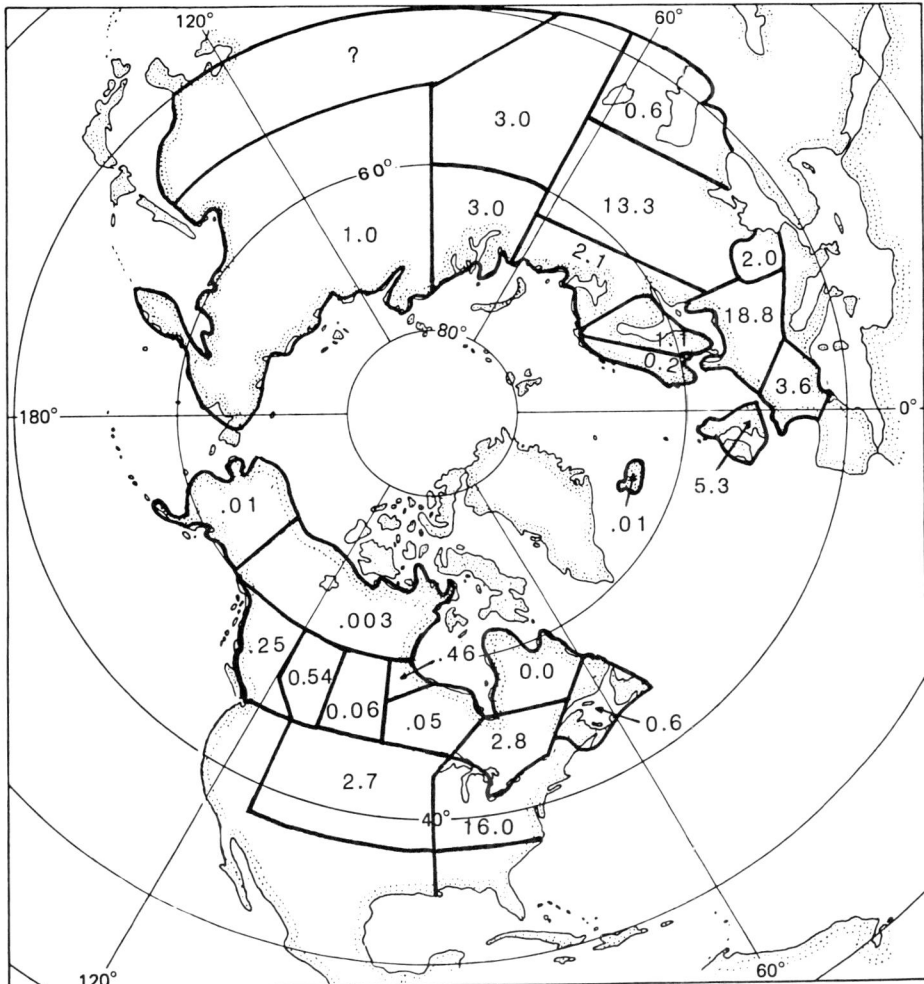

**Fig. 6.** Annual emission of $SO_2$ ($10^6$ tonnes) in regions of the northern hemisphere that influence the Arctic (from reference 8)

is stronger than the North American source. Their analysis found that outbreaks of arctic haze are preceded by surges of air out of Eurasia traveling northward. These in turn were associated with weather patterns that involved the entrance of a low-pressure system from the region of Iceland into the Murmansk area or across Europe; in either case the cyclonic system runs up against the Siberian high and the northward corridor of air between the high and low pressure system is maintained, typically, for a few days. It seems that the greatest fraction of midlatitude pollution that enters the Arctic does so during these events of northern-directed surges. A somewhat schematic representation of such a transport pathway carrying polluted European air to the Arctic is shown in Fig. 7.

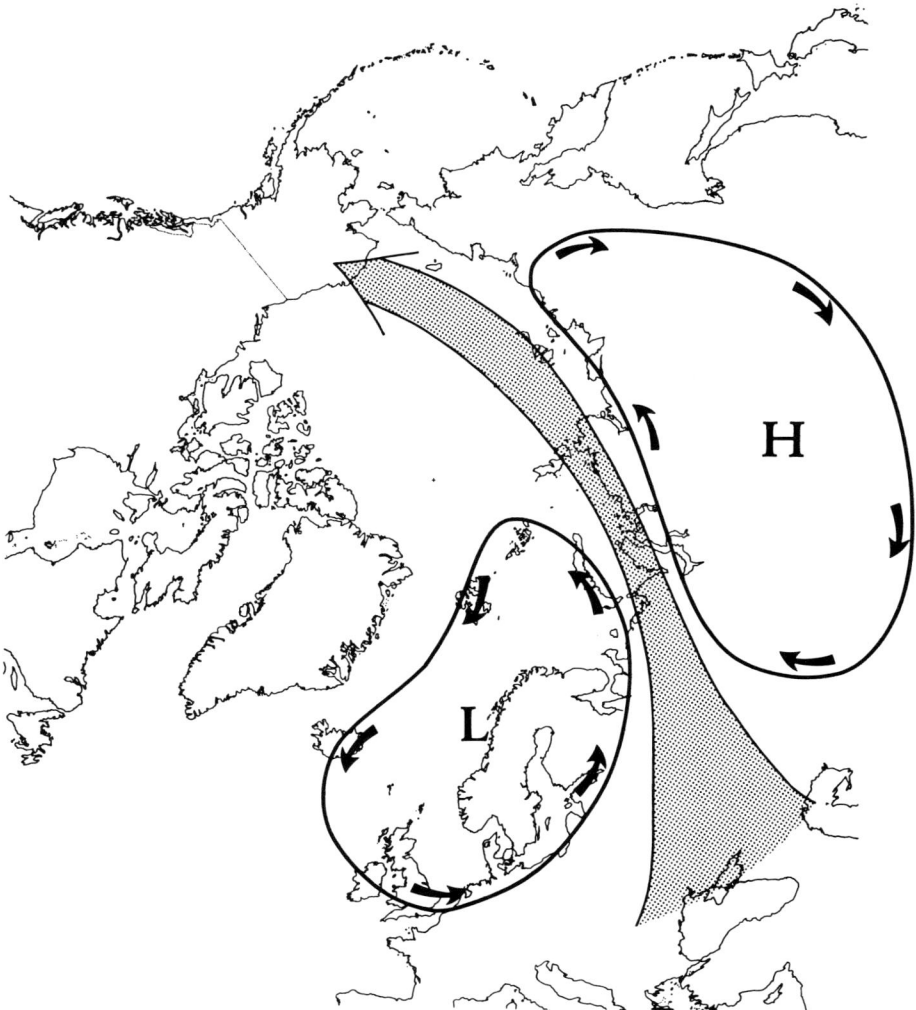

**Fig. 7.** One of the major transport pathways of arctic haze pollution (from reference 9)

## Phenomenology of the Arctic Haze

### Seasonal Variability

By the mid-1970s, from research at Barrow, it was realized that nearly every aerosol constituent (e.g., total mass or mass loading of individual elements, light-scattering coefficients, etc.) underwent strong seasonal variation, peaking strongly in late winter, around April, and nearly disappearing in summer. Similar trends were suspected from the sparse aerosol data available from other locations in the Arctic, but it really wasn't until a couple of years after the establishment of

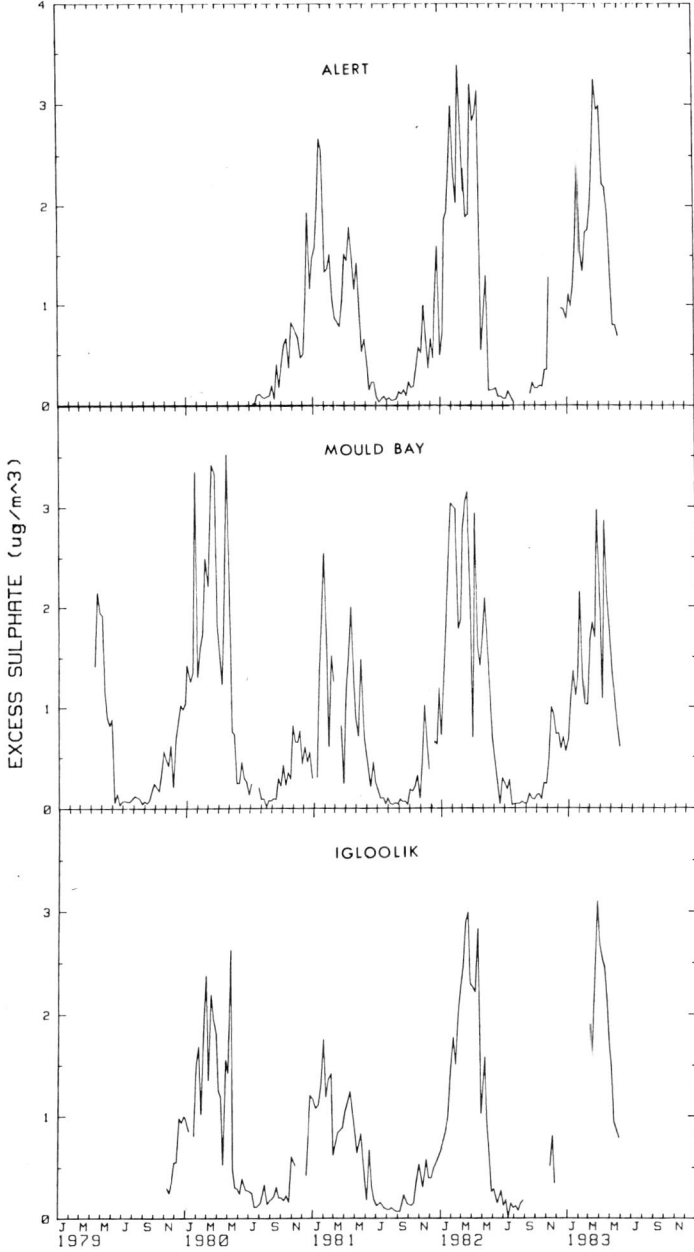

**Fig. 8.** Temporal variation of weekly mean sulfate in the atmosphere at three locations in the Canadian Arctic (from reference 31)

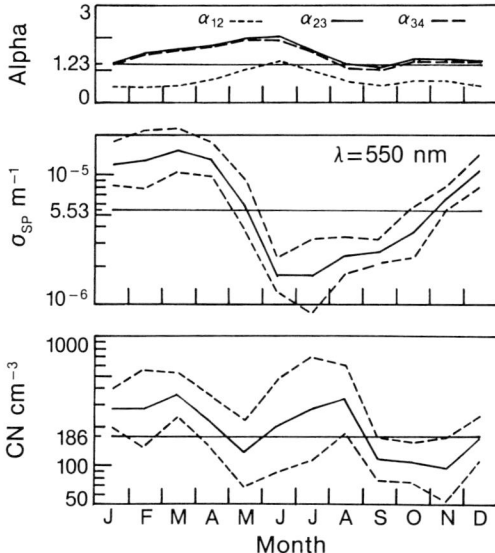

**Fig. 9.** Geometric means by month of CN concentrations (lower), $\sigma_{sp}$ at 550 nm (middle), and Angstrom exponent (upper), at Barrow Dashed lines: one standard deviation (from reference 90)

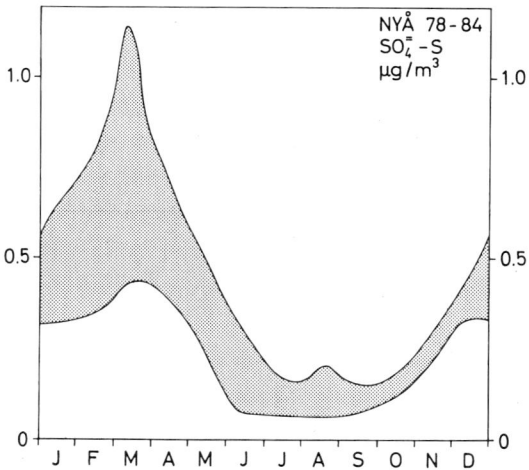

**Fig. 10.** Annual variation of sulfate as $SO^{2-}$–S on Spitzbergen (hatched) and sky-cover in % in the European Arctic (after reference 91)

the informal arctic network that it became apparent the seasonal variation was an Arctic-wide phenomenon.

During the 1980 symposium on arctic air chemistry at Rhode Island, the existence of the "spring maximum" and "summer minimum" were amply confirmed from a variety of measurements all over the Arctic, and all subsequent work has strengthened this finding. Its cause, however, has never been adequately explained, though there have been surmises.

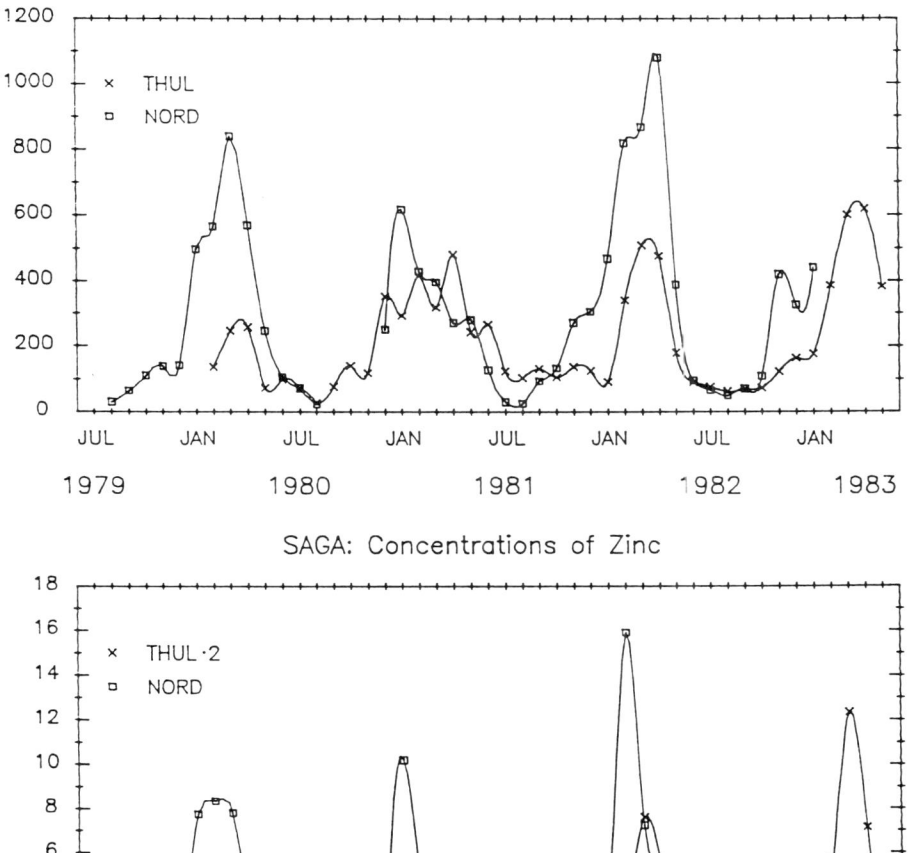

**Fig. 11.** Sulphur, zinc, monthly geometric mean concentrations (in ng m$^{-3}$) on the Greenland Ice Cap (reference 26)

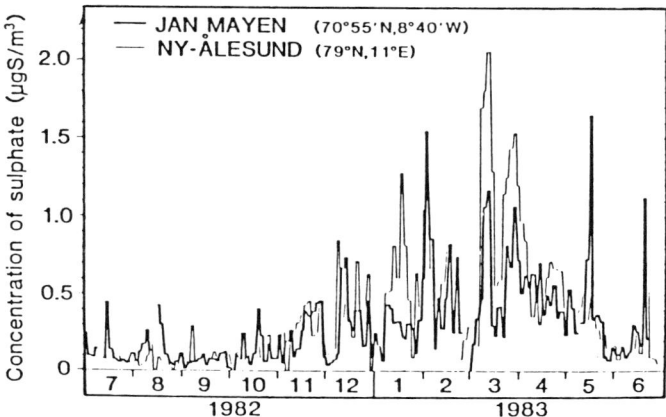

**Fig. 12.** Concentration of sulfate in atmospheric aerosols, Jan Mayen and at Ny-Ålesund, Norwegian Arctic. Values are means for a 2+2+3 days' sampling sequence applied each week (from reference 92)

The temporal variation of sulphate in the atmosphere at three locations in the Canadian Arctic is demonstrated in Fig. 8, from which it can be seen that all locations exhibit spectacular and nearly identical rhythmic oscillations, with $SO^{2-}$ rising in the winter months and abruptly falling off around May, after which it disappears for several months. The persistence of more or less the same pattern year after year, at all stations, is striking and made all the more impressive by bringing in data from other locations in the Arctic. For example, the light-scattering coefficient at Barrow (Fig. 9) and the sulfur at Spitzbergen (Fig. 10) vary in nearly an identical fashion, as does S, Zn, Mn and Ca on the Greenland Ice Sheet (Fig. 11).

A more detailed presentation of research results, covering the spring maximum at Jan Mayen and Ny Ålesund in the Norwegian Arctic, is shown in Fig. 12 to illustrate the good correlatability of the aerosol haze record at the two stations.

**Vertical Morphology**

Some rather limited information is available about the vertical structure of the arctic haze at around the time of the spring maximum. Measurements of the in situ optical extinction coefficient with airborne sun photometers [4] are in quantitative agreement with a number of vertical profiles of optical scattering coefficient for dry air made with nephelometry or with other forms of light-scattering measurement devices [10, 11, 12]. The light-scattering coefficient values as a function of altitude shown in Fig. 13 are representative. The haze is normally confined to the lower half of the troposphere (beneath 4 to 5 km) and is frequently strongly banded, as shown in the figure. Barrie [8] computed the mean vertical profile of 23 samples: Barrie's mean profile showed concentrations decreasing approximately exponentially with increasing elevation to about 2 km,

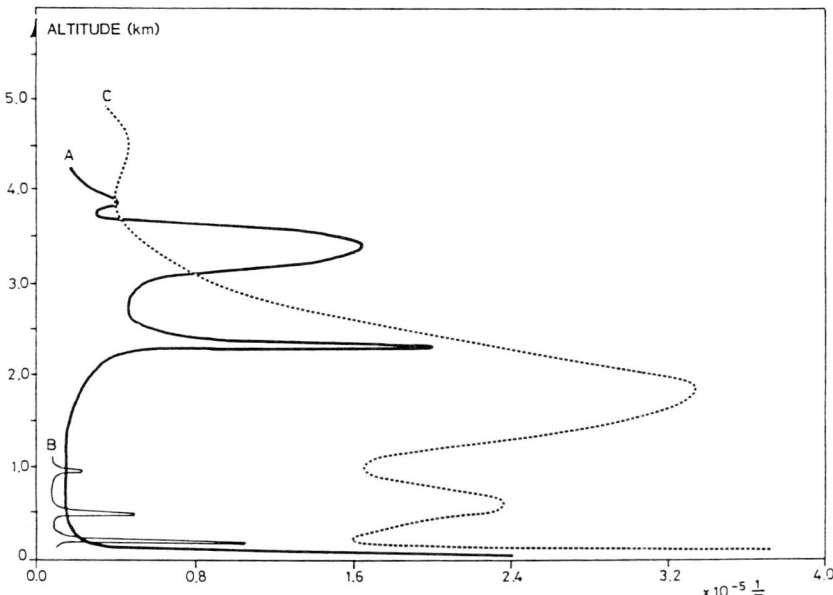

**Fig. 13.** Light scattering coefficient, $\sigma_{sp}$, vs. flight levels in the Norwegian Arctic on (A) 18 August 1983; (B) 25 August 1983; and (C) March 1984 (from reference 12)

above which the haze concentration was roughly constant to 5 km, then fell off to negligible values in the upper half of the troposphere.

### Size Distribution of Arctic Haze

The size distribution ar "probability distribution" or microscopic particles of the arctic haze can provide important insight into the sources of the particles and into nucleation and removal processes in the atmosphere.

Most measurements of the number size distribution indicate a rapoid fall-off of particle concentration with increasing particle diameter [13, 14, 15]. The smaller aerosols, sometimes referred to as Aitken particles, are much more variable in space and time.

Though the size distribution of arctic haze undergoes complex variations, two frequently observed distribution types are illustrated in Fig. 14. In the figure, the top box refers to the number size distribution, while the bottom box shows the distribution of aerosol mass or volume. The heavy line represents distribution typically found in Alaska when air is flowing in from the northern Pacific; such air is frequently depleted in larger particles, as shown, but charged with numerous very small particles (up to several thousand particles cm$^{-3}$). The small Aitken particles are presumably produced by photalysis in the sunlit regions of the oceanic regions south of Alaska.

Arctic air masses (the finer line curves in Fig. 14) typically contain significantly larger concentrations of particles around a tenth to a few tenths of a micron

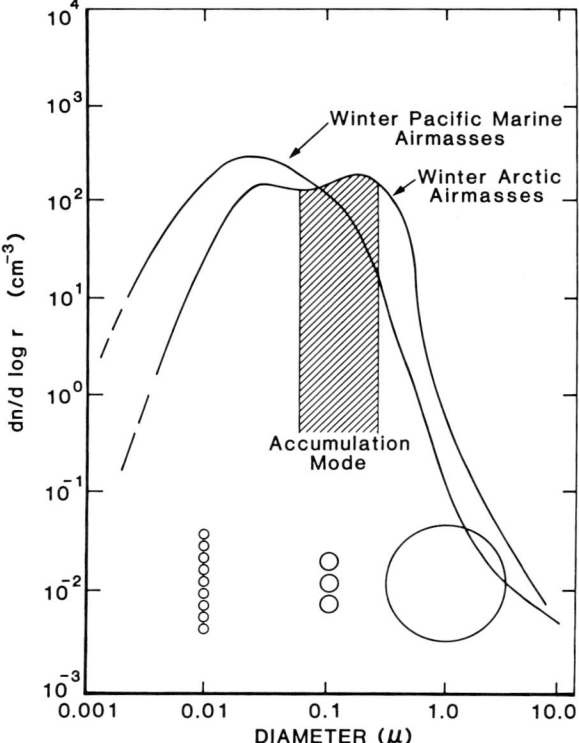

**Fig. 14.** A synthesis of aerosol size distribution (top) for cold, arctic-derived and warm, Pacific-derived marine air masses in Alaska. Note that cold air is depleted in CN, but enhanced in larger particles

diameter, but fewer of the small Aitken particles. As a result, the total particle number concentration in arctic air masses is substantially lower (i.e. 350 cm$^{-3}$) than it is for the northern Pacific air mass systems, which sometimes contain up to several thousand particles cm$^{-3}$, but of very small sizes.

The mass of particulate material is concentrated in a mode a few tenths of a micron in diameter and, as seen, this mode is much stronger in arctic-derived systems than in cleaner air masses representative of the northern Pacific air mass systems. It is this latter mode, sometimes called the accumulation mode, that contains the majority of particle mass and that dominates the chemistry and interacts most strongly with radiation [15, 16, 17].

In addition to the accumulation mode one also sometimes finds a larger mode of particles several microns in diameter (e.g. Fig. 15). Occasionally, "giant" (<10 μ) particles have been found, also, in arctic haze [18].

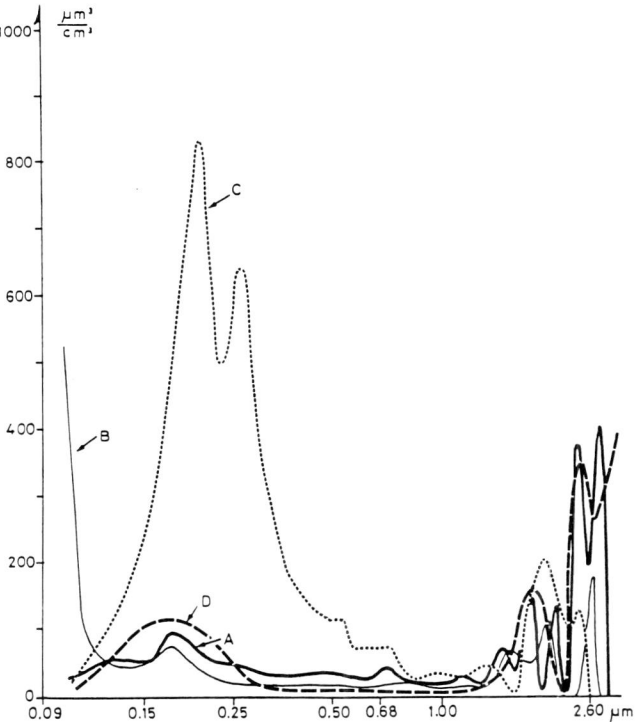

**Fig. 15.** Volume concentration of particles vs particle size: (A) at 3,300 m on 18 August 1983; (B) at 900 m on 25 August 1983; (C) at 1,800 m on 3 March 1984; (D) at 3,500 m on 3 March 1984 (from reference 12)

**Volatility-Hygroscopicity of Haze Particulates**

One of us (G. E. S.) has conducted studies of the volatility of the aerosols in arctic air masses during winter and spring. Some 90 to 98 percent of the accumulation mode aerosols (centered at $0.12 \pm 0.04$ μm mass mean radius) evaporated upon exposure to a temperature of 200 °C.

The larger-mode aerosol, centered at 1 to several micron radius, was virtually unaffected by passage through a temperature field up to 450 °C. Examination with optical and electron microscopes on impacted samples showed that the larger particles were mainly composed of crustal aggregates and individual particles, frequently with plate-like structure and concoidal fractures. X-ray maps with the scanning electron microscope images did indicate however, in addition to the usual Si, Al and Ca signatures, fairly strong S lines suggesting that these larger particles were coated with a sulfur-bearing substance, which, because of its ubiquity and low vapor pressure, was probably $H-SO_4$. During strong flow of Pacific marine air into central Alaska, marine compounds were identified, but these were not detected in arctic air masses.

Passage of the aerosol-laden air through pipes surrounded by wicks of excess salt solutions (to control the relative humidity) showed that the accumulation mode aerosols were hygroscopic and grew rapidly in size in elevated humidity; the experiments were not accurate enough to determine the hygroscopicity parameter, however.

Attempts to filter out and pass only the larger-mode particles and grow them in elevated humidity were unsuccessful.

To summarize, the arctic haze aerosol probability distribution seems to be best described by two mass modes, an accumulation mode of volatile particles centered around 0.12 μm and a much more variable, large crustal mode.

## Aerosol Chemistry

### Chemical Tracers

The chemistry of arctic haze is obviously tied to the nature of its sources. To complicate matters, however, there are significant chemical fractionation processes that go on in the atmosphere during the long transport to the Arctic. There is, in addition, an unusual photochemical situation, since sunlight is nearly constant in summer but absent in winter in the Arctic and subarctic (e.g., Fig. 2).

In spite of the complicating factors, the chemistry has provided important tracer information which has shed light on the source regions and atmosphere-transport pathways. One often overlooked characteristic of the arctic haze is that it occurs over a region free of particulate sources, and thus there is the resulting simplification that aerosols of nearby origin are not mixed in with aerosols transported in from afar. Nevertheless, it should be kept in mind that it is by no means a trivial matter to deduce the source of haze events, since the process of meteorological hindcasting (back trajectories) is a very uncertain exercise over the vast distances involved.

The first use of chemistry as a tracer was in the work at Barrow in 1976, which was described earlier. This first study indicated an origin in sandstorms over the Asian deserts. Subsequent samples were found to contain a variety of enriched (with respect to crustal material using aluminium as reference) elements which are known to be produced by industrial pollution. A tracer system to help elucidate sources on a geographic basis was proposed by Rahn [19, 20], and used the Mn/V ratio, from which it was possible to partition source-transport pathways to the high Arctic from Europe and the U.S.S.R. Raatz and Shaw [9] went on to use Rahn's time series of Mn and V data at Barrow to infer source regions for haze episodes and constructed a series of principal synoptic configurations and transport pathways from North America, Europe and Asia, which were remarkably consistent with those sources deduced from the elemental data. Arguments were made in the early 1980s, based on the chemistry, that the U.S.S.R. must be a major source region of the haze [21, 22].

A considerable improvement in tracer ability based on the chemistry was put forward by Lowenthal and Rahn [23] who introduced a 7 element tracer scheme, using elements that primarily are found on submicron aerosols and which, accordingly, tend to be preserved during long transport (large particles are removed rapidly near their sources). With the use of this more sophisticated tracer system, Rahn and Lowenthal deduced that Eurasia was the dominent source of aerosol to Barrow during winter 1979–1980: only 5%–10% of the aerosol could be attributed to North American sources. Within Eurasia, the central Soviet Union contributed an estimated 75% of the tracer elements, with the remainder comming from Europe.

Pacyna and Ottar [24] and Heidam [25, 26] employed crustal enrichment factors along with synoptic analysis to argue that chemistry is variable for haze episodes in the Greenland and Norwegian Arctic and depends on source regions; for example, metals tend preferentially to come from the industrial belt alongside the Urals.

Davidsen et al. [27], on the basis of chemistry, presented evidence that the Dye 3 station on the Greenland ice cap receives material from Eurasia in an over-the-pole transport pathway.

Using a somewhat different chemical approach, Sheridan and Musselman [28] determined the composition of individual particles collected during the AGASP (Arctic Gas and Aerosol Sampling Program) flights in the Alaskan Arctic. Sulfuric acid predominated in the submicron particles, while larger particles included Zn-Cu-Pb, graphitic carbon, coal and fly ash, and were correlated with trajectories from industrial belts in the Soviet Union [29, 30].

By coupling meteorology with observation of $^{210}$Pb, ions and cations and elements in aerosol at stations in the Canadian Arctic, Barrie and Hoff [31] and Graustein and Barrie [32] were able to construct major sources of individual species, including
1. Anthropogenic constituents Cr, Cu, Mn, Ni, Pb, Sr, V, Zn, $SO_4$, $No_3$;
2. Halogens (excepting Cl);
3. Sea salt elements Na, Mg, Cl; and
4. Soil constituents Al, Ba, Ca, Fe, and Ti.

The soil and sea salt constituents tend to reside on supermicron sized particles.

**Bromine**

Bromine concentrations in the arctic troposphere are the highest of any non-urban region ever found from mid-February to mid-May [33, 34]. Some portion, perhaps up to nearly half, is in the form of gaseous bromoform [34, 35], but there remain considerable questions about possible bromine gas-artifacts on filters. The source of the Br is unknown, but a likely source is the Arctic Ocean. Shaw [36] found that Br was the most independent variable of a suite of 13 elements measured in central Alaska and deduced a possible pollution source for this element.

**Sulfate**

On the basis of its mass, nonmarine sulfate is the most important species in arctic haze, constituting more than half the mass of the arctic pollution aerosol. It is found at most locations in the Arctic in concentrations ranging from about half

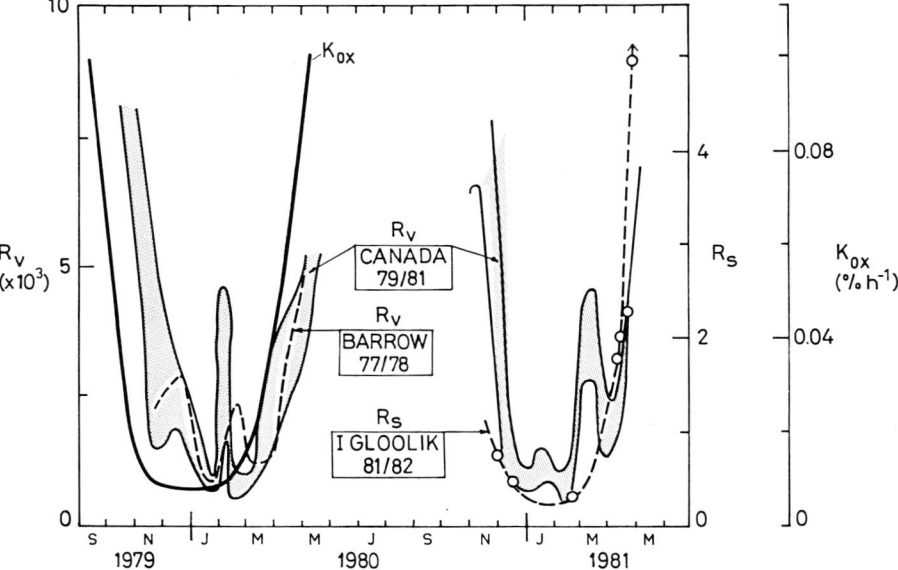

**Fig. 16.** Summary of the observed temporal variation of $R_v$ (E–S$^{2-}$/E–V) and $R_s$ (E-SO$_4^{2-}$/SO$_2$) in the North American Arctic as well as of Kox, the mean SO$_2$ oxidation rate in the lower kilometer of the atmosphere at 55°N predicted by Altshuller (reference 39) (from reference 38)

to several μg m$^{-3}$ SO$_4^{2-}$. Ever since the earliest days of research into the phenomenon of arctic haze, it has been realized that there is more SO$_4^{2-}$ than can be accounted for by simply transporting particulate air pollution from its source to the Arctic intact. A large fraction of the SO$_4^{2-}$ is produced into the particulate phase by the conversion from a gaseous predecessor, presumably from photolyzed pollution-derived SO$_2$. However, this is not as simple as it might at first appear, since Zalabsky and Twomey [37] have shown that ultraviolet radiation alone is incapable of producing embryos of SO$_4^{2-}$ from trace SO$_2$ and water vapor unless the radiation is very short wavelength (e.g., 185 nm): there are virtually no photons of this wavelength in the troposphere. The conversion of SO$_2$ to SO$_4^{2-}$ is complex and poorly understood.

By considering time series of both particulate nonmarine SO$_4^{2-}$ and gaseous SO$_2$, along with noncrustal vanadium, which can be used as a tracer of primary pollution, Barrie and Hoff [38] constructed a model with which they were able to deduce the oxidation rate of SO$_2$ in the arctic atmosphere. In this very important paper, the authors also derived the mean transport time of polluted air from the mid-latitudes to the Canadian Arctic. The essential features of the model are illustrated in Fig. 16 which shows nonmarine or "excess" sulfate expressed two different ways: 1) in terms of the molar ratio to vanadium, Rv, and 2) in terms of (to the ratio) sulfur dioxide, Rs. Notice that both Rv and Rs have deep minima in the months January to March, implying that during these dark months there is relatively little converted secondary sulfate. This makes sense because of the near total absence of sunlight (e.g., Fig. 2). In the sunlit months the excess $SO_4^{2-}$ (expressed in terms of V, Rv or Rs) rises considerably.

There is an interesting and important asymmetry with respect to the seasonal change of solar illumination. Solar illumination is directly related to the mean $SO_2$ oxidation rate in the lower kilometer of the atmosphere, as calculated by Altshuller [39] in a model involving kinetics of OH, $HO_2$ and $CH_3O_2$, which is shown as the curve KOx in figure 16, over 2–5 weeks. Barrie and Hoff showed that 1) selected scavenging, 2) transit time of pollutants from source and 3) seasonal variations of Rv were negligible in comparison to 4) the $SO_2$ oxidation rate en route, and the latter could be derived from the shape of the Rv curve. To do this, Barrie and Hoff used a simple Lagrangian model that considered conversion of $SO_2$ to $SO_4^{2-}$ with a rate constant K and loss of $SO_4^{2-}$, V and $SO_2$ by dry deposition en route from source to receptor. They concluded that the lower estimates for the mean $SO_2$ oxidation rate between Eurasian mid-latitudinal sources and the North American Arctic is 0.1% $h^{-1}$ in early December, 0.4% $h^{-1}$ in late February and 0.1%–0.2% $h^{-1}$ in early April. If the oxidation takes place in the first 5 days of travel, as would be expected for photochemical mechanisms, the oxidation rates would be 2.3–3.3 times higher. Photochemical $SO_2$ oxidation mechanisms involving OH, $HO_2$ and $CH_3O_2$, and radicals are insufficient by an order of magnitude to explain those deduced by Barrie and Hoff; they surmise that reactions involving hydrocarbons and aerosols, perhaps those containing graphitic carbon, increase the sulfur dioxide conversion rate.

**Size-Dependent Chemistry of Arctic Haze**

There have been a number of investigations of the size dependence of chemical composition, mainly in terms of elemental tracers since these can be easily derived by analyses of aerosols size segregated with multi-stage impactors [40, 16, 41, 17]. The results have not been substantially different from the size-dependent aerosol elements compiled by Rahn [42] for aerosols from a wide variety of locations. Na, Mg, Al, Mn, Ca, Ti, Fe and Si are preferentially in the supermicron coarse mode, whereas As, Si, V, Cr, Pb and Br are preferentially located on particles in the submicron mode. The accumulation mode aerosols have a cation-anion budget balance mainly of $H^+$, $NH_4^+$ and $SO_4^{2-}$ [41].

**Light-Absorbing Carbon**

Graphitic carbon has been identified as being present in arctic haze from Raman scattering (e.g., Fig. 17). It is the predominant species responsible for optical absorption in the visible region of the optical spectrum where the majority of solar radiation lies. The graphitic carbon, therefore, is of concern because it "colors" the haze and, when placed over a reflecting polar ice sheet, would lead to a heating of the earth's atmosphere system.

The carbon has been deduced to be congregated in the finer of the two aerosol mass modes. Graphitic carbon normally consists of agglomerates of spheres, each around a hundredth of a micron, the resultant agglomerate being in a mass mean mode around a tenth of a micron.

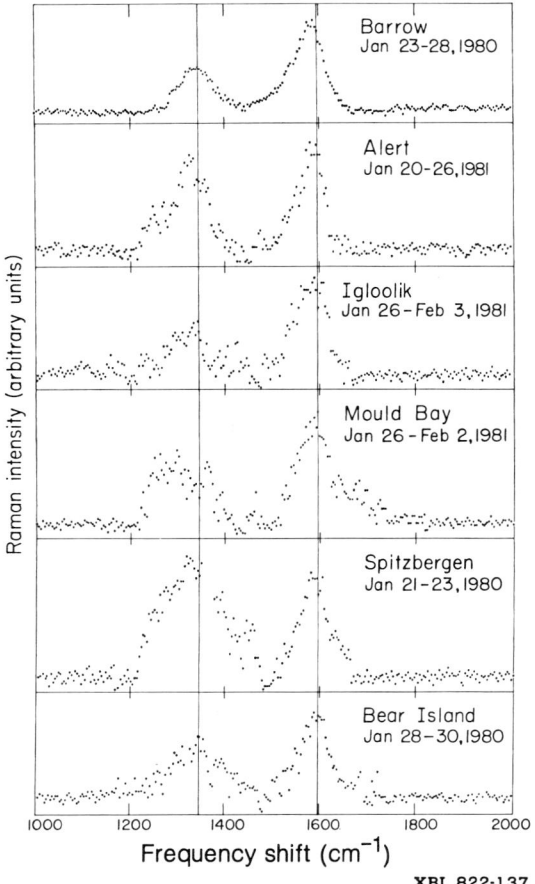

**Fig. 17.** Raman spectra of particles collected in Alaskan Arctic (Barrow) compared with samples collected in the Canadian Arctic (Alert, Igloolik, Mould Bay) and the Norwegian Arctic (Spitzbergen, Bear Island) (from reference 64)

Figure 18 shows that the graphite carbon tends to have a similar vertical distribution in the atmosphere as that of the mass or light-scattering properties of arctic haze, mainly a tendency to occur in the lower atmosphere layers (but above the near-surface inversion) of about 2 km. The profile in Fig. 18 was taken on one of the AGASP (Arctic Gas and Aerosol Sampling) experiments.

## Trace Gases and Arctic Haze

It is convenient to divide the trace gases related to arctic haze into three categories: (1) trace gases in the Arctic that do not cause arctic haze, but serve as indicators (tracers) of the types of sources that may contribute to arctic haze or

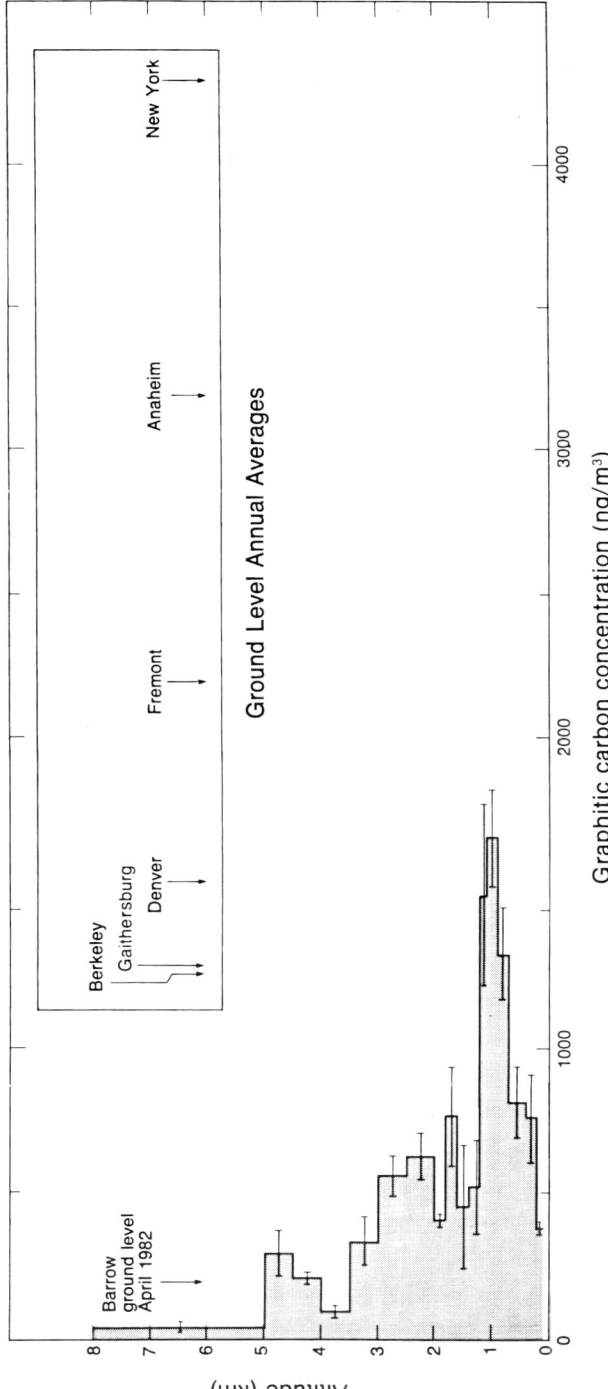

**Fig. 18.** Vertical profile of graphitic carbon concentrations (ng/m$^3$) on 31 March 1983 at about 74°N, 25°E. Shown for comparison are the annual average ground-level graphitic carbon concentrations at US urban locations, and the mean April 1982 ground-level values at the NOAA-GMCC observatory, Barrow (from reference 64)

where such sources may be located [43]; (2) trace gases that can disturb arctic atmospheric chemistry or the radiation balance; and (3) gases that may be precursors of some of the arctic haze aerosol. The tracers have been studied more extensively than the other classes and are the main subject of this section.

If arctic haze or its precursors orginate from anthropogenic sources, then trace gases characteristic of massive industrial or agricultural activities should be present in unusually high quantities in the haze and in the Arctic during the seasons when arctic haze is seen. The first indications of this occurrence were documented by Khalil and Rasmussen [43]. They showed that at Barrow, Alaska, concentrations of chlorofluorocarbons ($CCl_3F$ (F-11), $CCl_2F_2$ (F-12), $CHClF_2$ (F-22), chlorocarbons ($CH_3CCl_3$, $C_2Cl_4$, $C_2HCl_3$, $CH_3Cl$) and carbon monoxide were higher during the winter and early spring compared with other seasons. This behavior is approximately the same as excess sulfate and vanadium at Barrow that were the original indicators of the anthropogenic origins of arctic haze [44]. For most gases, highest concentrations were observed in February. The amount of excess is inversely proportional to the average atmospheric lifetimes of these gases as shown in Fig. 19 [43]. These results established the criteria and methodol-

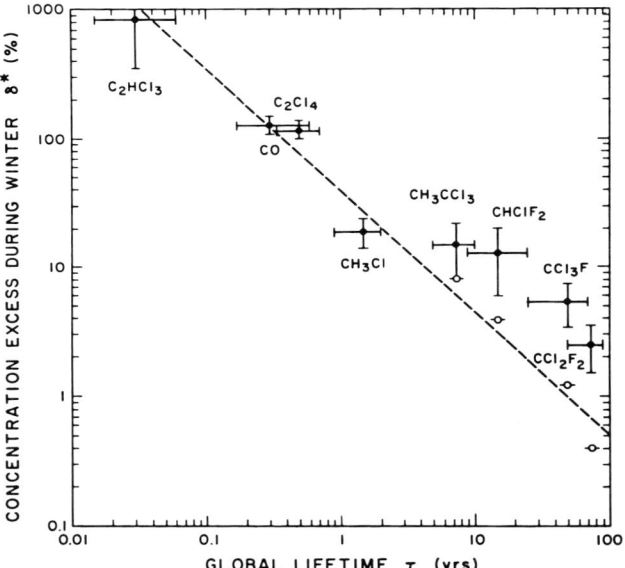

**Fig. 19.** The relationship between relative concentration variation of trace gases in the arctic atmosphere ($\delta^* = 100$ $(C_{max} - C_{min})/C_{min}$) and the atmospheric lifetime $\tau$. Error bars are approximate 90% confidence limits of $\delta^*$. The inverse relationship between $\delta^*$ and lifetime is explained by the seasonal variations of production, loss and transport of tracer gases from distant origins to the Arctic. The background concentrations ($\sim C_{min}$) of long-lived trace gases are large and hence more difficult to perturb, reducing $\delta^*$. A more complete explanation is offered in the text. The best tracers of arctic haze are gases with large, measurable winter excesses $\delta^*$. The figure demonstrates that such conditions require that the tracers either be short-lived ($\tau \leq 5$ years) or their concentrations be far from equilibrium with current annual emissions. The open circles are values of $\delta^*$ after the effects of long-term annual increases of sources and background concentrations have been removed from the observed $\delta^*$.

ogy for using trace gases as indicators of both the regions where arctic haze originates and the type of sources that may be involved. The chlorocarbons mentioned above, F-22, and CO are removed from the atmosphere by reactions with OH radicals. Because OH concentrations undergo large seasonal variations at high latitudes the concentrations of these gases are likely to be highest during winters compared with summers. However, the observed variations exceeded those expected from the OH cycle. Moreover, the chlorofluorocarbons F-11 and F-12 do not react with OH, and their cycles are likely to be caused entirely by more air pollution reaching the Arctic during the winters. Samples collected from a small commercial aircraft soon established that high concentrations of man-made trace gases, particularly nonmethane hydrocarbons, occurred in layers presumably also containing aerosol particles responsible for arctic haze [45].

The high concentrations of man-made trace gases in the haze layers was documented in a comprehensive set of experiments called the Arctic Gas and Aerosol Sampling Program (AGASP) in which fully instrumented research aircraft were used to determine the physical and chemical characteristics of arctic haze [46]. The existence of haze layers was determined by condensation nuclei counts and light scattering. Flask samples collected within the haze layers showed significantly higher concentrations of dozens of man-made gases compared with concentrations outside the haze layers [47, 48]. These results are summarized in Fig. 20 [47]. There are a number of significant features in these results and in similar studies. First, the relative excesses of concentrations in and out of the haze layers were much greater for shorter-lived species compared with longer-lived gases as shown in Fig. 21. Khalil and Rasmussen [47] explained "... that the background concentration of short lived gases is generally very small, especially at remote locations such as the Arctic. During the winter and spring, when arctic haze occurs, these gases may have long lifetimes, expecially if they are removed primarily by OH radicals, which are expected to be at much lower concentrations during winter than at other times of the year. If these gases arise from human activities, then they can survive the long distance transport to the Arctic during the winter and spring in the same layers of polluted air that contain the particles responsible for arctic haze... Since the background concentrations are small, delta (the relative excess) will be large." Second, as shown in Fig. 2, excesses of gases such as $N_2O$ and $CO_2$, as well as the $C_2$–$C_6$ hydrocarbons and other gases, suggest that high temperature combustion sources contribute to arctic haze. High concentrations of $CO_2$ in arctic haze were also reported by Conway et al. [49]. Halter et al. [50] analyzed episodes of sporadic high $CO_2$ levels at Barrow during the winter and spring of 1979–1980. They found that high concentrations of $CO_2$ were related to arctic haze since they occurred at the same times as high concentrations of excess vanadium and excess sulfate. Moreover, back trajectory analyses suggested that the these polluted air masses often originated in western U.S.S.R. or Europe brought by air streams within about 10 days. Third, Kahil and Rasmussen [47] found that most of the man-made gases were highly correlated with each other. No distinct clusters were found suggesting that the signatures of specific sources may be erased by transport, mixing, and chemical processes. It is noteworthy, however, that trace gases such as the fire extinguishing compound $CF_3Br$ and the solvent F-113, both of which are used in the high

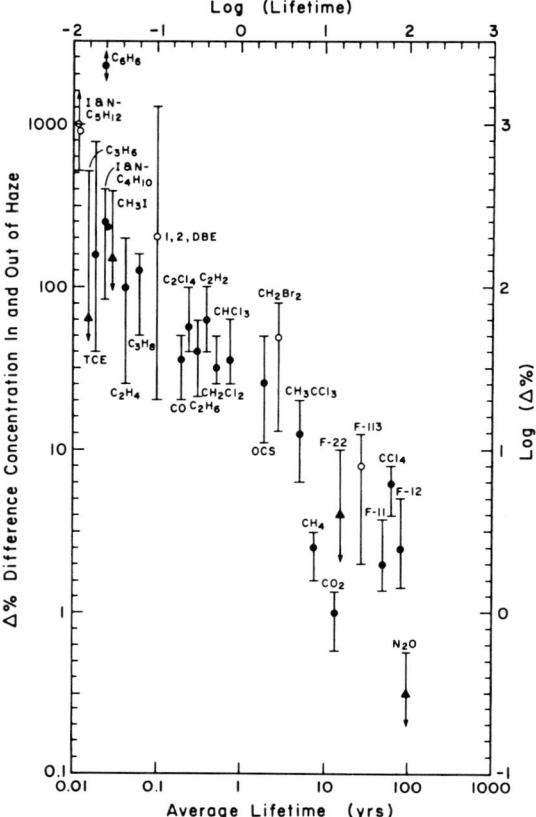

**Fig. 20.** The comparison of trace gas concentrations in the regions of arctic haze and in clean air. The regions of haze and clean air are as shown in Figure 1. The points represent estimates of the percent difference, $\Delta\% = 100\,[C(\text{background})/C(\text{haze})-1]$, and the vertical bars are estimates of the approximate 90% confidence limits of $\Delta\%$. The horizontal axis is the average atmospheric lifetimes. The estimated lifetimes of the trace gases are approximate. In a few cases there was insufficient information to determine the lifetimes. Such gases are represented by open circles. The gases represented by triangles have $\Delta\%$'s that are only marginally greater than $0$ ($\alpha = 0.05$).

technology industries of the United States, were not significantly elevated in the haze layers compared with background levels in the Arctic. These results suggested that the U.S. was not a contributor to arctic haze, at least in 1983 when this experiment was conducted. From these studies it was concluded that the Soviet Union and Europe were the likely regions where these trace gases originated because these regions have large populations and industrial development at latitudes above 50°N. In the same context, measurements taken in Spitzbergen (Norway) showed that during summers the concentrations of man-made trace gases were about the same as at Barrow, Alaska. But in spring, concentrations of man-made trace gases were significantly higher at Spitzbergen than at Barrow (Fig. 21). Spitzbergen is near the European side of the Arctic

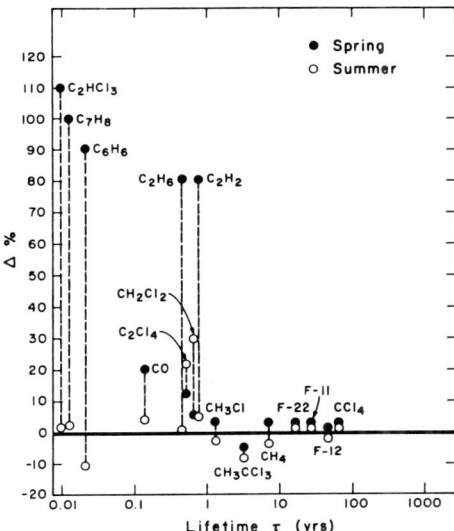

**Fig. 21.** Spatial variation of the gaseous tracers of arctic haze. The relative difference $\Delta$ of concentrations C of trace gases observed at Spitzbergen, Norway, and at Barrow are shown as a function of the average atmospheric lifetime of the trace gases. The open circles show $\Delta$ during summer (August–September); the dark circles represent $\Delta$ during spring (March–April). $\Delta\% = 100 \times [C(\text{Spitzbergen}) - C(\text{BRW})]/C(\text{BRW})$.

(about 20°E by 78°N) and is therefore much closer to the Soviet Union and the industrial regions of Europe than is Barrow, which is on the North American continent on the other side of the pole (about 157°W by 71°N). This may explain the greater pollution at Spitzbergen during the season of arctic haze if it originates from Eurasia, particularly in view of some of the transport pathways discussed by Shaw [51] and Raatz and Shaw [9]. High springtime concentrations of organic trace gases in the Norwegian Arctic and their origins from the Soviet Union have been also been reported by Hov et al. [52] and of polychlorinated hydrocarbons (pentachlorobenzine, hexachlorobenzine, alpha and gamma-hexachlorosyslohexanes) have been reported by Oehme & Ottar [53]. The findings of these studies do not pinpoint where the pollution in the Arctic comes from, but they suggest that it comes from Eurasia, or perhaps from even nearer sources and that combustion processes contribute significantly. These results are in agreement with analyses of the chemistry of the arctic haze aerosol [81].

The other categories of gases mentioned above have not been studied as extensively as the tracers. Chlorine- and bromine-containing gases, for instance, may destroy ozone in the arctic atmosphere [54, 55]. Berg et al. [33] reported particulate (excess) bromine levels of 70–100 $\times \text{ng/m}^3$ at Barrow, Alaska, during the springs of 1976–1980. During the same period even higher amounts of gaseous bromine were measured with an average concentration of about 118 pptv and gas-to-particle ratios of 7 to 18 by mass [33]. Subsequent work on identifying specific gaseous bromine species has shown much smaller seasonal variations for the man-made bromine-containing trace gases at least for the species that have been measured and lower particulate bromine levels as well [35, 56]. Rasmussen & Khalil [35] reported time series of $CH_2BrCH_2Br$, $CBrClF_2$, $CH_2BrCl$, $CH_2Br_2$, and $CH_3Br$ at Barrow and concentrations of the same gases in the arctic haze layers. The total bromine in these gases, excluding $CH_3Br$, varied by about 10% between the season of peak concentrations in winter or spring compared with the lowest values during summer. Of these gases only $C_2H_4Br_2$, $CH_2BrCl$, and $CH_2Br_2$ were more concentrated in the haze layers than in adjacent clean air. These gases also undergo the largest seasonal variations at Barrow with highest concentrations during the period of arctic haze [35]. Berg et al. [34] reported the concentrations of some of the same gases. Their study included concentrations of bromoform ($CHBr_3$) at about 15 pptv on average with a modal value of around 6 pptv which shows the non-gaussian distribution and the highly variable nature of the observed concentrations. Khalil & Rasmussen [57] also found about 15 pptv of $CHBr_3$ at Barrow in subsequent experiments. Bromoform at about 45 pptv of bromine, therefore, makes up more than half of the total gaseous bromine in the arctic troposphere (reported to be about 80 pptv as bromine [57]). However, bromoform is believed to be of natural origin and not much, if at all influenced by human activities. Therefore, it is possible that the catalytic destruction of ozone involving the BrOx cycle in the lower arctic troposphere is dominated by bromoform and thus of natural origin [55]. The longer lived man-made bromine-containing trace gases are likely to have a greater effect on arctic stratospheric ozone.

Perhaps $SO_2$ is of greatest significance among the gases that may directly affect the arctic haze aerosol because some $SO_2$ converts to particulate sulfate ($SO_4^{2-}$) causing some of the arctic haze. This conversion may take place near the source regions, along the transport pathways or in the Arctic when springtime brings sunshine to the region. High concentrations of $SO_2$ are found in the Arctic during wintertime (Rahn et al., 1980). In reporting the measurements of $SO_2$ and $SO_4^{2-}$ in the Canadian Arctic Barrie and Hoff [58] write "... that the variation in the conversion of $SO_2$ to $SO_4^{2-}$ is controlled mainly by the (seasonal) oxidation rates of $SO_2$... The net effect is to maintain high levels of arctic haze for a longer time than if sulfates originated solely from primary particulate emissions."

In summary then, gases in the Arctic and in the arctic haze itself may serve as tracers of the regions where the haze or its precursors originate and as tracers of the types of sources that contribute to the haze. The likely regions are parts of Europe and the USSR or the activities of the USSR in or around the Arctic. Sources appear to related to high temperature combustion processes. Not as well studied are the effects of some gases on the arctic atmospheric chemistry. It is probable that both natural and man-made bromine and chlorine gases, brought to the Arctic with the haze, affect tropospheric and stratospheric ozone concentrations. Finally, gases such as $SO_2$ may convert to $SO_4^{2-}$ causing some of the arctic haze.

## Optics and Radiation

### Radiation Budget

Shaw and Stamnes [59] developed a simple monochromatic radiative transfer model to estimate the climatic forcing due to haze over the Arctic. They predicted a 1 °C/d heating (20–30 wm$^{-2}$) of the earth's atmosphere system during haze maximum around vernal equinox; the model also predicted a slight cooling (5 wm$^{-2}$) at the surface and therefore had the effect of leading to increased atmospheric stability. Porch and MacCracken [60] and Cess [61] employed more complete spectral models, and reached similar conclusions. All these models neglected atmospheric thermal radiation because of uncertainties in modeling the problem and the expectation of negligible effects to low infrared aerosol opacity.

Blanchet and List [62] extended the model to include the infrared radiative effects, along with the radiative properties of the soot aerosol contained in the snow, as modeled by Wiscombe and Warren [63]. They considered dry and high humidity (hazy) conditions to determine the effect of moisture on the radiative equilibrium. They solved the equations of radiative transfer with a Delta-Eddington approximation, and employed 490 wavelength intervals covering solar and terrestrial radiation spectral regions. This complete and sophisticated model indicated that the earth atmosphere system would heat up, but by about a factor of two lower than had been predicted in the earlier models. The previously expected surface cooling can be offset by direct to diffuse conversion in the haze and by absorption of sunlight in the contaminated snow.

According to Blanchet and List's model, deliquescent aerosol material in the temperature inversion can increase upwelling thermal infrared and therefore lead to a significant cooling term (although the solar heating in day predominates). Significant net warming can take place in dry optically thin aerosol layers, like those observed about the arctic boundary layers, provided the aerosols contain elemental carbon, as measurements by Rosen and Hansen [64, 65, 66] and Heintzenberg [67] indicate to be the case.

All these modeling exercises, of course, have to appeal to experimental measurements of the optical parameters. Measurements of optical depth were summarized for the Arctic as a function of seasonal wavelengths [68] and as a function of altitude [4] for the Alaskan Arctic. The optical depth undergoes a strong seasonal variation, with a maximum in spring and a minimum in summer, which once and for all puts an end to speculations that the summer troposphere might be filled with pollution, except just near the surface. In fact, the summer optical depth values in the Arctic are similar to those found on the antarctic ice sheet. The entire atmospheric column is very clear in summer.

During the spring maximum, the optical depth reaches levels similar to the rural midwestern United States and is about a factor of three higher than the seasonal mean at Tucson, Arizona.

Freund [69] developed a method by which a "weighted" optical depth of haze layer can be derived from climatological all-wave radiation measurements (global, reflected and diffuse solar and total net radiation). In addition, one can estimate the aerosol single-scattering albedo with the method. McGuffie et al.

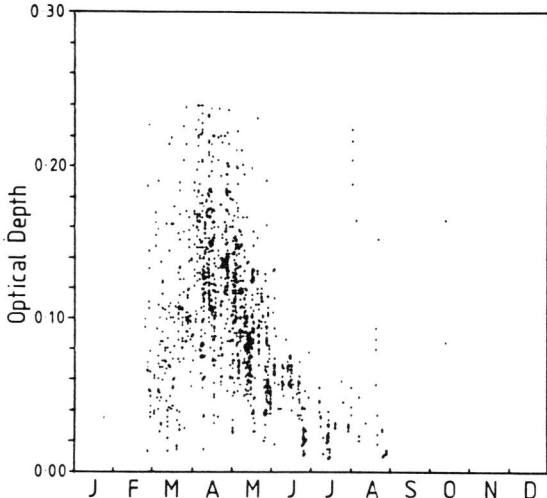

**Fig. 22.** Scatter diagram of aerosol optical depths derived by delta-Eddington direct method (individual measurements, 1970–1979) (from reference 70)

[70], using Freund's method on a ten-year data set, derived climatologically representative aerosol optical depths for Resolute (75°N, 95°W). The optical depth variation (Fig. 22) is very similar in shape and magnitude to that reported by Shaw [68] for the Alaskan Arctic. Note the presence of the strong spring maximum and the deep summer minimum. Spring maximum value for aerosol single-scattering albedo was deduced to be $0.85+0.02-0.04$, while that in June was $0.95+0.02-0.04$.

The parameters of interest for estimating radiative properties are the albedo of single scattering, $\omega$ (ratio of scattering to scattering plus optical absorption), and the asymmetry factor, g, which is a measure of the ratio of forward to backward scattering. Both parameters depend sensitively on the particle size distribution, and while g is slightly dependent on the particle composition (refractive index), it is crucial to have an accurate knowledge of light-absorbing graphitic carbon to estimate $\omega$. Total measurements of graphitic carbon can and have been made [63, 64, 66] but there are still rather major uncertainties in $\omega$ due to a lack of knowledge of whether the carbon particles are imbedded in or coated with films of sulfuric acid or are independent particles coexisting with droplets of $H_2SO_4$ (externally mixed).

In addition to atmospheric aerosols (and gases) disturbing the radiation balance, soot deposited in the snow also can have a significant effect, even if the soot is under the snow surface [62]. Clarke and Noone [71] measured black carbon in snow from a variety of locations in the Arctic; the concentrations were sufficient to reduce the snow albedo by from one to several percent, which is comparable to the perturbation caused by aerosol. Blanchet and List's model incorporated soot in the snow.

Arctic Haze

Valero and Ackerman [72] carried out measurements of the radiation field from an aircraft, including spectral measurements of upwelling and downwelling hemispheric radiation, during AGASP flights. Radiative heating was found within the haze layers, of the magnitude predicted with a radiative transfer model.

**Satellite Sensing**

An interesting possibility of sensing haze aerosol over the arctic basin by satellite was suggested by Shine et al. [73]. They analyzed low resolution (6 km) imagery from the broad band DMSP sensor, using the contrast between open water in polynyas and snow or fast ice. The ratios of system albedo of snow to water measured with the satellite (Fig. 23) are significantly smaller than would be the case for an aerosol-free atmosphere, and the effect is believed to be related to scattering and absorption of light by haze aerosol. Contrast is lowered by light scattering into the low albedo areas, while light both scatters out and is absorbed by the atmospheric path leading down to the lighter areas. As Fig. 20 indicates, the method is potentially quite sensitive and could in principle be extended to monitor the buildup and development of arctic air pollution over the entire polar cap.

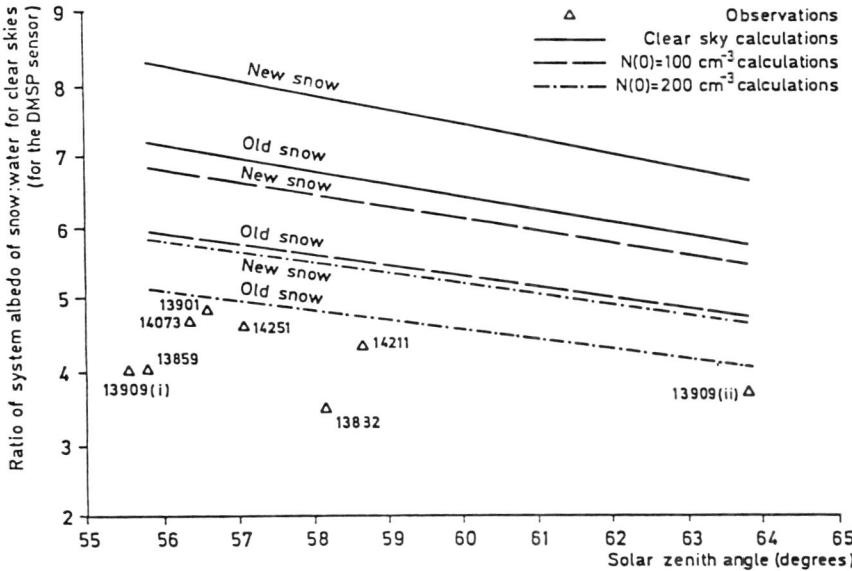

**Fig. 23.** Variation of ratio of the system albedo over snow to system albedo over water. Observations are marked by triangles and numbers refer to the identification number. Radiative transfer calculations for clear skies and two degrees of aerosol loading are given by the lines (from reference 73).

# Deposition

## Physics of Particle Removal in the Arctic

Arctic haze is such a large-scale (semiglobal) phenomenon just because the material remains suspended and does not get removed very rapidly. One can make an analogy to a situation in hydrostatics, where a head of water is developed, even in the event of only a trickle entering, by simply turning down the output leak. The more usual situation that we think of in air pollution, of course, is a large flow entering into the reservoir, but this isn't the case in arctic air pollution. The central paradigm of arctic air pollution, in fact, might be stated to be "lack of removal." A corollary is that removal by deposition must be small, as indeed most studies have found to be the case except for certain instances of sea salt material and crustal material generated near the sampling location. But if one considers the flux of imported pollution material down to the surface, the depositional rates are very small indeed.

Before going on to consider the experimental data on deposition, it may be useful to sketch out the major physical mechanisms of how material is removed from the atmosphere. There are four mechanisms for removal:

(1) Inertial removal: Particles slip across aerodynamic streamlines and impact on hydrometeors, surface obstacles or on other particles. Once contact ensues, Van der Waal forces assure adhesion.

(2) Diffuse removal: Particles are envisioned as heavy molecules (of tens or hundreds of thousands of atomic mass units) executing Brownian motion within their carrier gas. Thermodynamic equilibrium is assumed to hold. Einstein's relation $D = kTB$ relates mechanical mobility, $B$, to a diffusion coefficient, $D$, ($cm^2 s^{-1}$) (k and T are Boltzmann's constant and temperature). One considers the particles as diffusing to hydrometeors or surface obstacles. This process is more efficient at the smaller particle sizes and effectively mitigates against the concentration of extremely small particles, i.e., 0.001 μm, building up in the atmosphere.

(3) Water nucleation: Equilibrium vapor pressure of water is higher over a curved than a flat surface because of capillarity (the Kelvin effect), and the equilibrium saturation vapor pressure rises as the radius of the droplet decreases. Thus, when humid air in the free atmosphere cools to its dew point, no condensation occurs, but as the air mass slowly supersaturates, larger and larger particles serve preferentially as surfaces of condensation. As the process continues, the supersaturation eventually passes through a maximum at the point where the collective surface of nucleated droplets is sufficiently large to condense water at the rate at which it is produced in excess by further cooling or lifting of the air parcel. The result is a water cloud, with droplets in concentration of typically $10^2$–$10^3$ $cm^{-3}$. All the activated aerosols (cloud condensation nuclei, CCN) larger than those which can nucleate at the maximum supersaturation ($S_{max}$) are incorporated in droplets, which can proceed to coagulate and, in one out of every ten or so situations of cloud formation, to precipitate. Thus there is a transfer of aerosol mass from small to larger sizes and net removal from the atmospheric system down to the surface. Water-soluble particles of about one order of mag-

nitude smaller than insoluble particles are nucleated because of the decrease of equilibrium vapor pressure over a solution (Raoult's Law).

(4) Ice nucleation: In below-zero temperatures, particles with an atomic lattice structure similar to ice serve as nuclei, this being most probable at temperatures around $-15\,°C$. Since vapor pressure over an ice surface is slightly lower than vapor pressure over a water film, the more rapidly growing ice crystals, if present in sufficient concentration, may prevent smaller aerosols serving as CCN.

Particle removal rates from inertial mechanisms increase with particle radius, r, roughly as $r^2$, whereas diffusive removal increases roughly as $r^{-1}$ to $r^{-2}$. Both acting in concert therefore give rise to minimum particle removal, which turns out to be in the submicron regime. This phenomenon has variously been called the "Greenfield Gap," the "accumulation mode," "the removal resistant debris," etc. Thus if one were to inject an extremely heterodisperse population of particles into the complicated turbulent atmosphere, the fundamental processes of inertial and diffusive removal alone would eventually deplete all but a minority of the particles centered in a mode in the submicron particle range, the exact size depending on the original polydisperse size distribution and on the relative strengths of the inertial and diffusive processes, which are determined by the distribution of obstacles and energy in the atmosphere.

The additional transfer of particles onto or into water droplets or ice crystals by nucleation represents, in the long run, a preferential erosion of the larger sized particles, though in some cases with rapid cooling rates (strong upward convection), the process of nucleation could incorporate particles as small as a few hundredths of a micron.

**Particle Removal – Evolution Sequences for Arctic Air Pollution**

Pollution by-products injected into the atmosphere in the mid-latitudes typically have high concentrations of Aitken particles. The concentrations are in fact sufficiently high so that significant interparticle coagulation occurs on scales of kilometers and over times of hours. Upon undergoing transport to the Arctic, additional Aitken particles might appear from gas-to-particles production (e.g. organics condensing from "essential oils" or terpenoids or productions of sulfates from sulfur dioxide), but this process takes place at the sunlit latitudes. During the first few days of transport, then, the particle size distribution and the particle chemistry undergo rapid and violent alterations, not only because of coagulation and particle production, but also because of the high amount of turbulence, and the relatively high probability of cloud formation. Though this mixture is extremely complex, there is growing evidence that the size distribution effectively winnows down into two mass modes, one consisting primarily of refractory particles, presumably of primary crustal and sea salt debris, and another, smaller, mode of volatile, water soluble material. The smaller mode is centered at a tenth of a micron diameter and the larger mode is centered at 1 to several microns diameter and is the more variable of the two mass modes.

What seems likely for arctic air pollution is that the dynamics of the particle size distribution settles down in comparison to the rather violent evolution that took place in the first thousand or so kilometers of transport at or near mid-latitudes. The submicron accumulation mode of mainly pollution-derived material and a super-micron mode of crustal and sea salt debris are all that are left and these, presumably, are rather stable over large scales of distances and time while the air mass remains in the Arctic.

This material – congregated in the two modes – has the important property of being removal resistant. Thus one expects deposition to the surface to be low and in particular to be low in comparison to the concentrated material in the atmosphere. Indeed this expectation is fulfilled for dry deposition, especially for particles in the smaller of the two modes.

**Experimental Data on Deposition**

One way to express particle deposition is to relate it to a first-order rate constant (k) using the mass concentration within the atmosphere, thus the mass flux F, (g $cm^{-2}s^{-1}$) on the surface is $F = ky$, where y is the mass concentration in the atmosphere (g $cm^{-3}$). We expect k to be small for aerosol material over the arctic basin for reasons outlined above. The constant k has dimensions of velocity (cm $s^{-1}$), and hence is sometimes referred as "deposition velocity," a highly misleading term since k has nothing to do with the speed with which material is carried on to the surface from above.

Deposition is broken down into "dry deposition", that which occurs in the absence of precipitation, and "wet deposition," that which is brought down in water precipitation; the latter, furthermore, is sometimes broken up into below-cloud scavenging (particle attachment to falling hydrometeors by inertial or diffusive mechanisms) and within-cloud scavenging.

Precipitation scavenging conveniently is expressed as a scavenging ratio, W, with the motivation to relate concentration of aerosol material in snow (or rain) to that in the air. W is simply the ratio of mass loading per gram of water to mass loading per gram of atmospheric carrier gas, $W = Cs\ p_a/y$ where Cs is mass mixing ratio of aerosol material in melt water (g aerosol/g water) and y is, as before, mass mixing of aerosol suspended per volume of air (g aerosol $cm^{-3}$); $p_a$ is the density of air (g $cm^{-3}$), and W is dimensionless.

Davidson et al. [27] calculated W for conditions over the Greenland ice sheet, by sampling and analyzing freshly fallen snow, and simultaneously sampling air and determining its chemistry. Since aerosol concentration would be higher or lower at cloud level than at the surface (where the sampling was carried out), the estimates of W are somewhat uncertain, but probably by no more than a factor of two. Results in Table 1 indicate that precipitation scavenging of crustal non-enriched elements are higher than enriched elements. Pollution-derived elements As, Ag, Pb and pollution-derived $NO_3^-$ and $SO_4^{2-}$ are known to be mainly in the smaller of the two modes and, accordingly, are removed inefficiently from the atmosphere. Davidsen et al. point out that the somewhat larger W for nitrate compared to sulfate (both of which are believed to be mainly on removal-resistant submicron particles) may be an artifact caused by scavenging of $HNO_3$ vapor by snow. This has implications in the stratosphere for ozone depletion.

It probably can be assumed that below-cloud attachment is negligible in comparison to in-cloud scavenging [74]. Furthermore, there is evidence that crustal

**Table 1.** Scavenging ratio, W, for elements on the Greenland Ice Sheet (after Davidson et al., 1985)

| | W (ng g$^{-1}$ H$_2$O/µg g$^{-1}$ air) |
|---|---|
| Crustal elements | |
| Al | 1300 ± 130 |
| Fe | 1700 ± 400 |
| K | 1200 ± 620 |
| Mg | 1800 ± 1100 |
| Mn | 1600 ± 860 |
| Na | 2000 ± 1200 |
| Enriched elements | |
| Ag | < 250 |
| As | < 490 |
| Cd | 1100 ± 520 |
| Cu | 1800 ± 860 |
| Pb | 160 ± 70 |
| Anions | |
| NO$^-$ | 980 ± 780 |
| SO$_4^{2-}$ | 180 ± 120 |

Data represent average from 3 snowstorms during April–May 1988 at Dye 3

material, especially micas and clays [75, 76, 77], are effective as ice nuclei, which may be responsible for the larger scavenging ratios of crustal or crustal-like material in Table 1.

An interesting feature of Davidson et al.'s computation of scavenging ratio is copper, which, though is usually present predominantly on smaller submicron-sized particles [e.g. 48], has a higher W than other submicron material like SO$_4^{2-}$, Ag, Pb, etc. Davidson et al. suggest that the abnormally large value may be explained by the rather high abundance of Cu in clays compared with the lower Cu abundance of other crustal material, as was pointed out by Rankama and Sahama [78].

Dry deposition rates to the surface were also derived by Davidson and colleagues on the Greenland Ice Sheet by monitoring the buildup of aerosol in snow between precipitation events. This process has potential problems, such as the unknown deposition of small quantities of ice crystals from clear air displays that tend to contain larger aerosols as nuclei than normal [76] and possible sublimation of snow or drifting. The depositional velocity, k, for crust material ranged from about 0.1 to 0.3 cm s$^{-1}$, while that for submicron sulfate was $0.03 \pm 0.01$ cm s$^{-1}$. Interestingly, k for copper was also very small, being $0.08 \pm 0.04$. Calculations of dry deposition using particle size distributions representative of those in the polar atmosphere from fundamental principles [e.g. 79] are more or less in agreement with the measurements, though they tend to predict somewhat smaller values.

## Arctic-Wide Deposition

If we make the assumption that dry deposition values reported by Davidsen et al. are representative of the arctic basin at large, one can compare the mass budget of sulfur as sulfate deposited with that emitted. The results of the computation are in Table 2 and indicate that dry plus wet deposition in the polluted arctic air mass (assuming that these are apportioned in the ratio 1:3 as reported by Davidsen et al., [27]), account for only a few percent of total anthropogenic sulfur. Furthermore, the acidity of 10 µeq l$^{-1}$ is, within a factor of two, in agreement with that measured in snow at Poker Flat Rocket Range in Alaska and in the Agassiz ice sheet on Ellesmere Island [80]. These estimates would suggest that environmental degradation (from acid snow) would be unlikely, at least for the arctic basin as a whole.

**Table 2.** Mass budget of sulfur for the Arctic Basin

| | |
|---|---|
| Area of arctic air mass * | $44 \times 10^6$ km$^2$ |
| Mean concentration of sulfur in the atmosphere ** | 470 µgm$^{-3}$ |
| Deposition velocity, h | 0.03 cm s$^{-1}$ |
| Arctic dry deposition | .19 Tg S yr$^{-1}$ |
| Arctic wet deposition | .57 Tg S yr$^{-1}$ |
| Northern hemisphere emission | 30 Tg S yr$^{-1}$ |
| Fractional deposition in Arctic | 2.5% |
| Snow acidity assuming H$_2$SO$_4$ and water precipitation rate of 10 cm H$_2$O yr$^{-1}$ | 10.3 µ eq l$^{-1}$ |

\* Area is taken to be that within the boundary of the arctic as depicted in figure 3 and represents 9% of the earth's surface. This represents an area about 30% larger than the African continent.
\*\* Mean annual from data of Barrie et al. [78]

## Additional Remarks on Wet Scavenging

The process of wet deposition in the Arctic is not very well understood. For simplicity, assume that the entire removal is by nucleation and that below-cloud scavenging is negligible. The main problem comes about in the assumption made of writing a scavenging ratio as a linear relationship between the variables $C_s$ and y, where $C_s$ is the concentration of a given species in precipitation and $\gamma$ is the concentration of that species in air, albeit that air in which cloud forms. The implication is evidently one in which if air concentration were to double, the snow concentration would follow suit. There is no particularly compelling reason to believe that this would be the case, as cloud nucleation and growth are highly nonlinear processes. Moreover, the precipitation efficiency of a cloud also varies in a complex manner in response to the cloud droplet size distribution.

The key simplifying factor to decide the validity of the assumption (for clouds involving condensation and supercooled liquid water) is the slope of the cloud condensation nucleus spectrum, K, where it is assumed that an expression of the form $N_c = \text{const } S^k$ is adequate to describe the relationship between the number concentrations of nuclei activated, $N_c$ (cm$^{-3}$), wehen air charged with CCN is subjected to supersaturation S.

A theory relating cloud droplet concentration, Nc, maximum supersaturation encountered during cloud formation, $S_{max}$, to K was put forward by Sean Twomey in 1959 (Twomey activation theory) and predicts that there is a critical value of K=2 above which the number of cloud droplets, Nc, is independent of CCN concentration. For K≪2, the number of cloud droplets becomes approximately proportional to the number of aerosol particles [81]. Unfortunately we don't know the value of K for arctic air masses, so it is not possible to say very much about the wet scavenging mechanism. It is probable, however, that the introduction of pollution aerosol into what had been a more or less pristine environment might increase wet scavenging by creating clouds with a larger number of finer droplets.

## Trends

Ice cores taken from glaciers in or near the Arctic have provided records of past deposition. Most of the sampling locations have been at elevations 2–5 km above

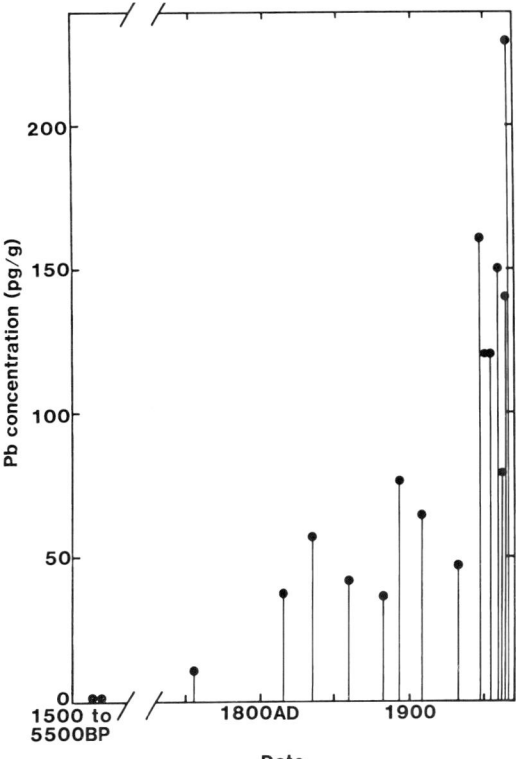

**Fig. 24.** The increase of anthropogenic lead pollution in Greenland snow (from reference 93)

sea level, so they may not be representative of arctic haze, which is predominantly contained in the lower atmospheric layers.

Ice cores have been collected from Mt. Logan, northwest Canada [82], on the Greenland ice sheet [83], and at northern Ellesmere Island [84, 80]. The record of electrical conductivity and hydrogen ion concentration in the melt water from the Agassiz Glacier, Ellesmere Island, are well correlated and undergo a rhythmic seasonal variation like that seen in air samples at arctic locations. The conductivity and acidity are rather constant until the mid 1950s, after which they increased, nearly doubling by the early 1970s. Ice core electrical conductivity increased by 75%; the water acid activity increased from 7–9 μeq l$^{-1}$. Trends in sulfate and nitrate deposition in Greenland at Dye III are also upward (Fig. 24) as is anthropogenic lead pollution in Greenland snow. All the records show evidence of increasing pollution at the high latitudes since the industrial revolution.

## Conclusion

According to the records so far recovered from glaciers in or near the Arctic, pollution has been on the increase ever since the industrial revolution. Such trends are not so readily apparent in the Antarctic.

Much of what is known about the so-called arctic haze has been learned in the last decade and a half and although great progress has been made, there are still many unanswered questions. The occurrence, trends and extent of pollution in the Soviet sector of the Arctic and the coupling of radiative models to general circulation climate grid models to determine climatic impacts are two areas that especially deserve attention.

There is of course a growing need to develop and extend transport and deposition models if we are to be able to assess intelligently environmental impacts of the growing industrialization at high latitudes. A continued development and verification of chemical tracer methods [85] needs to be encouraged, as the method has great potential for inventorying pollution impacts from regional source areas.

An area that so far has received practically no attention, but which is very important, is the connection between arctic air pollution and the microphysical properties of clouds [86]. There may well be significant alteration in cloud albedo over the Arctic from the imported cloud condensation nuclei.

## Acknowledgements

The authors would like to acknowledge funding under grants ATM86-11438, DPP86-20395 and DPP83-0286 from the U.S. National Science Foundation. Additional support was provided by Andarz Co. and by Biospherics Research Corp.

Funding from Office of Naval Research under Contract N00014-86-K-0055 is gratefully acknowledged.

# References

1. Mitchell, J.M. (1957): J. Atmos. Terr. Phys. (Special Supplement) 195–211
2. Raatz, W.E. (1984): Tellus *36B:*126–136
3. Holmgren, B., Shaw, G.E., Weller, G. (1974): AIDJEX Bull. *27:*135
4. Shaw, G.E. (1975): Tellus *27:*39–49
5. Shaw, G.E., Wendler, G. (1972): Conference proceedings on atmospheric radiation, Fort Collins, Colorado, Boston, American Meteorological Society, 181–181
6. Rahn, K.A., Borys, R., Shaw, G.E. (1977): Nature *268:*713–715
7. Rahn, K.A., McCaffrey, R.J. (1979): Proceedings of the WMO Symposium on the Long Range Transport of Pollutants and its Relation to General Circulation including Stratospheric/Tropospheric Exchange Processes, Sofia, 1–5, Oct. WMO No. 538 , 25–35
8. Barrie, L.A. (1986): In: Stonehouse B. (ed) Arctic Air Pollution. Cambridge University Press, Cambridge, 326 pp
9. Raatz, W.E., Shaw, G.E. (1984): J. Clim. Appl. Met. *23:*1052–1064
10. Schnell, R.C., Raatz, W.E. (1984): Geophys. Res. Lett. *5:*369–372
11. Hoff, R.M., Trivett, N.B.A. (1984): Geophys, Res. Lett. *11:*389–392
12. Ottar, B., Pacyna, J.M. (1986) In: Stonehouse B. (ed) Arctic Air Pollution. Cambridge University Press, Cambridge, pp 55–681
13. Radke, L.F., Lyons, J.H., Hegg, D.A., Hobbs, P.V., Bailey, I.H. (1984): Geophys. Res. Lett. *11:*393–396
14. Bigg, K. (1980): J. Appl. Meteor. *119:*521–533
15. Shaw, G.E. (1984): Geophys. Res. Lett. *11:*409–412
16. Heintzenberg, J. (1980): Tellus *32:*251–260
17. Pacyna, J.M., Vitols, V., Hanssen, J.E. (1984): Atmos. Environ. *18:*2447–2459
18. Bailey, I.H., Radke, L.F., Lyons, J.H., Hobbs, P.V. (1984): Geophys. Res. Lett. *11:*397–400
19. Rahn, K.A. (1981 a): Atmos. Environ. *15:*1447–1455
20. Rahn, K.A. (1981 b): Atmos. Environ. *15:*1457–1464
21. Barrie, L.A., Happ, R.M., Daggupaty, S.M. (1981): Atmos. Environ. *15:*11407–1419
22. Heidam, N.Z. (1981): Atmos. Environ. *15:*1421–1427
23. Lowenthal, D.H., Rahn, K.A. (1985): Atmos. Environ. *19:*2011–2024
24. Pacyna, J.M., Ottar, B. (1985): Atmos. Environ. *19:*2109–2120
25. Heidam, N.Z. (1984): Atmos. Environ. *18:*329–343
26. Heidam, N.Z. (1985): Atmos. Environ. *19:*2083–2097
27. Davidson, C.L., Santhanm, S., Fortmann, R.C., Olson, M.P. (1985): Atmos. Environ. *19:*2065–2081
28. Sheridan, P.J., Musselman, I.H. (1985): Atmos. Environ. *19:*2159–2166
29. Harris, J.M. (1984): Geophys. Res. Lett. *11:*453–456
30. Raatz, W.E., Schnell, R.C., Bodhaine, B.A., Oltmans, S.J. (1985): Atmos. Environ. *19:*2143–2151
31. Barrie, L.A., Hoff, R.M. (1985): Atmos. Environ. *19:*1995–2010
32. Graustein, W.C., Barrie, L.A. (1983): Presented at the Third Symposium on Arctic Air Chemistry
33. Berg, W.W., Sperry, P.D., Rahn, K.A., Gladney, E.S. (1983): J. Geophys. Res. *88:*6719–6736
34. Berg, W.W., Heidt, L.E., Pollock, W., Sperry, P.D., Cicerone, R.J., Gladney, E.S. (1984): Geophys. Res. Lett. *11:*429–432
35. Rasmussen, R., Khalil, M.A.K. (1984): Geophys. Res. Lett. *11:*433–436
36. Shaw, G.E. (1988): Chemical air mass systems in Alaska, Atmos. Environ. *22:*2239–2248
37. Zalabsky, R., Twomey, S. (1979): J. Rech. Atmos. 147–155
38. Barrie, L.A., Hoff, R.A. (1984): Atmos. Environ. *12:*2711–2722
39. Altshuller, A.P. (1979): Atmos. Environ. *13:*1653–1661
40. Heintzenberg, J., Hansson, H.-C., Lannefors, H. (1981): Tellus *33:*162–171
41. Hoff, R.M., Leaitch, W.R., Fellin, P., Barrie, L.A. (1983): J. Geophys. Res. *88:*10947–10956
42. Rahn, K.A. (1976): The chemical composition of the atmospheric aerosol, Tech. Report, Graduate School of Oceanography, Univ. of R.I., 276 pp
43. Khalil, M.A.K., Rasmussen, R.A. (1983): Environ. Sci & Technol. *17:*555–559

44. Rahn, K.A., McCaffrey, R.J. (1980): Ann. N. Y. Acad. Sci. *338:*468–503
45. Rasmussen, R.A., Khalil, M.A.K., Fox, R.J. (1983): Geophys. Res. Lett. *10:*144–147
46. Schnell, R.C. (1984): Geophys. Res. Lett. *11:*361–364
47. Khalil, M.A.K., Rasmussen, R.A. (1984a): Geophys. Res. Lett. *11:*437–440
48. Khalil, M.A.K., Ramsussen, R.A. (1984b): Geophys. Mon. Clim. Change *12:*141–144
49. Conway, T.J., Raatz, W.E., Gammon, R.H. (1985): Atmos. Environ. *19:*2195–2202
50. Halter, B.C., Harris, J.M., Rahn, K.A. (1985): Atmos. Environ. *19:*2033–2038
51. Shaw, G.E. (1985): Atmos. Environ. *19:*2025–2031
52. Hov, O., Penkett, S.A., Isaksen, I.S.A., Semb, A. (1984): Geophys. Res. Lett. *11:*425–429
53. Oehme, M., Ottar, B. (1984): Geophys. Res. Lett. *11:*1133–1136
54. Yung et al. (1980): J. Atmos. Sci *37:*339–353
55. Barrie, L.A., Bottenheim, J.W., Schnell, R.C., Crotzen, E.G., Rasmussen, R.A. (1988): Nature *334:*138–142
56. Hansen, A.D.A., Rosen, H. (1984): Geophys. Res. Lett. *11:*381–384
57. Khalil, M.A.K., Rasmussen, R.A., Gunawardena, R. (1988): Geophys. Monitoring Clim. Change *15:*123–125
58. Barrie, L.A., Hoff, R.M. (1984): Atmos. Environ. *12:*2711–2722
59. Shaw, G.E., Stamnes, K. (1980): Ann. N.Y. Acad. Sci. *338:*533–539
60. Porch, W.M., MacCracken, M.C. (1982): Atmos. Environ. *16:*1365–1371
61. Cess, R.D. (1983): Atmos. Environ. *17:*2555–2564
62. Blancet, J.P., List, R. (1987): Tellus *39B:*293–317
63. Wiscombe, W.J., Warren, S.G. (1980): J. Atmos. Sci *37:*2712–2733
64. Rosen, H., Hansen, A.D.A. (1984): Geophys. Res. Lett. *11:*461–464
65. Rosen, H., Hansen, A.D.A. (1985): Atmos. Environ. *19:*2203–2207
66. Rosen, H., Hansen, A.D.A. (1986) In: Stonehouse B. (ed) Arctic Air Pollution, Cambridge University Press, Cambridge pp. 101–120
67. Heintzenberg, J. (1982): Atmos. Environ. *16:*2461–2469
68. Shaw, G.E. (1982): J. Appl. Met. *21:*1080–1088
69. Freund, J. (1983): Atmos. Ocean *21:*158–167
70. McGuffie, K., Cogley, J.G., Henderson-Sellers, A. (1985): Atmos. Environ. *19:*707–714
71. Clarke, A.D., Noone, K.J. (1985): Atmos. Environ. *19:*2045–2053
72. Valero, F.P.J., Ackerman, T.P. (1986) in: Stonehouse, B. (ed) Arctic Air Pollution. Cambridge University Press, pp. 121–134
73. Shine, K.P., Robinson, D.A., Henderson-Sellers, A., Kukla, G. (1984): J. Climate Appl. Met. *23:*1459–1464
74. Junge, C.E. (1977) in: Isotopes and Impurities in Snow and Ice, 63–77 pp, IAHS publication
75. Mason, B.J. (1975): Cambridge University Press, Cambridge, U.K., 66–67 pp
76. Kumai, M. (1967) in: Proc. First Nat. Conf. on Weather Modification, Albany, N.Y., pp. 414–422
77. Pruppacher, H.R., Klett, J.D. (1980): Microphysics of clouds and precipitation. D. Reidel, Dordrecht, Holland
78. Rahnkama, K., Sahama, T.G. (1950): Geochemistry, pp. 697 and 700, University of Chicago Press, IL
79. Twomey, S. (1977): Atmospheric Aerosols, Pergamon Press
80. Barrie, L.A., Fisher, D., Koerner, R.M. (1985): Atmos. Environ. *19:*2055–2063
81. Twomey, S. (1980): J. Phys. Chem. *84:*1459–1463
82. Holdsworth, G., Peake, E. (1985): Ann. Glaciol. *7:*153–160
83. Herron, M.M. (1982): J. Geophys. Res. *87:*3052–3060
84. Koerner, R.M., Fisher, D. (1982): Nature *295:*137–140
85. Rahn, K.A., Lowenthall, D.H. (1985): Science *223:*132–139
86. Borys, R.D. (1983): Atmospheric Science Paper No. 367, Department of Atmospheric Science, Colorado State University, Ft. Collins
87. Shaw, G.E. (1986) in: Stonehouse, B. (ed) Arctic Air Pollution. Cambridge University Press, Cambridge, 328 pp
88. Benson, C. (1986) in: Stonehouse, B. (ed) Arctic Air Pollution. Cambridge University of Press, Cambridge, p. 69–84
89. Liljequist, G.H. (1970): Klimatologi Stockholm, General stabens Litografiska Anstalt

90. Bodhaine, B. (1986) in: Stonehouse, B. (ed) Arctic Air Pollution. Cambridge University Press, Cambridge, pp. 159–173
91. Heintzenberg, J., Hansson, H.C., Ogren, J.A., Covert, P.S., Blancet, J.P. (1986) in: Arctic Air Pollution. Cambridge University Press, Cambridge, pp. 25–35
92. Joranger, E., Ottar, B. (1984) Geophys. Res. Lett. *11:*365–368
93. Wolf, E.W., Peel, D.A. (1985): Nature *313:*353–540

# Air Pollution and Materials Damage

*Frederick W. Lipfert*

Brookhaven National Laboratory
Upton, NY 11973, USA

| | |
|---|---|
| Introduction | 114 |
| Background | 115 |
|   Pollutants and Atmospheric Processes | 115 |
|     Classification of Pollutants | 116 |
|     Atmospheric Pollutant Delivery Processes | 117 |
|     Pollutant Interactions | 121 |
|     Air Concentrations | 122 |
|     Other Atmospheric Factors | 124 |
|     Urban Effects | 126 |
|   Previous Reviews of Air Pollution Damage to Materials | 126 |
|     Descriptive Studies | 126 |
|     Economic Studies | 128 |
|     Recent Comprehensive Technical Reviews | 129 |
|   Separating Pollution-Induced Damage from Natural Weathering | 131 |
|     Requirements for Damage Functions | 132 |
|     Three-Step Damage Function Development | 133 |
| Air Pollution Effects on Specific Materials | 138 |
|   Selection of Materials | 138 |
|   Corrosion of Metals | 140 |
|     General Mechanisms | 140 |
|     Corrosion of Zinc | 142 |
|     Copper | 146 |
|     Carbon Steel | 147 |
|     Aluminum | 148 |
|     Other Metals | 149 |
|     Major Unknown Factors – Metal Corrosion | 150 |
|   Damage to Masonry | 151 |
|     $CaCO_3$ Damage Mechanisms | 151 |
|     Theoretically-Based Calcite Damage Function | 152 |
|     Measured Stone Loss Rates | 153 |
|     Experimental Determination of Carbonate Stone Damage Functions | 156 |
|     Damage to Concrete and Mortar | 161 |
|     Conclusions and Major Unknowns Regarding Masonry Deterioration | 164 |

Air Pollution Effects on Paints and Organic Coatings . . . . . . . . 164
　　　　　Components of Paint Systems . . . . . . . . . . . . . . . . . . 165
　　　　　Types of Damage to Paints . . . . . . . . . . . . . . . . . . . . 165
　　　　　Observations of Environmental Damage to Paints . . . . . . . . . 166
　　　　　Major Unknowns with Respect to Environmental Damage
　　　　　to Paints . . . . . . . . . . . . . . . . . . . . . . . . . . . . . 169
　Configuration Effects . . . . . . . . . . . . . . . . . . . . . . . . . . . 169
　　　Boundary Layer Concepts . . . . . . . . . . . . . . . . . . . . . . . 170
　　　　　The Atmospheric Boundary Layer . . . . . . . . . . . . . . . . 170
　　　　　Boundary Layers on Structures . . . . . . . . . . . . . . . . . . 172
　　　　　Application of Boundary Layer Theory to Buildings and Structures . 175
　　　　　Application of Boundary Layer Theory to Non-Buildings . . . . . 176
　　　Discussion of Boundary Layer Calculations and Results . . . . . . . . 177
　　　　　Deposition with Little or No Surface Resistance . . . . . . . . . 177
　　　　　Deposition with Surface Resistance . . . . . . . . . . . . . . . 178
　　　　　Other Configuration Effects . . . . . . . . . . . . . . . . . . . 178
　Concluding Remarks . . . . . . . . . . . . . . . . . . . . . . . . . . . 180
　Acknowledgments . . . . . . . . . . . . . . . . . . . . . . . . . . . . 180
　References . . . . . . . . . . . . . . . . . . . . . . . . . . . . . . . . 180

## Summary

This chapter reviews and summarizes the current state of knowledge concerning the effects of atmospheric pollution on deterioration of materials. The major groups of materials discussed are metals, masonry (primarily calcareous stones), and painted surfaces. Emphasis is placed on the deposition processes that relate air pollution to materials damage, as well as the roles of rain and rain acidity. Although some fundamental damage processes are discussed, quantitative damage relationships are emphasized, and some new data towards this end are presented for carbonate stones. Specialized topics that are included in this chapter discuss experimental and statistical methods for separating "natural" weathering from air pollution related damage and the problem of relating research findings to practical structures in urban environments.

## Introduction

Although the deleterious effects of impurities in the air on certain materials have been known virtually from ancient times, the scientific study of cause and effect mechanisms has begun only relatively recently. R. Angus Smith was perhaps the first (1872) to (qualitatively) link such damages to both wet and dry deposition of acidic compounds [1]; the study of acidic deposition effects in both Europe and North America now includes damage to materials as well as to ecological systems. There are two main motivations for such research: preservation of cultural resources and national treasures, and accounting for materials damage losses to national economies. The first concern tends to emphasize stone buildings and monuments, whereas the second deals with common construction materials. This chapter will echo these concerns in selecting topics for emphasis and literature for review.

Until serious consideration of air pollution abatement began in the latter half of this century, environmental damage to materials was implicitly considered as part of the price paid for an industrial society, i.e., a factor to be reckoned with in selecting materials and designing structures rather than an additional motivation for the abatement of air pollution. In order to properly estimate the benefits that might accrue in the form of lengthened materials service lives as a result of air pollution abatement, a detailed understanding of the underlying cause and effect relationships is necessary and a substantial economic and environmental data base is required. Even today, while it is recognized that materials damage may have large monetary effects on the pollution abatement cost/benefit equations, research on this topic receives a relatively small portion of the total budget for such research.

This chapter begins with some background material on the topic, including a discussion of the important atmospheric processes and a summary of previous reviews of materials damage. The problem of identifying the contributions of pollution damage to materials is addressed next, followed by detailed discussions of what is currently known about such damage with respect to specific materials. For each material grouping (metals, masonry, paints), mechanisms are discussed, the important research is summarized, and major unknowns are listed. An emphasis is placed on quantitative "damage functions" that relate rates of material degradation to the presence of specific pollutants or agents in the atmosphere. The final topic is a discussion of the problems of relating research results to service lives of actual structures in place.

## Background

### Pollutants and Atmospheric Processes

It is often convenient to discuss air pollution effects on materials according to the major categories of pollutants and their primary modes of deposition. For example, Nriagu discusses materials damage in the context of sulfur pollution alone [2]. It is axiomatic that pollutants must actually be deposited on surfaces in order to cause damage, and it has been only with the advent of a better understanding of these deposition processes that real progress has been made towards materials damage dose-response functions having a firm theoretical backing.

Both wet and dry deposition can be important to atmospheric deterioration of materials. Their relative importance depends on the type of material, its orientation on a structure, air and precipitation quality, and climatic factors, particularly rainfall amounts and frequency of high relative humidity. Other climatic factors that may contribute to materials degradation include freeze-thaw cycles, intensity of solar radiation, and wind conditions.

Dry deposition, which includes several different physical processes, has been defined as "the aerodynamic transfer of atmospheric trace gases and particles to the surface by mechanisms not associated with the fall of hydrometeors" [3]. Wet deposition is then the converse, those processes that involve hydrometeors (rain, snow, hail, sleet, etc.) but notably excepting dew, in part because of its connection with dry deposition.

## Classification of Pollutants

*Particles* are conveniently classified by average size and composition. Except for fog situations, large particles (>10 μm mass median aerodynamic diameter [mmad]) are usually alkaline, comprising soil materials and/or resuspended road dust. Deposition of such particles is largely controlled by gravitational settling, and thus horizontal surfaces are much more likely to be impacted than vertical ones. Marine aerosols may also be included in this particle size fraction.

Smaller particles (<2–3 μm mmad) may be more acidic. This size fraction is often dominated by sulfates, which can include free sulfuric acid and various ammonium salts. Note that even ammonium sulfate, which is nominally neutral, can behave as a weak acid. Free sulfuric acid is usually found at the smaller end of this particle size range (<0.8 μm mmad) and is usually a minor portion of the total aerosol sulfate loading. Deposition in this size range is strongly affected by fluid mechanics, including turbulence and mixing.

*Soiling* is a special case of air pollution damage to materials, since often the damage is caused more by removal of the offending particles rather than by chemical reactions directly resulting from particle deposition. Soiling usually has an operational definition: deposited material which is removed by cleaning. The cleaning processes may involve abrasion of the surfaces (sand blasting) or chemical processes. Carbonaceous particles (soot) are most likely to cause soiling; these are typically agglomerates of fine (<1 μm mmad) carbon particles whose deposition may be affected by a number of aerodynamic and physical processes. Soot deposits may also play a role in catalyzing surface reactions involving other deposited air pollutants, especially $SO_2$. Early demands for air pollution abatement in cities such as St. Louis and Pittsburgh often focused on soiling as a prime benefit, although it was often couched in more general terms such as "reduced materials damage." In extreme cases of soot deposition, soiling of personal clothing was often a consideration.

*Gases* of concern to materials damage include both acid-forming and oxidizing species. Acidic gases and those that form acids are of primary interest for inorganic materials. These pollutants include sulfur dioxide ($SO_2$), nitric acid ($HNO_3$), hydrochloric acid (HCl), hydrofluoric acid (HF), and formaldehyde (HCHO). The latter may form a complex with $SO_2$ (hydroxymethanesulfonic acid, HMSA). Nitric acid has been found to deposit very readily on virtually all surfaces, being limited only by aerodynamic effects, and in the absence of specific data it is presumed that HCl and HF would behave similarly. However, deposition rates for $SO_2$ or HCHO are strongly affected by surface conditions as well as by aerodynamic considerations. Wet surfaces absorb such gases by means of dissolution, and thus their Henry's Law characteristics become important as applied to the liquid surface layers. Surfaces may be wetted by either moisture adsorbed from the atmosphere (condensation) or by precipitation.

Oxidizing gases (ozone, $NO_2$, $H_2O_2$, PAN, etc.) can have both direct and indirect effects on materials damage, with direct effects largely limited to organic materials [4]. Such direct effects include changes to organic molecular structure with accompanying physical property changes such as embrittlement and subsequent cracking. Indirect effects involve oxidation in liquid surface layers, notably dissolved $SO_2$. Oxidation of sulfite to sulfate reduces the volatility of the acid and thus increases the duration of possible acidic surface attack, as well as facilitating the dissolution of additional deposited $SO_2$.

*Acidic precipitation* is the third pollutant category, in which the air pollutants of concern have previously been scavenged by either clouds or by falling droplets. The effects of acidic precipitation on wetted materials may be controlled by kinetic limitations and thus may be sensitive to configuration factors which affect residence time of precipitation contact with the surface. There is empirical evidence that atmospheric damage is maximized on horizontal or near horizontal surfaces, but since moisture condensation may be more frequent on such surfaces in addition to precipitation contact, both dry and wet deposition may be involved.

Acidic precipitation may have several different modes of action on surfaces. Direct attack may occur through exchange reactions as a result of contact with dilute acids. Indirect effects can occur as a result of washing action and removal of other substances. Such substances may include previously deposited particles (which could be either acidic or basic), corrosion products formed from dry deposition or "natural" weathering, or deliberately applied protective coatings. Such indirect effects may not depend only on the composition of rainfall, but more so on its duration and frequency.

*Indoor air pollution* can also damage materials, notably paper, art works, and electrical contacts. Such damage is limited to dry deposition under conditions of moderate relative humidity and essentially "dry" surfaces. However it has been shown that certain precursors of acidic precipitation, notably sulfate and nitrate aerosols, are also found in indoor environments and as deposits on indoor surfaces [5]. Such rates of deposition are extremely low when referred to outdoor air concentrations, in part because of the reduced indoor air concentration levels and in part because of relativity quiescent air flow. Reactive gases such as $SO_2$ tend to deposit on many types of indoor surfaces and thus air concentration levels are usually low. Baer and Banks present a useful summary of aspects of art conservation relative to indoor materials damage [6].

*Atmospheric Pollutant Delivery Processes*

The relative importance of wet versus dry deposition depends not only on the deposition pathways but also on the level of pollution and on climatic conditions. Other factors influence the relative effectivenesses of pollutant delivery pathways, including the building configuration and design features and the degree of sheltering from the rain. Useful test data are only now beginning to be acquired for material test specimens and real structures.

Wet Deposition

Wet deposition, which includes all forms of precipitation (e.g., rain, snow, sleet) and fog impaction can have indirect effects on many types of material surfaces and on materials performance. First, it can remove normally protective alkaline particles, which are more likely to have been deposited than acidic particles because of their larger average diameter; as the alkaline particles dissolve in the dew layer, they may help neutralize the acids formed by dry deposition of $SO_2$.

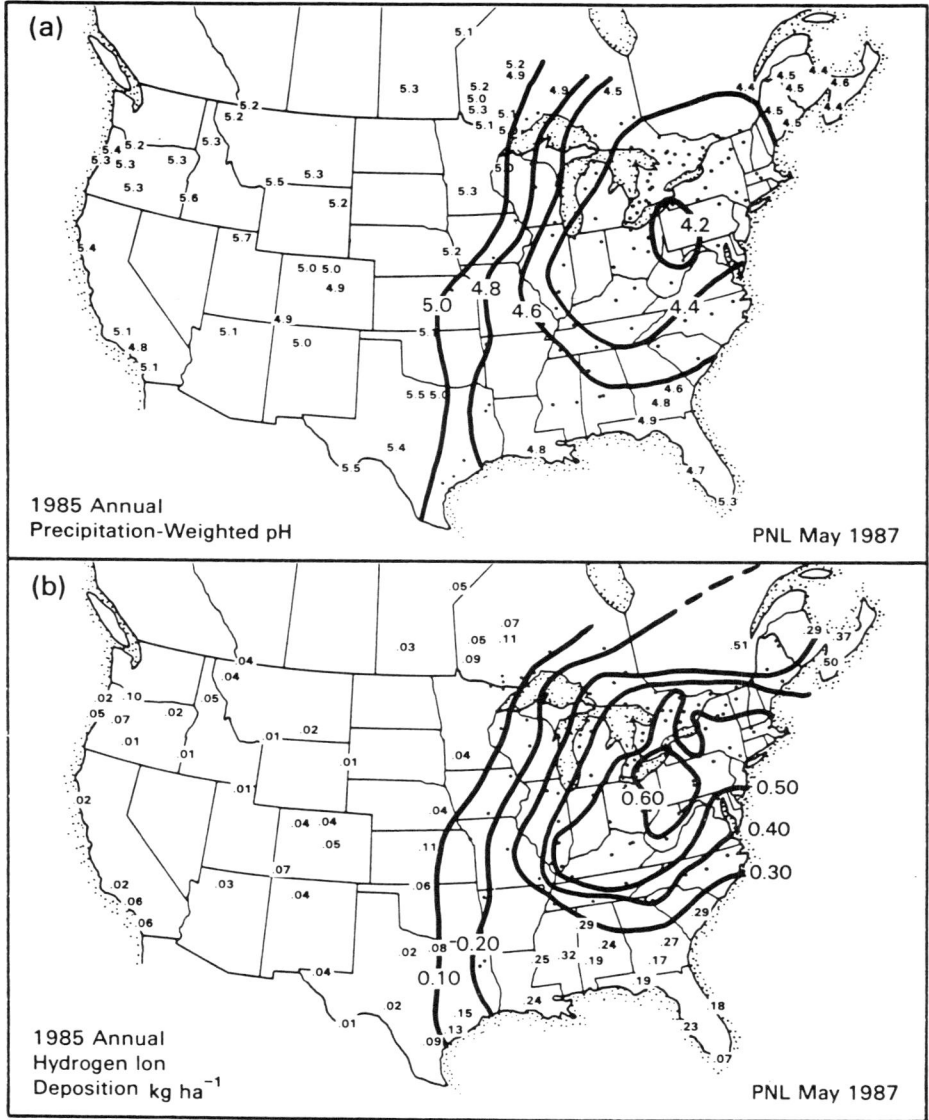

**Fig. 1.** 1985 Isopleths of precipitation pH and hydrogen ion deposition for the United States. Source: National Acid Precipitation Assessment Program, 1987 [54]

Second, acidic precipitation may dissolve protective (passivating) oxide or carbonate corrosion films, thus reexposing fresh surfaces of the underlying material to either subsequent wet or dry deposition of acids. This mechanism is important in zinc corrosion, for example, since the solubility of zinc carbonate increases with decreasing pH.

Of the various types of precipitation, rain is most likely to be important with respect to materials damage, because snow is not a good adsorber for $SO_2$ and frozen surfaces are less reactive. However, in many locations in the United States, wetting of surfaces may be dominated by dew because precipitation occurs only a small percent of the time and dew may form on practically every clear night. Dew is an important mechanism in dry deposition because it can adsorb gaseous pollutants such as $SO_2$.

The total amount of acidity that can be delivered to a surface by wet deposition is the product of the $H^+$ concentration of the rain and the volume of rainfall deposited on the surface. For dilute strong acids, the hydrogen ion concentration is the most important indicator of aggressiveness. Figure 1 shows typical isopleths of median pH and $H^+$ deposition over the United States, which provides some indication of the exposure of materials in the region. These data were obtained from monitors deliberately sited away from major emission sources and cities, and thus may be biased towards somewhat lower levels of deposition.

The available experimental evidence from field studies and special purpose networks indicates that urban areas can have considerable influence on precipitation chemistry [7, 8]. These effects include excess concentrations of both acidic and basic ions in various instances, but the net effect on the acidity of precipitation depends on the relative balance. The influence of the urban environment is germane to the materials damage issue because of the high density of exposed materials and cultural resources in urban areas.

Dry Deposition

*Controlling Factors for Gases.* Although rates of materials damage often are referred directly to air concentrations of pollutants, these agents must actually be deposited on the surfaces in question for damage to occur. The rate of deposition, the deposition velocity ($V_d$), must therefore be considered. Dry deposition processes can be complicated and may be affected by surface composition and chemical reactions, surface roughness and shape, wind speed, atmospheric turbulence, and surface wetness. These factors are discussed in more detail in the last section of this chapter.

$SO_2$ is of special interest because of its relatively high concentrations in urban areas and the sensitivity of many materials to deposited sulfur. $SO_2$ can deposit through either of two mechanisms. One is the dissolution of gaseous $SO_2$ in moisture films present on surfaces. In addition, as a potential second mechanism, Edney et al. [9] showed by laboratory experiment that a monolayer of $SO_2$ will deposit on a nominally dry zinc surface (no condensation present); this phenomenon also was shown for dry deposition of $NO_2$ [10]. However, Mikhailovskii's data [11] indicate that there may be up to several monolayers of water present on such a surface, depending on the relative humidity, and Spedding [12] has raised questions as to whether such thin films may be expected to follow the same laws as "bulk" water. This and the implied high ionic strengths indicate that use of Henry's law relationships to predict dry deposition of $SO_2$ at low relative humidities may be somewhat problematic.

For dissolved $SO_2$, the strong dependence of the effective Henry's Law constant on pH indicates that $SO_2$ uptake by water will be sharply curtailed as the pH decreases [13]. This mechanism will control the rate of $SO_2$ deposition to surfaces with sufficient water present. If the base material is sufficiently reactive, the moisture layer will be buffered by corrosion products, maintaining the pH level, and $SO_2$ will continue to be dissolved and thus to "dry" deposit. On nonreactive surfaces (e.g., glass or Teflon), much lower effective $SO_2$ deposition velocities would be expected as the moisture layer pH drops. This was the case with dew measurements obtained by Pierson et al. in rural Pennsylvania [15].

pH effects on deposition were demonstrated by Adema et al. [14] in wind tunnel experiments involving deposition to buffered solutions for $SO_2$ and $NO_2$; experiments also were conducted on $NH_3$ deposition. The deposition of $SO_2$ dropped markedly as the liquid pH decreased below about 5.0. With buffering from $NH_3$, this decrease did not take place. $NO_2$ deposition was about 5% of the (maximum) $SO_2$ values and was not dependent on solution pH. This finding is consistent with the low solubility of $NO_2$ in water as reported by Lee and Schwartz [16].

For reactive materials, such as steel, zinc, and calcareous stones, buffering of the dew layer will keep the pH sufficiently high to allow $SO_2$ to deposit continuously, as long as the aerodynamic conditions continue to deliver fresh reactant to the surface. Edney et al. [9] have demonstrated this process for galvanized steel in the EPA laboratory exposure chamber. Less reactive materials (e.g., some paints, asphalt) may not react fast enough to prevent this moisture layer from becoming too acidic to continue to absorb $SO_2$. Thus, the dry deposition velocity is in an indication of surface reactivity to $SO_2$ [17]. Some typical values from the literature are given in Table 1.

The oxides of nitrogen (NO, $NO_2$) differ from $SO_2$ in at least one important characteristic: their greatly reduced solubility in water [16]. Thus, for a given atmospheric concentration level, $NO_2$ is expected to deposit on a wet surface much more slowly than $SO_2$. Under photochemical smog conditions with production of the OH radical, some $NO_2$ will be converted to nitric acid. Normally a vapor at standard atmospheric conditions, $HNO_3$ has been observed to deposit readily on most surfaces [18, 19]. (Because the resulting nitrate salts usually are quite water soluble, they are not likely to be found on corroded surfaces or in corrosion product films exposed to rain washing.) Therefore, the principal harmful atmospheric nitrogen compound may be nitric acid vapor, rather than the more common oxides (NO, $NO_2$).

*Typical Values of $SO_2$ Dry Deposition Velocities.* The preceding discussion has dealt with the ability of a surface to absorb a depositing gas. However, there are two processes that are necessary for dry deposition to occur. The pollutant must be aerodynamically delivered to the surface, and the surface must be able to absorb the pollutant as it is delivered. In atmospheric models, these two processes are often considered as resistances in series. Both resistances must therefore be favorable for a substance to be deposited.

Table 1 displays some dry deposition velocities for $SO_2$ (taken from various sources in the literature) that illustrate the role of surface reactivity. Note that

**Table 1.** Typcial $SO_2$ dry deposition velocities (maximum, under daytime conditions)

| Surface | $V_d$ (cm/sec) |
| --- | --- |
| Cement | 1.6–2.5 |
| Lead sulfation plates | 1.4 |
| Zinc (wet) | 1.1–1.9 |
| Limestone | 0.3–1.3 |
| Paints | 0.1–0.6 |
| Wood (bare) | >0.1 |
| Asphalt | 0.04 |

most experimental determinations of $V_d$ measured the combined surface and aerodynamic resistances. However, in this case, many of the values shown (including the highest and lowest) were obtained from a flow reactor, which allowed the surface resistance to be determined independently [20].

Judeikis and Wren used this same apparatus to measure deposition velocities (i.e., surface resistances) for NO and $NO_2$, to soils and cement [21]: For cement, the results were 0.21 cm/s for NO and 0.32 cm/s for $NO_2$, essentially independent of humidity. Edney et al. (1986) found that $NO_2$ deposited to wet zinc plates was about 4% of the deposition rate for $SO_2$ [10]. Other investigators have shown that ozone deposition is actually inhibited by surface moisture because of its low solubility in water [22]. The solubility of an atmospheric trace gas together with typical ambient concentration levels may be used as indicator of importance for materials damage, assuming that chemical reactions following or associated with deposition in fact constitute damage to the material [17].

*Dry Deposition of Particles.* For particles, deposition velocities are controlled largely by physical processes associated with the effective aerodynamic particle diameter. Gravitational settling increases with increasing particle size, while diffusion controlled deposition decreases with increasing particle size. Although the exact shape of the deposition velocity-size curve varies according to experimental methods and receiving surface, the minimum $V_d$ occurs for particles approximately in the range 0.1–1.0 μm diameter, which includes may common secondary particles such as sulfates [23]. These particles are often acidic and usually hygroscopic, but their deposition velocities are so low that they are not believed to be a major factor in materials degradation. Freshly emitted soot particles, on the other hand, may be even smaller (in the range 0.01–0.1 μm) and thus may deposit more readily. Such particles will be retained by rough surfaces such as stonework. Particles larger than 1 μm, which settle gravitationally, may be more important with respect to horizontal surfaces (but only until the next rainfall).

*Pollutant Interactions*

The previous discussion focused primarily on individual atmospheric factors relative to materials damage. There also are interactions to consider. For example, in addition to creating soiling, soot can help catalyze the oxidation of $SO_3^=$ to $SO_4^=$, which may be especially important on non-metal surfaces; metallic surfaces may provide their own catalysis. Soot also is an essential ingredient in the black sulfate crusts that form on stonework in rain sheltered areas [24], although the particle sizes cited seemed large for soot as it is usually characterized [25].

With respect to other particle interactions, outdoor surfaces can be wetted at relative humidities less than 100% if salts are present on the surface with deliquescence points less than 100% (about 90% relative humidity for zinc sulfate, for example). This is one of the mechanisms by which sea or road salt particles can increase the corrosion of metals. All deposited particles may affect the dry deposition of gases by providing additional adsorption sites [26].

Oxidants (e.g., ozone, hydrogen peroxide) are thought to play only indirect roles in acidic deposition materials damage. Certain polymers (including paints) may be sensitive to oxidant degradation and subsequent cracking of the material could allow acidic deposition to penetrate to the substrate. Also, oxidants may be required in surface layers to oxidize the dissolved $SO_2$ from sulfite to sulfate, the more stable form. Ozone is not very water soluble, and thus is probably not as important as hydrogen peroxide in this regard.

Following the analogy for rain acidification, pre-acidification of surface moisture layers by other atmospheric acids ($HNO_3$, HCl) could retard the subsequent deposition of $SO_2$ if not buffered by reaction products. However, the reverse order, $SO_2$ followed by $HNO_3$, would result in additional dew acidification. Deposition of alkaline substances (particles, $NH_3$) could work in the opposite direction, but their alkalinity would neutralize at least a portion of the deposited acids. On dry surfaces, $NO_2$ may be more important in oxidizing $SO_2$ [27], since both gases will be more nearly equally deposited in the absence of moisture.

Formaldehyde (HCHO) has been shown to form a complex (HMSA) with sulfite [28]. The exact consequences of such complexing on corrosion rates are not completely understood, but it appears that the effect is antagonistic, in that the net material loss due to the complex is less than the sum of separate $SO_2$ and HCHO effects. Oxidation of HCHO to formic acid is expected to be slow, even in the presence of hydrogen peroxide.

*Air Concentrations*

Both air concentrations and deposition velocities must be considered in studying material sensitivities to specific air pollutants. While this chapter does not address pollutant source attribution per se, it should be noted that information on pollutant source type and spatial distribution may be important in determining deposition characteristics and exposure conditions for materials at risk.

With the implementation of the 1970 Clean Air Act, the spatial distribution of pollution sources changed in the United States, especially for $SO_2$. Whereas the distribution of emissions was previously weighted towards many smaller sources in urban areas, it is now dominated by a much smaller number of very large sources, primarily outside urban areas.

Table 2 presents some typical characteristics of urban air pollutants thought to be important for materials damage, together with estimated deposition velocities (to a zinc surface, for reference). Deposition velocity ($V_d$) values for

Air Pollution and Materials Damage 123

most of the gases shown are taken from chamber tests by Edney et al. [9] to provide an equivalent basis for comparison of relative pollutant fluxes ($V_d \times C$); values for $H_2S$ deposition velocity were taken from Judeikis and Wren [21]. The classification of important source types (area, point, or regional background) is intended to provide guidance as to the possible spatial distributions for each pollutant, in the context of the US. Area sources are agglomerations of small sources such as the network of a city; point sources are separately identifiable large sources usually with elevated release points. Regional background refers to air concentrations at distances greater than 50 to 100 km from emission sources.

Aside from water, Table 2 indicates the dry highest deposition fluxes (up to 95 g/m²y) for coarse particles, which would only be applicable to horizontal (or near horizontal) surfaces sinces these large particles tend to settle gravitationally. By way of comparison, normal rainfall in the eastern United States deposits $0.5–1 \times 10^6$ g/m²y of water, which is capable of removing this particle loading assuming the particles are inert. For the remaining pollutants which are thought to be capable of causing material damage, the highest surface flux is indicated for $SO_2$ (2.2 g/m²y), followed by HCHO, $HNO_3$, and HCl (0.3–1.2 g/m²y). This rank order, based on the quantitative data shown, is in contrast to that of Graedel and McGill [29], who assigned first priority for zinc corrosion to formaldehyde based mainly on qualitative corrosion rate data.

The deposition flux for sulfate particles (0.1 g/m²y) was substantially lower than for $SO_2$, which would indicate only minimal importance for sulfates even if

**Table 2.** Typical air pollutant concentration levels and source types (adapted from Baer et al. [29])

| Pollutant | Typical concentration (µg/m³) | $V_d$ to zinc (cm/s) | Flux g/m²y | Pollutant source types | | |
|---|---|---|---|---|---|---|
| | | | | Point | Area | Reg. backgr. |
| $SO_2$* | 30 – 40 | 0.25 | 2.4 – 3.2 | X | X | X |
| NO | 50 –130 | low | | X | X | |
| $NO_2$ | 50 – 65 | 0.03 | 0.5 – 0.6 | X | X | X |
| $O_3$ | 40 – 50 | <0.1 | <1.3 | | X | X |
| $HNO_3$ | 1.5– 2.1 | 0.7 –1.7 | 0.3 – 1.2 | | X | X |
| HCHO* | 12 – 25 | 0.08–0.15 | 0.3 – 1.2 | | X | X |
| HCl | 1.5– 6 | 0.7 | 0.3 – 1.3 | X | | |
| $H_2S$ | 0.5 | 0.02 | 0.003 | X | X | |
| $SO_4^-$ | 2 – 9 | 0.05 | 0.03 – 0.14 | X | X | X |
| Particles: | | | | | | |
| fine | 10 – 20 | 0.05 | 0.16 – 0.32 | X | X | X |
| coarse | 20 –100 | 1 –3 | 6 – 95 | X | | |
| $H_2O$ | | | | | | |
| rainfall | | | $0.5 – 1 \times 10^6$ | | | |
| dewfall | | | $0.025– 0.05 \times 10^6$ | | | |

* For the estimated deposition of $SO_2$ and formaldehyde (HCHO) to wet surfaces, their deposition velocities have been multiplied by 0.25 to account for the estimated time of surface wetness in an urban area.
Note: Air concentrations are based on US national and regional averages where possible and do not necessarily represent worst cases. $V_d$ for HCl is assumed equal to that for $HNO_3$ as a first order approximation.

they were all present as sulfuric acid. ($H_2SO_4$ mists from primary sources such as manufacturing plants could have severe local impacts, however, because of characteristically larger particles and hence higher deposition velocities.)

The rank order in Table 2 could be altered by variations in local sources and air concentrations as well as by different values for the fraction of wet surface time. Also, the deposition velocity for nitric acid has been found to exhibit large diurnal swings when measured in open country [30]; it is not clear that this would also happen in cities. The $HNO_3$ deposition velocities used in Table 2 are attempts at day-night averages. Uncertainties notwithstanding, it appears that the attention that historically has been focused on materials damage due to $SO_2$ may have been justified, especially because $SO_2$ levels were generally much higher in the past. For areas remote from fuel combustion pollution sources, other pollutants may be more important.

With respect to wet deposition of acidity, an average precipitation pH of 4.3 at 1 meter/year would result in $H^+$ wet deposition of 25–50 mmol/m²y. Assuming that the $SO_2$ levels of Table 2 were "dry" deposited to a surface wet from condensation and resulted in $H_2SO_4$ being formed, the resulting $H^+$ flux by this route would be 75–100 mmol/m²y, an amount which is comparable to the wet deposition level. Also, the ionic strength implied by dissolving the deposited $SO_2$ in the dewfall [15] is considerably higher than typical precipitation concentrations. The relative importance of wet vs. dry deposition also depends on reaction kinetics and precipitation residence time in contact with the surface. But because all strong acids are equivalent when diluted to the same pH, whether the $H_2SO_4$ is received through wet deposition or by dry deposition of an acidic gas or vapor is not an important factor, all other factors being equal.

*Other Atmospheric Factors*

In addition to air concentrations, there are many important atmospheric factors relative to materials degradation. Thus, damage predictions based on air concentrations alone are likely to be inaccurate.

*Time of Surface Wetness.* Prediction of materials damage from dry deposition requires knowledge of the time of wetness for exposed surfaces, which is usually based on the frequency of occurrence of relative humidities above some critical value. Because neither corrosion nor $SO_2$ deposition are expected to be important on frozen surfaces, wetness conditions should be further limited to hours with dew points above freezing.

Because of radiational cooling, dew may be expected to form most readily on surfaces that have a clear view of the night sky. This factor accelerates the rusting of galvanized roofs as opposed to vertical walls. Available data on the composition of natural dew [15, 31], usually obtained with horizontal inert surfaces, tend to show much smaller $SO_2$ deposition velocities than obtained by Edney et al. [9] in the EPA laboratory chamber with artificially induced dew. However, even if dew is not visibly present, high relative humidity enhances the adsorption of atmospheric water vapor [11]. Currently, no data are available on the relative amounts of surface water necessary for either corrosion or $SO_2$ deposition to proceed.

Statistical analysis of National Weather Service data [32] has shown that the degree of urbanization decreases the frequency of high relative humidities, apparently because of the urban heat island effect. Urban areas are characterized by having lower relative humidities and low or wind speeds than the surrounding region outside the city. In the US, east of the Mississippi River, frequency of high relative humidity increases with decreasing latitude and with proximity to large water or bodies or coast lines [32].

*Temperature.* In addition to time of wetness and relative humidity, ambient temperature may have effects on material deterioration processes. Reaction rates accelerate with increasing temperature, but dissolution of gases increases at lower temperatures. Calcareous stone ($CaCO_3$) dissolution, for example, is increased by lower temperatures primarily because of increased $CO_2$ solubility [33]. As the previous night's dew layer is dried by solar heating, its chemical concentrations will increase. The complete ramifications of these competing thermal processes are poorly understood.

*Wind and Meteorological Factors.* The aerodynamic resistance to dry deposition is inversely proportional to wind speed. In the case of low surface resistance, dry deposition velocity should be nearly proportional to wind speed, as demonstrated by Sickles and Michie [34] in calibrating sulfation plates; their data also demonstrate the influence of the local flow regime on deposition. Thus, the wind speed dependence of dry deposition flux ($V_d \times C$) will depend on the relationship between wind speed and air concentrations.

For area sources, air concentrations at a given height are inversely proportional to wind speed [35]; thus there should be little or no dependence of the deposited flux on wind speed. However, as a function of height, there is some evidence that the air concentration drops off more slowly when the wind speed increases, resulting in an increase in deposition with height, as observed, for example on the Cologne Cathedral by Luckat [36]. This relationship also will depend on the nature of the nearby pollution sources. Downwind of buoyant point sources (1-20 km), the maximum concentrations are nearly independent of wind speed because of the tradeoff between plume rise and dilution [37]. Thus, the increase in deposition velocity with wind speed should result in a net increase in deposited material. In remote areas, changes in pollution concentrations are associated with the passage of polluted regional scale air masses. Since air concentrations will be largely independent of wind speed, deposition should increase with increased wind velocity.

In addition to wind speed, the stability of the atmosphere plays an important role with regard to both deposition velocity and pollutant concentration. Deposition velocities (aerodynamic component) tend to be increased by greater atmospheric turbulence, which tends to occur during the midday hours (on non cloudy days). Whether an increase in material corrosion will follow depends on the simultaneous behavior of air concentration and surface wetness, both of which will have their own typical diurnal patterns.

## Urban Effects

Much of the extant data on rates of dry and wet deposition has been derived from rural measurements, intended to provide "regional representativeness" for ecological studies. Since the highest densities of materials at risk to air pollution tend to be located in cities, it is important to understand any urban-rural differences that might affect rates of deposition.

Air concentrations of many air pollutants tend to be higher in cities, because of local sources. Exceptions may include ozone (because of titration by NO) and aerosol acidity (because of neutralization by basic particles and ammonia). In the US, some of the highest concentrations of $SO_2$ now occur in the vicinities of rural point sources, rather than in cities. As a result, seasonal and diurnal concentration patterns of $SO_2$ have changed as well. Wet deposition of sulfur is also higher in some cities, but as discussed above, deposition of $H^+$ in cities depends also on the presence of neutralizing agents such as $Ca^{++}$ and $NH_4^+$.

Deposition of air pollutants depends on several other atmospheric factors besides air concentration. Cities can affect the deposition climatology relative to rural areas as follows:
- wind speeds are higher at night and lower in daytime
- urban atmospheres tend to be less stable
- high relative humidities occur less frequently (times of surface wetness will be lower).
- urban precipitation amounts are slightly higher.

All of these factors must be considered when extrapolating materials service lives from urban to rural situations, or vice versa.

## Previous Reviews of Air Pollution Damage to Materials

As might have been expected with so venerable a topic, the technical literature is relatively rich in discussions of atmospheric effects on materials. Many of the previous reviews have been rather qualitative and only since the advent of serious consideration of air pollution controls to abate acid precipitation has a comprehensive research effort been mounted to fully understand cause-and-effect relationships and the resulting economic ramifications. The following literature review is necessarily subjective, but includes many of the important reference works. In this chapter, the emphasis in selection of literature is on comprehensive reviews and studies developing quantitative relationships between materials damage rates and environmental factors, as opposed to detailed studies of damage mechanisms. Information on the latter may be obtained from any of the excellent bibliographies referenced below.

## Descriptive Studies

Stern's reference work on air pollution includes a chapter on materials damage [38]. The chapter emphasizes $SO_2$ effects, presents some empirical data and photographic examples of damage, and includes discussion of metals, building materials (masonry), paint, leather, paper, textiles, dyes, elastomers and rubber, glass and ceramics, as well as brief discussions of economics. However, the relevant atmospheric processes and effects of acidic precipitation were not included in this chapter.

Most (metallic) corrosion reference works and handbooks, such as that by Schreier [39] contain chapters on atmospheric corrosion, with emphasis on qualitative descriptions and selection of appropriate engineering materials to yield adequate service under specified general conditions (such as "industrial" or "marine" atmospheres). Barton's book on metallic corrosion [40] adds considerable theoretical insight, with emphasis on the European literature. Consideration of precipitation effects is also largely lacking in all of these works. Although Barton and virtually all other authors emphasize the role of atmospheric moisture in providing the electrolyte necessary for electrochemistry to proceed (galvanic action), they have not recognized the additional role of surface moisture with respect to facilitating surface deposition of $SO_2$ and other water soluble gases. This recognition has largely come from the atmospheric sciences community.

Sulfur effects on various materials are discussed by Nriagu [2], including both mechanisms of damage and some of the empirical damage functions that had been proposed at that time. This is one of the first such reviews to discuss the role of acidic precipitation in removing protective corrosion films. The chapter also includes effects of sulfates in aqueous systems and of high temperature corrosion in industrial processes involving sulfur-bearing gases.

Flynn et al. [41] prepared a very comprehensive bibliography listing over 1300 references concerned with atmospheric damage to metals, masonry, organics, and other materials. However, of these citations, the number providing quantitative information was much smaller, as was the number of references that considered precipitation chemistry as a factor.

The North Atlantic Treaty Organization (NATO) has considered materials damage through its Committee on Challenges to Modern Society (NATO/CCMS) [42]. Their document includes national reports from various contributing countries:
- Germany, emphasizing native stone types, examples of degradation, and weather influences
- the Netherlands, emphasizing empirically-based sulfur oxides effects
- Norway, presenting quantitative metals corrosion data from a 4-year test program
- United States, a general review of the topic.

A summary of air pollution damage effects on materials was prepared by Benarie for the Commission of the European Communities [42a]. This report predates most of the newer research motivated by acid rain concerns and attempts to place the data available at that time into a theoretical physico-chemical framework aimed at postulating damage functions. The work emphasizes metal corrosion, but also discusses damage to paints, textiles, elastomers, marble and concrete and includes some ozone and $NO_x$ effects. An appendix discusses air pollution effects on stress-corrosion cracking (catastrophic failures).

Graedel and McGill [4] present a largely qualitative review of atmospheric effects on materials, in matrix format. They present data on a number of pollutants not previously discussed, including water, ammonia, carbonyl sulfide, hydrochloric acid, organic acids, and hydrogen peroxide. The information on pollutants includes estimates of concentration trends. Materials included by Graedel and McGill which are not often discussed by others include solder, polymers, and several indoor materials. Matrices of severity of effects are presented in six different contexts: urban atmospheric gases, rain in the Northeastern US, atmospheric particles in the central US, dew in Detroit, fog in Los Angeles, and indoor air in a mobile home. Unfortunately, the matrix presentation is quite complex and requires a full color reproduction in order to grasp all the information available. Criticisms of this review include the neglect of nitric acid and the assignment of water effects on carbonate stone as "low".

The most recent comprehensive literature review on atmospheric damage to materials was prepared by Jorg et al. [43] for the West German government. In addition to qualitative discussions of damage mechanisms and a review of published materials damage "dose-response" functions, it lists 998 references, with emphasis on European sources, especially the "gray" literature. References are cross-classified by material as well as by pollutant, including acid precipitation. A number of polymeric and organic materials are included, including paints.

*Economic Studies*

Various types of economic studies of the costs of atmospheric corrosion have appeared in the literature, beginning with early smoke abatement studies which emphasized soiling. Economic studies are important for damage to common construction materials since the primary concern is in the aggregate, i.e., damage summed over some suitably large area rather than to an individual building or group of buildings. However, the reliability of such economic estimates depends strongly on knowledge of damage processes at the individual building or component level.

Salmon [44] made estimates of the annual materials damage costs for the US, based on production quantities, economic life, and estimated fractions of material quantities exposed to air pollution. Extrapolated to today's dollar values, his results would amount to several billion (US) dollars annually for damage to galvanized and painted surfaces.

Fink et al. [45] addressed a similar objective, estimating annual corrosion costs for painted and galvanized steel at $ 1–45 billion in 1970; the largest portion was associated with industrial roofing and siding. This analysis was based on observed differentials between "clean" and "polluted" areas. Both the Fink et al. and Salmon estimates are based on environmental conditions prior to large scale air pollution abatement in the US and do not explicitly consider the effects of precipitation acidity on rates of materials damage.

A more recent estimate of the cost of all metallic corrosion processes in the US was presented by Meredith [46]: 4% of gross national product or $ 122 billion in 1982. Meredith also compared this figure to previous estimates for other countries at different times, which ranged from 1.8 to 3.5% of GNP. However, of this total amount, only about 14% was considered "avoidable" damage costs and about 9% of the total was associated with automobiles.

In West Germany, the Umweltbundesamt estimated 1980 costs at DM 1.5 billion for damage to buildings (excluding cultural objects), 2.0 billion for corrosion, and 1.0 billion for soiling, totaling DM 4.5 billion not including the embedded costs for corrosion prevention [47]. It is not clear how much of this total would be considered "avoidable", but this estimate corresponds to US$ 44 per capita.

A more limited economic assessment was performed by Murray et al. [48] for the Los Angeles air basin. The materials included were painted surfaces, galvanized steel, and outdoor statuary. Pollutants included $SO_2$, $NO_2$, ozone, and particulates; neither nitric acid nor acid precipitation were included because of a lack of suitable damage functions. The resulting economic estimates were dominated by damage to paints and totaled about US$ 4/per capita. Soiling cots would add a like amount, but there was some uncertainty as to whether adding soiling damage to the acidic pollutant damage would constitute double counting.

On a per capita basis, these various estimates of "avoidable" air pollution damage costs to materials have ranged from $ 2 to $ 30 annually (1980 US dollars), corresponding to between 0.10 and 0.53% of GNP. Manning gives estimates for the UK of between US$ 9–27, or 10% of total corrosion costs [49]. The corresponding figure for the 1985 NAPAP assessment (see below) was US$ 32 per capita, with a range from US$ 8.50 to US$ 105 per capita (these figures were later revised downward by about 20%).

*Recent Comprehensive Technical Reviews*

*Symposia.* The effects of acid deposition on materials damage were specifically addressed in two US conferences in 1985: The American Chemical Society's Symposium "Materials Degradation Caused by Acid Rain" (R. Baboian, ed. 1986 [50]) and the Electrochemical Society's Symposium "Corrosion Effects of Acid Deposition" (Mansfeld, Kucera, Haagenrud, Haynie, and Sinclair, eds. [51]). Both conferences included accounts of ongoing work in the various national research programs in the US and Europe, although only the ACS Symposium included work on non-metallics (masonry, painted wood, woody plants, and nylon).

Both conference proceedings include brief accounts of the US National Acid Precipitation Assessment Program (NAPAP) economic assessment efforts (the NAPAP "1985 Assessment"), which have never been officially released by NAPAP. The pros and cons of this effort were recently summarized, based on a 1986 technical symposium by the Air Pollution Control Association [52]. The 1985 Assessment estimated annual damages of $ 2.2 billion, due to all acid deposition, with upper and lower bounds of plausibility (not statistical confidence limits) of $ 0.7 and $ 6.7 billion. These estimates were based on ca. 1980 conditions and did not attempt to estimate what portion of the total could conceivably be saved through air pollution abatement. The estimated US costs due to acid deposition are seen to fall within the ranges for other countries, summarized above.

The International Conference on Acid Rain, held in Lisbon in September 1987, featured a session on materials damage. The papers presented dealt with damage to stone, metals, and paints (risk assessment methods).

*Reports.* A summary of the state of the art of acid deposition damage to materials was prepared for the Electric Power Research Institute (EPRI) [29]. This report includes sections on atmospheric processes and mechanisms of damage, as well as on ongoing research activities. No attempts are made at economic analysis.

In Canada, a summary report [53] was also recently produced on the topic. However, it dealt mainly with older data and not at all with damage done by precipitation per se. The Canadian report also discussed differential damage rates on different parts of a structure, a topic largely neglected by others.

*The NAPAP Interim Assessment* [54]. This four volume document represents the US Federal Government's first official comprehensive "assessment" publication under the Acid Precipitation Act of 1980 (P.L. 96-294, Title VII) and was released in September 1987. As such, it responds to the Congressional mandate for "economic assessments of the environmental impacts caused by acidic precipitation on . . . aesthetic resources and structures . . . " The accompanying Federal press release stated "acidic deposition can accelerate the natural weathering of materials. The degree of the acceleration is an open question, and is the subject of current research within NAPAP."

Volume I of the Interim Assessment, the Executive Summary, contained a one page synopsis dealing with materials damage, emphasizing dose-response research conducted by NAPAP at five field exposure test sites and in laboratory chambers. Three classes of materials were emphasized: galvanized steel (zinc), carbonate building stone, and surface coatings (paints). Effects on cultural resources were not specifically mentioned. The synopsis listed a number of research topics which need to be addressed before a "credible economic assessment" can be made. The use of data from concurrent and previous research by others was apparently not considered in this regard.

Volume III of the Assessment (Atmospheric Processes) contains a brief mention of dry deposition of pollutant gases to material surfaces (Chapter 4) and the importance of surface wetness in this regard. Chapter 5, which discusses air quality and deposition data, presents little information on urban levels, with the exception of Dupuis' analysis of wet deposition in the New York City metropolitan area [7].

Volume IV presents more details on the NAPAP materials damage research (Chapter 9) and the research results achieved as of early 1987. The chapter is oriented towards the needs of economic assessments and contains sections on failure modes, selection of materials for study, weathering processes, development of dose-response functions (for metals, stone, and painted surfaces), distributions of materials, and application to real structures. The research results presented are limited to dose-response function development and include:

- a 1:1 molar relationship between $Zn^{++}$ and $SO_x^=$ in artificially induced dew, based on $SO_2$ exposure chamber tests of galvanized steel.
- a proportional relationship between $Ca^+$ in precipitation runoff from marble test slabs and incident $H^+$ loading from precipitation (note that the effect of precipitation per se was not isolated).
- a review of theoretical and experimental findings with regard to air pollution damage to paints.

The chapter concludes that:
- a qualitative understanding of the effects of wet and dry deposition on galvanized steel has been obtained.
- a direct relationship between carbonate stone surface recession and hydrogen ion loading has been demonstrated.
- previous economic estimates which "related large costs for repainting due to pollutant damage were premature" (this refers to the 1985 NAPAP Assessment, mentioned above).
- significant information gaps for an economic assessment remain.

Recommendations include improved environmental monitoring at exposure test sites, continuing the experimental exposure work, considering additional materials, defining damage effects to real structures, and improving the data bases on spatial distributions of materials.

**Separating Pollution-Induced Damage from Natural Weathering**

All materials degrade to some extent in the natural environment. Examples include dissolution of carbonate stones (marble, limestone) by rain containing dissolved $CO_2$, rusting of iron, ultra-violet radiation effects and oxidation of paint films. In order to ascertain how much improvement in service life could result from reduced air pollution, careful theoretical and experimental work is required, especially since pollution-induced deterioration is often manifested as an acceleration of the "natural" processes rather than as a distinct observable impact of its own.

Prior to the identification of the need for such differentiation, materials scientists and engineers were largely content to test candidate new materials under various reference atmospheric conditions, such as "industrial" (often on the premises of the material's production facility), "marine" (a specified distance from the sea), or "rural" (remote from cities or industrial facilities). However, such testing has never emphasized urban atmospheres per se. The American Society for Testing Materials (ASTM) has been conducting such tests for over 75 years [55]. However, during that time, conditions in the US have changed dramatically for the better in most cities, due to reduced levels of primary air pollutants, but often for the worse in Northeastern US rural areas due to transported air pollution including acid rain. Thus site to site comparisons over time have been confounded. ASTM-type testing at a single site generally has the goal of comparative evaluation of several candidate materials, often exposed until failure occurs. While such tests are useful in terms of identifying better material performance, they cannot identify which specific agents in the atmosphere may have contributed to the relative performances observed. Quantitative mathematical expressions for this purpose, relating material damage rates to concentrations or deposition rates of specific environmental agents, are termed dose-response or damage functions.

The purpose of a damage function is to identify the factors that control corrosion or materials degradation, *in an average sense*. To the extent that other factors exist at a specific time or location that are not included in a damage function (such as variations in material composition or configuration, other pollutants or atmospheric variables), the absolute value of damage to a specific structure or component may not be predicted accurately. However, as long as these factors do not interact with the factors that are included in the damage function, the relative effects of the factors that are included should still be predicted satisfactorily. One should remember that the intended purpose of such damage functions often is to make estimates of economic losses aggregated over a large number of structures and locations. One thus hopes that the uncertainties for individual applications will average out.

There have been many attempts to derive empirical damage functions for materials degradation, notably for metals corrosion, but relatively little attention has been given to systematic experimental designs for this purpose, from first principles. A recent paper by Haynie and Lipfert [56] was intended to provide guidance for such an idealistic experimental program design.

For the purposes of environmental assessment or cost benefit analysis, "damage" must be defined in terms of loss of function. Often, this may involve both a structural and an aesthetic element; processes leading to such "damage" may include loss of mass, discoloration, loss of surface finish (including paint film failure), or structural failure. Actual costs may only be incurred when some critical damage level has been reached requiring maintenance or replacement. However, the processes leading up to these critical levels constitute damage as well, and the economic evaluations must account for the time required to reach critical levels, assuming such time does not exceed the useful lifetime of the object in question.

*Requirements for Damage Functions*

In order to be useful, damage functions must be credible and widely applicable. Credibility requires a number of elements in order to satisfy potential users:
- consistency with underlying theory
- verification under realistic ambient conditions, encompassing the conditions under which the functions will be used
- statistical robustness of the verification data.

The applicability of a damage function is related to its completeness, i.e., the extent to which it characterizes atmospheric conditions. Ultimately, damage functions must be verified under realistic field exposure conditions in order to be credible, and thus must account for all of the natural and manmade forces in the atmosphere that may influence rates of degradation. This can be a very long list that includes:
- various air pollutants, including both suspended and deposited dusts and precipitation-borne contamination
- various forms of atmospheric moisture, including rain, snow, dew, frost, fog, etc., and the attendant variations in atmospheric and material surface moisture
- thermal variations, including sunlight, ultraviolet (uv) radiation, freeze-thaw cycles, and thermal lag with respect to the atmosphere.

For a cost/benefit analysis, only the pollutants that are subject to possible future controls need be included. However, interactions of many pollutants are often subtle and difficult to detect with field experiments of limited scope. For example, $SO_2$ and suspended particulate matter (TSP) are often covariant because of common sources, as are wet and dry deposition of sulfur oxides. A damage function that fails to separate the effects of these covarying agents will not be universally applicable, for example in areas of high $SO_2$ but low TSP or vice versa. Since there are many interacting agents in the environment, this requirement, which is related to statistical robustness of field data, can lead to a need for expanded field verification programs in order to obtain sufficient observations of the important damage agents, taken one at a time.

*Three-Step Damage Function Development*

This section discusses an optimal way to combine theoretical and experimental approaches to damage function development. As a first step, theory can be used to postulate the appropriate mathematical forms for damage to specific materials, leading at best to a complete theoretical model requiring only field verification of coefficients. At the very least, theory may be used to predict which environmental agents are likely to be important and what forms of damage might result (pitting, erosion, cracking, etc.). It is also important to establish the proper mathematical form (linear, exponential, etc.) as a framework for statistical analysis of field data. For many materials, theoretical information may be limited and laboratory experiments may be needed as a second phase in damage function development. Environmental simulation chambers are perhaps the most cost effective way to rapidly screen for the most important direct and interacting pollutant effects. Often, it is necessary to sacrifice some environmental realism in order to obtain first order estimates of the main effects. These quick answers may considerably reduce the scope of subsequent field exposure verification campaigns.

The third step in the ideal damage function development process is the use of field exposure tests to verify theoretical expectations and to supply data to be used to estimate the coefficients for the damage function models. The nature of such field tests is highly dependent on the type of damage that is to be predicted and the time history of its development. For example, the process of dissolution or erosion of certain materials may take place quickly enough and be sufficiently linear to allow average rates to be determined from analysis of the chemistry of precipitation runoff. In such cases, causality may be inferred from time series analysis of material losses versus pollutant input. However, for forms of damage requiring substantial time to develop, such as paint film failure by corrosion of underlying metal substrates, or when the large numbers of potentially important variables may confound the analysis of field test data, preliminary investigations in chambers or by theoretical means can have a large payoff.

Derivation of Theoretical Functional Forms

By its very nature, degradation is a process which increases with time. However, the process of degradation may not be observable in some instances until some specific amount of time has passed. In those cases, time to failure becomes a measure of the rate of degradation. The rates at which physical and chemical processes occur control the rates at which measures of degradation can be observed.

*Temporal Rate Considerations.* One of the most generally applied rate functions for metals corrosion has the form [57, 58, 76]:

$$C = At^b \tag{1}$$

where C is a measure of corrosion loss, t is time, and A and b are empirical coefficients. The exponent, b, usually has a value between 0.5 and 1.0. This empirical form is consistent with competing mechanisms of formation and dissolution of protective corrosion product films [59]. When a protective film is present through which atmospheric pollution must diffuse in order for damage to proceed, the equations are parabolic and $b = 0.5$. When the corrosion products are highly soluble and thus a fresh surface is more or less continually exposed, the system is linear and $b = 1.0$. Thus it can readily be surmised that the acidity of precipitation may play a role in determining the operational value of the exponent b.

*Other Theoretical Considerations.* The uniform erosion of the surface of a single component reactive material, such as zinc or a dense carbonate stone, is a relatively simple system to describe theoretically. However, other processes such as stress-corrosion cracking or crevice corrosion for metals or intergranular deterioration processes for calcareous stones or mortar are much more difficult. Additional factors must be considered and degradation processes may be discontinuous.

The development of theoretical damage functions for painted surfaces is complicated by the fact that there are many possible mechanisms for paint failure. A theoretical model must thus include potentially competing mechanisms for failure. Underlying all failure mechanisms is the steady erosion of material from the paint film. For paints containing carbonates as fillers or extenders, this erosion process is sometimes referred to as "chalking." In an ideal situation, this is the intended mechanism of ultimate paint failure. However, under real world conditions of application and surface conditions, other processes may interfere and result in premature failures of the paint coating system. These may include delamination from previously painted surfaces, cracking, checking, peeling, etc. In general, it will be difficult to model the hierarchy of possible paint failure modes, although it may be possible to provide some theoretical guidance as to causal mechanisms for each individual failure mode.

*Damage Function Models.* In addition to the time dependence relationship (Eq. 1), damage functions are usually based on mass balance, in which corrosion products result from described chemical reactions based on deposited pollutants. This requires knowledge of a number of elements:
- rates of pollutant deposition
- system of operational reactions
- degree to which chemical and physical equilibrium is achieved.

Each of these factors will be specific to combinations of pollutants and materials (including multi-pollutant mixtures), atmospheric conditions, and configuration of the exposed surface. Theory can only be used to describe these processes in a general sense, especially where kinetic limitations may exist. The details must be provided through experiments. Statistically designed controlled environment experiments (chamber studies) are used to separate the possible interacting effects of variables. Field studies are needed to confirm that the damage mechanisms are applicable under the more complex and dynamic real world conditions.

Experimental Methods for Damage Function Development

Methods for measuring atmospheric damage to materials include:
- Gravimetric: changes in test sample weight before and after some exposure period; most useful for testing metals. Substances that form during corrosion (corrosion products) must be removed prior to final weighing. Non-uniform corrosion such as pitting must be accounted for.
- Corrosion film composition: analysis of the chemical composition of corrosion products found on exposed surfaces to infer the operative reactions. For samples exposed to rain, this method can be confounded by loss of water soluble species.
- Precipitation run-off chemistry: comparing the composition of precipitation before and after it contacts the test specimens to determine the amount of soluble material removed. However, this method cannot measure corrosion products left on or stored in the material, such as in stone pores, and must account for dry deposits between precipitation events.
- Qualitative inspection: paint damage may have to be inferred from differences between exposed and control specimens in terms of qualitative measures such as fading or discoloration, loss of gloss, presence of surface cracks, peeling, etc. Some of these attributes may be amenable to quantitative surface characterization techniques as well.

Ideally, one would like to combine various methods in the same experiment, in order to have confidence in the mass balance and in the different roles of the various agents in the atmosphere. In any event, converting these damage measurements into quantitative damage functions is a statistical analysis task and can be hindered by the usual data analysis problems of confounding or unmeasured variables, too few observations, experimental error, etc. Also, if the damage mechanism is nonlinear in time, use of short term exposure data to infer time to failure may be misleading. Both outdoor exposures to the natural environment and tests under controlled laboratory conditions may be required in order to expeditiously derive meaningful damage functions.

*Chamber Studies.* Controlled environment experiments can be performed in enclosed systems in which the atmosphere is simulated by means of flowing gases. Useful information obtained from such experiments is optimized when they are statistically designed so that both direct and synergistic effects of factors that are expected to be important are observable and separable. In such experiments, no attempt is made to duplicate real world conditions per se although important factors found in the real world should be included in the design. For example, it is important that an adequate flow velocity be provided through the chamber in order to simulate realistic atmospheric deposition processes.

One advantage of chamber studies is that levels of environmental factors can be used that can accelerate the rate of deterioration with regard to real world conditions. Less time is thus required before measurable damage occurs. Another advantage is that exposure levels can be changed and held constant allowing experimental variability to be measured. Often, mechanisms of deterioration can be observed more readily in a laboratory than in the field.

There are two major problems that must be considered with chamber studies. First, an important factor present in the real world might be overlooked in the design of the laboratory experiment. Such an excluded factor may not necessarily produce a direct effect but could produce a synergistic effect with some factor that was included. Second, in an effort to accelerate deterioration by changing levels of factors from real world levels, the mechanism of deterioration could also be changed. Excessively low rain pH levels may provide such an example; changes in the factors controlling dry deposition of various species may be another.

*Field Exposure Studies.* Most of the existing data on the atmospheric corrosion of metals have been obtained from field exposure studies. The experimental technique has been to expose specimens for different periods of time at different locations using standard practices to measure differences in weight change, pit depth, tensile strength, per centage of rusted area on coated steel substrates, etc. Most ASTM experiments have not measured environmental factors but classified them into general categories such as urban, industrial, marine, and rural. As discussed above, within the last twenty five years there has been a growing awareness of the need to measure environmental parameters in field studies in order to interpret the observed variability in corrosion behavior. However, the list of required measurements is growing as knowledge of atmospheric processes improves.

The purpose of field exposure tests is to quantify materials performance under real world conditions. This is an essential and often difficult step in the damage function development process. Chambers often cannot replicate atmospheric turbulence levels and thus the factors that influence rates of dry deposition. In addition, it is difficult to simulate atmospheric particulate loadings and precipitation processes in closed chambers.

Realistic exposures can be obtained by testing at a single "representative" exposure site. However, in order to reliably distinguish the separate effects of the many operational atmospheric/predictions factors requires careful statistical design. Typically, field exposure experimental designs are based either on cross-sectional comparisons or time series analysis.

*Cross-Sectional Statistical Analysis.* One of the traditional ways of determining a damage function is through site to site comparison of the performance of materials in relation to environmental parameters. The statistical technique used for such an effort is referred to as cross-sectional analysis. This technique has been used to estimate damage functions for several experimental programs using multiple linear regression analysis [58, 60, 61]. The traditional ASTM test site classifications of urban, industrial, rural, marine, etc., are not useful for this purpose unless the detailed environmental characteristics and deposition levels at each site are available.

When site to site comparisons are used to deduce damage functions, two different types of potential errors must be considered. First, if important environmental characteristics have been overlooked and not measured, some portion of the observed damage may be ascribed to the wrong pollutant. Secondly, it will be quite difficult to separate the effects of pollutants having similar spatial patterns.

To a certain extent this is the case with hydrogen ion ($H^+$) and $SO_2$ (Figs. 1 and 2, respectively) in Eastern North America. Both of these problems may be overcome by considering a large number of test sites with diverse characteristics [58].

*Time Series Analysis.* For a material with nearly immediate response to pollutant deposition, an alternative method of estimating a damage function involves correlating increments of damage with the time series of deposits, which will in general be quite variable due to the stochastic nature of the atmosphere. This method avoids the site to site collinearity problems discussed above, but requires careful measurement of the time histories of all pertinent environmental factors, not just the long term averages. This may be difficult for dry deposition, for example. Again, multiple regression analysis is usually required in order to identify the various time varying effects.

The easiest way of determining a time series of materials degradation responses is by analysis of precipitation run-off chemistry, in comparison to incident precipitation and run-off from a similar but chemically inert surface (blank). This technique has been used to infer calcite dissolution rates for marble and limestone at several of the NAPAP field exposure sites [62]. Disadvantages of this

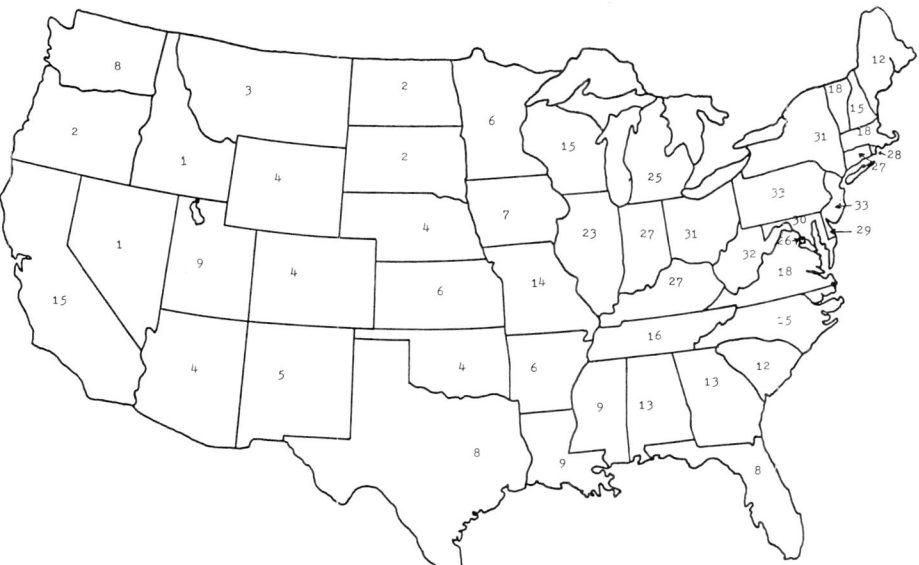

**Fig. 2.** Population-weighted $SO_2$ air concentrations, based on regional long range transport model calculations ($\mu g/m^3$)

technique include unsteady effects involving pollutant or corrosion product storage in the material (which may be quite porous in the case of stone) and cumulative effects due to progressive roughening of the exposed surfaces. A damage function resulting from such runoff studies represents only material dissolution; thus there may be an underestimate of the total damage if other loss mechanisms are also important. For stone, these may include loss of large particles through dissolution of the calcite binder, freeze-thaw cycles, etc. Nevertheless, assuming that the important damage processes are linear in time, the technique has promise in that results may be achieved within a fraction of the ultimate service life of the material.

*Semi-Controlled Field Experiments.* Recently some of the factors of gaseous and wet deposition have been physically separated by using semi-controlled field experiments [63]. At Research Triangle Park, NC, an automatic movable covering device excludes rain on half of a set of samples of galvanized steel and weathering steel. As seen by comparing the two sets of samples, rain accelerates the corrosion of galvanized steel by dissolving the corrosion product film. The corrosion rate of weathering steel is decreased by rain, which apparently washes away acids that form and accumulate during periods of dew. This experimental technique is superior to the use of fixed position rain shelters, which can affect (reduce) rates of dry deposition while sheltering from rain.

An alternative methodology is in use in the UK, involving outdoor $SO_2$ fumigation facilities [64]. The material samples are exposed to elevated, controlled $SO_2$ levels, tailored to match a predetermined pollutant frequency distribution. Preliminary results on limestone weight loss indicate that such a facility results in an $SO_2$ – weight loss dependency that is substantially stronger than what is observed under natural conditions [65]. This may be a result of the specific test site characteristics (near the coast) or perhaps of perturbing the natural air concentration – dry deposition relationships, by virtue of the $SO_2$ air concentration frequency distribution selected. Use of measured $SO_2$ deposition to the exposed surfaces (rather than air concentration) as a damage function parameter might help explain these results.

## Air Pollution Effects on Specific Materials

### Selection of Materials

Since each combination of pollutant and material comprises an unique case to be considered for research (Fig. 3), the number of cases may become quite large. It is necessary to invoke some practical constraints in order to focus the discussion on the cases of practical importance [66].

Damage by sulfur dioxide has been high on the list for many years, beginning with the early laboratory experiments on metals by Vernon [67]. Burdick and Barkely summarized tests by a number of investigations using very high $SO_2$ concentrations on cement, limestone, zinc, copper, steel, nickel, paints, leather, paper, and cotton fabrics [68]. These early chamber tests provided only qualitative indications of $SO_2$ damage since appropriate atmospheric rates of delivery

Material types (in order of estimated decreasing abundance)

| Pollutants (in order of increasing regionality) | Materials | | | | | | | | | | | | |
|---|---|---|---|---|---|---|---|---|---|---|---|---|---|
| | Concrete | Painted wood | Mortar | Painted steel | Aluminum | Galvanized steel | Plastics | Building stone | Copper | Weathering steel | Nickel | Bare carbon steel | Lead |
| Cl⁻ | | | | | | | | | | | | | |
| H₂S | | | | | | | | | | | | | |
| Coarse particles | | | | | | | | | | | | | |
| HCl | | * | | * | | * | | * | * | | | | |
| HCHO | | | | | | * | | | | | | | |
| HNO₃ | | * | | * | | ○ | | ○ | * | | | | |
| SO₂ | * | ○ | * | ○ | * | ○ | | ○ | * | * | * | * | * |
| NO₂ | | | | | | | | | | | | | |
| O₃ | | ○ | | ○ | | | * | | | | | | |
| Fine particles | | | | | | | | | | | | | |
| Precip. acid (H⁺) | * | ○ | * | ○ | * | | | ○ | | * | * | * | * |
| Precip. (H₂O) | | ○ | | ○ | | | | ○ | | | | | |

○ Full economic assessment    * Qualitative discussion    ☐ No data or not applicable

**Fig. 3.** Material-pollutant selection matrix. Source: Lipfert [66]

(dry deposition) were not simulated. However, they did establish the critical role of relative humidity in conjunction with SO₂ in shortening the lives of most of these materials.

Two separate concepts should be considered in selecting specific materials, either for research or for inclusion in a comprehensive discussion: abundance in use, and sensitivity to air pollution attack. Abundance refers to the amounts of surface area exposed to air pollution and thus requires data on the distribution of materials in place. Such data are difficult to acquire in sufficient detail [52, 69]. Sensitivity refers to the amount of pollution deposition required to effect a significant decrease in the useful life of the material. Sensitivity thus involves both the concept of reactivity (a chemical concept) and the amount of material loss that may be tolerated in a given situation or application. Galvanized steel is a sensitive material because zinc is reactive (to deposited acids, for example) and the amount of material loss that may be tolerated is small since the zinc is applied as a thin coating. In contrast, carbon steel railroad track may be reactive but will not be sensitive (because of the ample stock thickness). Marble or limestone may be sensitive when used as carved decorations, but probably not as window lintels or as steps. Materials that are both reactive and expensive to replace may also be deemed sensitive (such as carved stonework).

Over the years, consumers have learned to select materials that minimize maintenance expenses. Hence, in residential use in the more polluted parts of the US, galvanized steel rain gutters (sensitive) are seldom used and aluminum siding (insensitive) is often prevalent. We find that no single material or material grouping is both sensitive and abundant; as a result, economic assessments must often consider a range of types of materials. The discussions that follow will include metals, masonry (including concrete), and paints. For each group of materials, important mechanisms of damage are listed, potentially damaging pollutants are discussed, some of the significant research results are given with emphasis on damage functions, and some of the remaining unknowns are listed.

**Corrosion of Metals**

*General Mechanisms*

Metals corrode when their corrosion products are more stable than the base metal; this requires a decrease in the free energy when the corrosion product is formed. Since metals are good electrical conductors, practically all metallic corrosion is electrochemical, involving the flow of electric current through an electrochemical cell, entering the cell through the anode and leaving through the cathode. The exact nature of the cell will differ according to the situation, which may be classified for convenience as:
- direct corrosion, i.e., direct chemical attack, where electrical effects occur in the immediate vicinity of the reaction.
- electrochemical corrosion, in which the current flows an appreciable distance through parts of the metal in which corrosion is not occurring.
- oxidation, where a variety of electrochemistry processes take place, but all involving oxygen in the absence of moisture. In this case, the diffusion of metal ions and electrons is through the oxide film rather than through a liquid electrolyte.

In the first two instances, atmospheric moisture creates the necessary liquid electrolyte, and sufficient moisture will be present on the surface when relative humidity exceeds about 30%. A typical *direct corrosion* reaction for zinc might be:

$$Zn + 2H^+ \rightarrow Zn^{++} + H_2 \text{ (gas)} \qquad (2)$$

in which an acid solution dissolves the surface uniformly (no protective layers are formed). The acid solution might be either formed on the surface by the dissolution of $SO_2$ in surface moisture or deposited directly from either precipitation or perhaps nitric acid vapor.

**Table 3.** Examples of electromotive series [70]

| In sea water | Standard conditions |
|---|---|
| Anodic (corroded) end | |
| Magnesium | Magnesium |
| Zinc | Aluminum |
| Aluminum | Zinc |
| Cadmium | Chromium |
| Aluminum alloys | Iron |
| Steel | Cadmium |
| Wrought iron | Cobalt |
| Cast iron | Nickel |
| Stainless steel | Tin |
| Lead | Lead |
| Tin | Copper |
| Nickel | Silver |
| Brass | Platinum |
| Copper | Gold |
| Cathodic (protected) end | |

*Electrochemical corrosion* frequently involves two dissimilar metals (but not always), and is common in sea water environments, for example. This mechanism is also sometimes referred to as galvanic action. In such cases, two reactions take place, one at the anode and another at the cathode. Anodic reactions involve metal ions going into solution and thus loss of (solid) material. Cathodic reactions are essentially reductive, and thus generally do not involve the loss of metal since most metals cannot be further reduced. In order to determine which of the two metals is cathodic and thus protected, the concept of the *electromotive series* is useful. This series is a list of metals arranged in decreasing order of their tendencies to ionize by losing electrons (see Table 3 for example). This order will depend somewhat on the environment and cannot be predicted a priori with certainty for metals that are near one another in the series.

The metal that is higher in the series will be sacrificed in favor of the one that is lower, when connected by an electrolyte. While this concept is extremely important in marine environments, it may be less so for atmospheric corrosion. For example, a galvanized steel surface will corrode first by direct chemical attack. A galvanic cell will only be set up through imperfections in the coating (including cracks, edges, etc.), such that the underlying steel is exposed to moisture, or at the point when enough zinc has been worn away to expose steel (in which case there may be direct attack on the steel and electrochemical acceleration of the zinc loss). In contrast, nickel and chromium plating are used primarily for appearance (rather than galvanic protection of the underlying steel), and tin plate, while cathodic to steel in the presence of air, is anodic when exposed to nearly air-free food products [71]. Tin plate was also used for roofing (terne plate) many years ago, and might still be found in some historic buildings. Lead coatings, on the other hand, will protect steel only by means of excluding contact with the atmosphere, and thus will fail where imperfections exist in the coating. If copper and steel come into contact in the presence of an electrolyte, the steel will be sacrificed. This was the case with the internal structure of the Statue of Liberty in New York Harbor, because of failure of the insulating material originally placed between the copper skin and its steel supports.

It is also possible for galvanic cells to be set up on the surface of a single metal by means of concentration cells, because of slight inhomogeneities in the metal or its immediate environment. Particle deposits may be one such mechanism; joints and crevices are another. Severe localized corrosion can result, sometimes as pitting.

The *oxidation* corrosion mechanism (which is not intended to include the rusting of iron because of the importance of moisture to that process) can produce either oxide films (less than about 0.3 µm thick) or scales. The presence of a film generally decreases the rate of oxidation, but the thicker scale layers may be either protective or not, depending on whether the volume of the oxide is at least as great as the volume of the base metal which it was formed. Aluminum oxide (film) is almost completely protective, and anodized aluminum, in which a thicker oxide coating is applied electrolytically, provides even better corrosion resistance. The formation of aluminum oxide is probably one of the reasons that aluminum-zinc alloy coatings (with sufficient Al content) protect steel better than galvanizing alone.

The fate of the reaction products determines the rate of further corrosion. According to the well-known relationship for loss rate as a function of time (which was discussed above):

$$C = At^b \tag{1}$$

The value of the exponent b determines whether the rate of loss C will continue at a constant rate into the future ($b = 1$) or whether it will slow down because of the protective action of the corrosion products that are formed ($b > 1$). If the process becomes limited by the rate at which ions or electrons can diffuse through the layer of corrosion products on the surface, $b = 0.5$. Many practical situations lie between these two extremes, depending on environmental conditions and the age of the exposed material. It is necessary to consider the action of the environment with respect to the stability of these corrosion products; if they are easily removed by precipitation, the protection will be lost.

Most metal oxides are relatively insoluble, even in dilute acids, which is one of the reasons that aluminum is resistant to both dry and wet acidic deposition. In contrast, most nitrates and some sulfates are highly soluble, and it is unlikely that nitrate would be an important component of corrosion product films remaining on surfaces. Since no extant field measurement programs have monitored ambient nitric acid, its contribution to atmospheric corrosion rates has not been quantified. Standard solubility tables [72] may be consulted for solubility data for other potential corrosion products.

## Corrosion of Zinc

*Mechanisms.* Many empirically-based damage functions have been proposed for zinc or galvanized steel based on the results of field testing, but a reasonably complete understanding of the actual mechanisms of damage has only recently begun to emerge. As discussed above, since zinc is higher (anodic) than steel in the electromotive series (Table 3), in a galvanic cell it is sacrificed and the steel base metal becomes the cathode and is protected. This is an important mode of zinc loss once the integrity of the coating on steel has been breached, due to scratches or local chemical attack.

In a clean atmosphere, the products of corrosion may be zinc oxide, zinc hydroxide, or zinc carbonates (from atmospheric $CO_2$). To the extent that these compounds are insoluble (like ZnO, for example) they will passivate the surface and tend to protect it from the atmosphere. In this case, the exponent b in Eq. 1 would take on the value 0.5, representing the case for which diffusion through the corrosion product film controls the process. As anecdotal evidence for this protective mechanism, galvanized steel still appears to be intact in dry atmospheres in remote places in the Western US after 50–100 years of exposure.

However, zinc carbonate is weakly soluble in water and the solubility increases with decreasing pH. This is one of the mechanisms for acid precipitation to accelerate the loss of galvanized steel coatings, in addition to the mechanism of direct chemical attack. However, even "clean" rain can dissolve some corrosion product, a factor which has only recently been included in damage functions [73].

*Important Pollutants.* The action of $SO_2$ on zinc corrosion is essentially limited by the rate of dry deposition. Both chamber tests [9] and a statistical analysis of extant field tests [58] have shown a stoichiometric relationship for deposited sulfur, i.e., one mole of sulfur deposited results in the loss of one mole of zinc. This situation occurs because of the reactivity of zinc, in that the moisture layers which facilitate the deposition of $SO_2$ (through dissolution) are buffered by the soluble reaction products, such as $Zn(OH)_2 \cdot ZnSO_4$, such that the pH of the layer remains high enough for $SO_2$ to continue to be taken up (around $pH = 6.5$, independent of incident rain pH [74]), as discussed above. The presence of surface moisture, i.e., the time of wetness, is thus an important factor in establishing the effective annual average deposition velocity. For this reason, the effective $SO_2$ air concentration with respect to damage to zinc and many other metals is the air concentration multiplied by the fractional time of surface wetness. With respect to damage function formulation, use of $SO_2$ deposition flux would be a better parameter, since it accounts for both aerodynamic and surface resistances.

If other contaminants in the atmosphere are present, modifications can occur to these simple processes. For example, salt (NaCl) particles are hygroscopic and can increase the time of wetness, in addition to supplying chloride ion to the reaction. Alkaline particles, such as most ordinary soil dusts, can provide neutralization capacity for both wet and dry acidic deposition. Particles on the surface can also increase the effective surface area for $SO_2$ dry deposition, at least until the next precipitation event washes them off. This phenomenon can lead to differences in corrosion rate between rain sheltered and rain exposed surfaces beyond the direct effects of precipitation, with the nature of the rain effect depending on the acidity of the particles.

Oxidants can affect the rate at which dissolved $SO_2$ ($SO_3^=$) is oxidized to the less volatile form $SO_4^=$, which is more likely to persist on the surface once formed;

ozone is probably less important in this regard than hydrogen peroxide ($H_2O_2$) because of the latter compound's higher solubility in water.

Other atmospheric acids may also be important for zinc corrosion. Nitric acid ($HNO_3$), normally present in the atmosphere as a vapor, deposits quite readily to most surfaces, wet or dry. Since the corrosion product $Zn(NO_3)_2$ is readily soluble, this could be an important loss mechanism because it operates all the time, not just when surfaces are wet. However, nitric acid concentrations are thought to be much lower than $SO_2$ in most locations, limiting the importance of this mechanism. Formic acid (HCOOH) results from the dissolution and subsequent oxidation of formaldehyde (HCHO) in surface moisture, and has been shown to attack zinc in much the same way as $SO_2$ [10]. This could be a source of indoor corrosion (unrelated to acidic deposition) because of the common indoor sources of formaldehyde. Outdoor formaldehyde is a component of photochemical smog. Excess acidity on material surfaces from such sources could also, in theory, act to limit $SO_2$ dry deposition by lowering the pH of surface moisture layers and thus reducing the effective Henry's Law coefficient [13]. Such interference makes the task of interpreting field test data quite difficult, especially when soluble corrosion products are involved and precipitation runoff is not captured for analysis.

*Damage Functions.* Many authors have postulated damage functions based on corrosion tests from individual test programs [60, 61]; see for example the 1980 conference proceedings edited by Ailor [74]. However, much less effort has been devoted to validating candidate damage functions from one program against data from another, and little attention has been given to precipitation factors, in general.

Lipfert et al. [58] attempted to address these difficulties by combining data from 9 major test programs comprising 72 sites (256 observations) and estimating precipitation acidity where necessary. The pooled data were found to be reasonably consistent with a number of different damage function models, all loosely based on Eq. (1) and following Benarie's original development [76]. Eq. (3) below was selected by the authors as providing the best fit of the data consistent with the above physical and chemical principles:

$$M = \{t^{(0.779 + 0.0456 \ln H^+)}\}\{4.2 + 0.55(f \cdot SO_2) + 0.019 Cl^- + 0.03 H^+\} \quad (3)$$

where M is the cumulative mass loss of zinc ($g/m^2$), t is the exposure time (years), $H^+$ is the wet deposition of acidity ($meq/m^2 y$), f is the fraction of time the surface is wet (deduced from relative humidity statistics), $SO_2$ is the annual average gaseous concentration ($\mu g/m^3$) and $Cl^-$ is the estimated chloride ion particle deposition rate ($mg/m^2/day$). This relationship explained over 85% of the observed variance in zinc corrosion. Dustfall deposition data were available for this analysis, but this variable was not statistically significant, perhaps because of the opposing aspects of the alkalinity of the particles per se and the increased surface area for $SO_2$ adsorption. This could have been tested in the regression by including an interaction term, $SO_2 \times$ dust.

The exponent for exposure time ($0.779 + 0.0456 \ln H^+$) indicates the tendency for a linear temporal relationship to be achieved (no surface passivation) as precipitation acidity increases; according to this relationship, the growth in corrosion with time would be linear (exponent = 1) for $H^+ = 128$ (1.28 m of rain per year at pH = 4 or 1 m/y at pH = 3.9). However, according to this relationship, the temporal exponent will never drop below about 0.8, indicating only a modest degree of surface passivation even in the absence of precipitation acidity.

The $f \cdot SO_2$ term indicates the role of surface wetness in controlling deposition; the empirical constant 0.55 is consistent with a stoichiometric relationship and minimal surface resistance to deposition to a wet surface. A further refinement of this damage function might involve a wind speed dependence in the deposition velocity relationship. The small coefficient for chloride indicates the relative unimportance of this mechanism, after all the other factors (including time of wetness) have been accounted for. The constant 4.2 represents factors not included in the damage function, which could be the "clean" rain effect or the corrosion caused by $CO_2$ or by other unmeasured pollutants at the sites from which these data were taken. There could also be a dependence on ambient temperature, which affects the $CO_2$ content of rain and also reaction rates. Many of these concepts are discussed by Haynie [59].

The zinc damage function (Eq. 3) is represented in terms of $SO_2$ and precipitation pH (based on 1 m of precipitation annually) in Fig. 4. When combined with the geographic distribution of $H^+$ (Fig. 1) and $SO_2$ (Fig. 2), this function maps into a service life distribution shown in Fig. 5, assuming that service life is defined by the appearance of 5% rusting and that the original coating was 381 g/m². These environmental data are based on non urban conditions and indicate a service life for galvanized steel of about 20 years at the worst location. Note that many US manufacturers of metal buildings now offer a 20-year guarantees, but that these buildings are usually constructed from materials more durable than galvanized steel, such as aluminum coated steel.

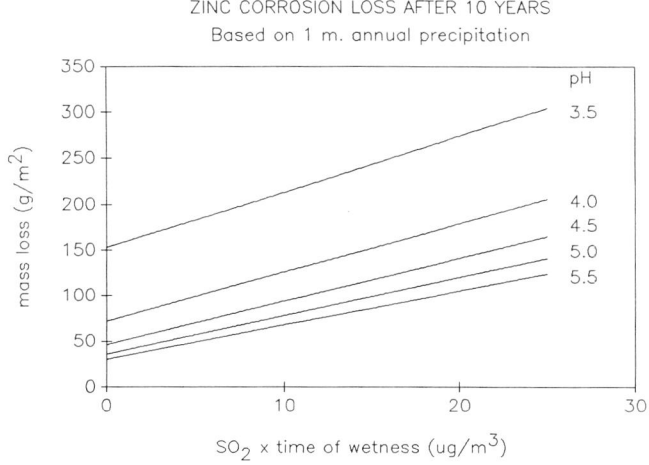

**Fig. 4.** Parametric plot of zinc (galvanized steel) damage function. Corrosion loss is shown for 10 years exposure based on 1 m. annual precipitation. Adapted from Lipfert [78]

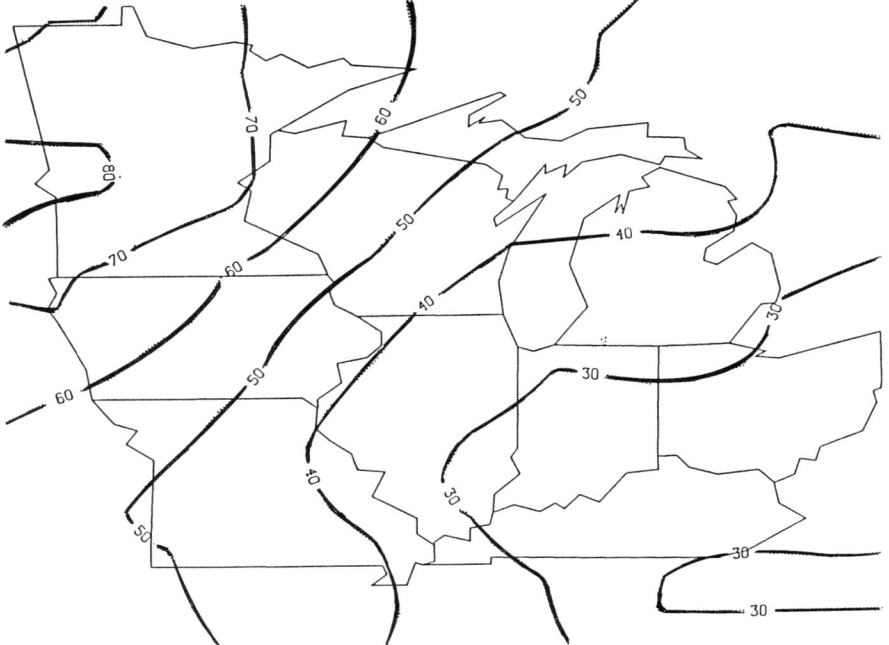

**Fig. 5.** Estimated service life of galvanized steel (yrs). Source: Lipfert [79]

Strictly speaking, these estimates are for small flat test samples. Fence wire and the like would be expected to deteriorate more quickly and large flat panels (roofing or siding) somewhat more slowly [77] as discussed below.

*Copper*

Copper differs from zinc in that its oxide is less soluble and the sulfate salts formed tend to be basic and also less soluble. Under the proper conditions, the corrosion products formed on copper constitute an attractive green "patina" which is stable over time. Also, copper is normally used as a solid material rather than a coating, which provides more stock thickness and precludes galvanic action (unless installed in contact with a dissimilar metal). Fundamental copper corrosion mechanisms are discussed by Nriagu [2] and by Mattson and Holm [80].

Unlike zinc, which corrodes nearly linearly with time, copper shows decidedly nonlinear relationships, apparently because of the interplay of various oxides and sulfur compounds and their relative solubilities. The short term copper exposure data (<1 yr) analyzed by Lipfert et al. [58] showed no statistically significant effect due to $SO_2$, whereas the longer term data did show such an effect. The long term regression results are given in Eq. 4 and plotted parametrically in Fig. 6. Chloride, while important in the damage function, is not included as a parameter on the plot since in general it is not a controllable pollutant.

$$M = 2.2 t^{(0.80)} \{0.49(f \cdot SO_2) + 0.33 Cl^- + 0.016 H^+\}^{0.64} \qquad (4)$$

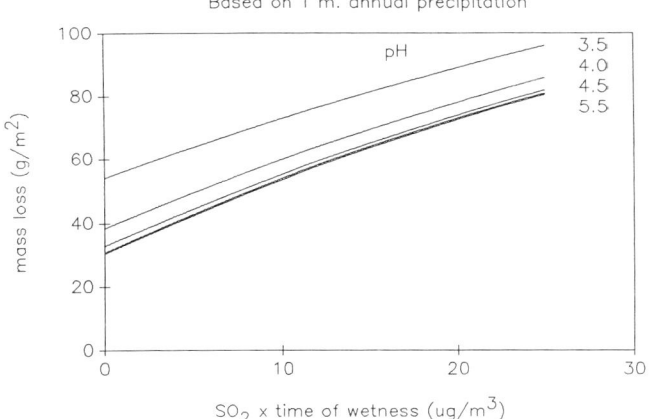

**Fig. 6.** Parametric plot of copper damage function. Corrosion loss is shown for 10 years exposure based on 1 m. annual precipitation. Adapted from Lipfert [78]

Although the copper corrosion rates are somewhat similar to those of zinc, the theoretical service life of copper is perhaps one to two orders of magnitude larger because copper is usually used as a solid (sheet) material rather than as a coating. The indicated service life for 26 gauge sheet based only on corrosion is of the order of 1000 years, but as a practical matter component failure will usually occur much sooner because of localized corrosion, loss of section and strength and often subsequent mechanical failure.

As an approximate check on the validity of the copper damage function given by Eq. 4, the corrosion loss was estimated for the skin of the Statue of Liberty in New York Harbor. The result was 0.09 mm, whereas based on actual measurements the National Park Service has estimated a loss of 0.13 mm (out of the original 2.5 mm skin thickness) [81]. This was considered as encouraging agreement, given the uncertainties in the estimated long term average environmental conditions.

*Carbon Steel*

The corrosion of ferrous metals is a very complex subject; the rates of corrosion are an order or magnitude higher than most other metals and the mechanisms are quite different. Vernon showed the importance of $SO_2$ in conjunction with moisture, a nonlinear relationship [67]. The effect of atmospheric sulfur is also an order of magnitude higher than simple stoichiometry would indicate; a molecule of deposited sulfur appears to result in the loss of many molecules of iron. Various mechanisms have been advanced to explain this phenomenon.

Perhaps the most common explanation is one of regeneration of $H_2SO_4$ by hydrolysis of $FeSO_4$ [70]. Evans [82] argues for a combined electrolytical and electronic process operating only in the presence of surface moisture; the process is unique to iron because of the conductivity of the oxide $Fe_3O_4$ (magnetite).

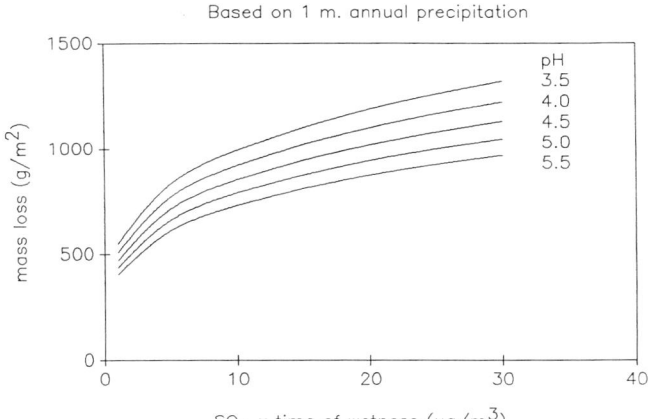

**Fig. 7.** Parametric plot of carbon steel damage function. Corrosion loss is shown for 10 years exposure based on 1 m. annual precipitation. Adapted from Lipfert [78]

Evans also argues that differences in oxide conductivity may explain the success of certain alloying elements (such as Cu) in slowing the rates of iron corrosion [80]. Other important factors in atmospheric corrosion of steel may include prompt oxidation of dissolved $SO_2$ to $SO_4^=$ because of iron catalysis and the fact that the critical relative humidity for the $Fe\text{-}FeSO_4\text{-}H_2O_{(g)}\text{-}O_2$ system is below the relative humidity over a saturated $FeSO_4$ solution. Mikhailovskii and Sokolov [83] discuss the role of $SO_3^=$ as a corrosion stimulator; this mechanism deemphasizes the need for aqueous phase oxidation of dissolved $SO_2$.

The statistical analysis of Lipfert et al. [58] estimated a damage function for steel which was nonlinear both in $SO_2$ and in exposure time, in accordance with the early work of Vernon [67]. Chloride is also a very important corrosion predictor for steel, again in part because of the hygroscopic properties of salt particles. Figure 7 is a parametric plot of steel corrosion loss as a function of $SO_2$ concentration and precipitation pH, showing the nonlinear behavior [58].

Since carbon steel is not normally used in an unprotected state, these corrosion data are only of somewhat academic interest. However, corrosion of the base metal is important as a reference for deterioration rates of steels with various protective coatings, or for weathering steel, for example.

*Aluminum*

The key factor in the durability of aluminum in most atmospheres is the insolubility of its oxides. The buildup of an oxide coating helps protect the base metal from other agents in the atmosphere. As an example of the low reactivity of $SO_2$ with aluminum, Sydberger and Vannerberg [84] found that aluminum adsorbed $SO_2$ at about 15% of the rate of zinc, at 90% relative humidity. One of the most important environmental agents with respect to aluminum is chloride, which has been shown to cause localized pitting [85]. Summit and Fink analyzed aluminum aircraft maintenance records and showed that chloride (proximity to the sea) was more important than regional $SO_2$ or precipitation acidity [86].

The aluminum data used in Lipfert et al.'s statistical analysis employed weight loss as a measure of corrosion, rather than pit depth [58]. In general, the data showed a weight gain in the early years of exposure, presumably due to the buildup of an oxide coating (which was not removed by the solvents used in the normal ASTM corrosion product stripping procedure). These gains were treated as "tares" against the succeeding weight loss data, but these weight loss data must be regarded as only crude indicators of the relative importance of various pollutants. The regression analysis of Lipfert et al. showed that times of surface wetness were associated with higher relative humidities than other metals, and that when a linear model was used, the regression coefficient for $SO_2$ implied an $SO_2$ deposition velocity about 20% of that for zinc.

Aluminum pit depth data indicated growth proportional to $(time)^{1/3}$, and since the weight loss data were roughly linear with time, the implication is one of constant pit shape so that pit depth is proportional to $(mass\ loss)^{1/3}$ [87, 88]. This would be a very rough way of converting the aluminum weight loss damage function to a pit depth basis. Other test programs have used reductions in specimen tensile strength as a measure of corrosion damage, but have usually not analyzed the contributions of environmental factors [88].

*Other Metals*

The metals discussed above are used chiefly as common construction materials. Many other metals have been used through the years and may now be of some interest with respect to historical preservation. In the absence of detailed data and analyses for these other metals, one way to rank their atmospheric corrosion performance is to compare their loss rates to the loss rate for a "reference" material having similar properties, under the same conditions. This is consistent with Nriagu's classification of metals into four groups based on oxidation, corrosion, and passivation characteristics [2]. Ailor's conference proceedings is an excellent source of such information for many different metals [75]. Since zinc corrosion is relatively well understood, it may be a useful candidate for such a (nonferrous) "reference" material.

Table 4 presents estimated cumulative corrosion losses for a number of metals in terms of thickness loss or penetration depth (mm); typical uses are also indicated [89]. These data are based on arbitrary reference environmental conditions, probably approximating a "worst case" for historic exposures in heavily polluted cities (before the advent of pollution control laws): 100 years at 500 µg/m³ $SO_2$ with (metal) surface wetness 20% of the time and 1 m precipitation per year with pH as shown in the table. For the table entries with more than one reference, the corrosion loss estimates were based on the ratio of loss data for the metal shown to that for a similar reference metal for which environmental effects data were available from Ref. 55, as described above. This procedure is quite approximate, but is used here to indicate the relative ranking in sensitivities to long term extreme environmental conditions. The blank entries in the table represent cases for which insufficient data were available to make even this type of approximate estimate.

**Table 4.** Sensitivity of various metals to acid deposition (Estimated cumulative thickness loss in mm for 100 years exposure to 500 µg/m³ $SO_2$ and rain acidity as shown @1 m/yr.)

| Material | Typical use | Loss at pH = 5.6 | Loss at pH = 4.3 | Ref. |
| --- | --- | --- | --- | --- |
| Aluminum* | Ornamentation | 0.10 | 0.11 | 58 |
| Bronze | Statuary | 0.09 | | 58, 80 |
| Copper | Roofing | 0.13 | 0.13 | 58 |
| Lead | Roofing, window cames | 0.05 | | 58, 92 |
| Monel* | Ornamentation | 0.10 | | 58, 90 |
| Steel+ | Structures | 0.58 | 0.75 | 58 |
| Cast iron+ | Ornamentation | 0.31 | 0.40 | 58, 93 |
| Wrought iron+ | Ornamentation | 0.38 | 0.49 | 58, 93 |
| Stainless steel* | Ornamentation | 0.002 | | 58, 90 |
| Zinc | Roofing, ornamentation | 0.29 | 0.66 | 58 |

\* may also be subject to local pitting
+ assumed to be unpainted

Of the common metals used in historic structures, the ferrous metals tend to be the most sensitive to atmospheric corrosion, followed by zinc. Although short term steel corrosion rates for plain carbol steel tend to be an order of magnitude higher than for zinc, the passivating effect of the rust layer on ferrous materials slows down the process so that the differences become much smaller at long exposure times. However, one should keep in mind that a loss of about 0.5 mm corresponds to complete consumption of 25–26 gauge sheet stock but might well be in the acceptable range for ornamental metal work or statuary (although clearly undesirable, at least because of the subsequent staining from corrosion products, if for no other reason).

Lead, copper, aluminum, and nickel alloys tend to have lower long term corrosion losses because of the lower solubilities of their oxides and/or corrosion products. However, the corrosion resistance of lead is related mainly to the low solubility of lead sulfate; it is more readily attacked by nitric and organic acids, which may be present in contemporary urban atmospheres in substantial amounts. Urban rooftop testing in South Africa showed lead corrosion rates only slightly less than zinc, for example (no detailed atmospheric composition data were reported) [90]; thus it is possible that the low corrosion rate shown for lead in Table 4 is an underestimate and that the rate would increase substantially with rain acidity. Terne-coating, an alloy of tin and lead has been used on historic buildings (including Thomas Jefferson's Monticello), and is now being promoted for use with modern buildings [91].

All of these corrosion rate data reflect arbitrary (flat) test configurations. It should be expected that damage rates would be at least a factor of two higher on carved details [77].

*Major Unknown Factors – Metal Corrosion*

The preceding discussion gives the impression that environmental damage effects are understood best for zinc (galvanized steel). While this is undoubtedly the case, a comprehensive, validated damage function has not yet been published.

The effects of nitric acid, atmospheric particles, and organic acids have not been fully quantified. The most difficult fundamental aspects of zinc corrosion deal with interactions of pollutants: other acids with $SO_2$, particles with gases, effects of ozone and other oxidants, wet with dry deposition. In addition, the effects of environmental interactions with actual structural configurations are not well understood.

While the scientific community has been busy studying galvanized steel, the industrial community has been busy developing better materials, notably aluminum-zinc alloy coatings, about which much less is known. Also, there are much less quantitative data available on weathering steel, statuary bronze, and various aluminum alloys. With respect to all structural metals, little is known about specific environmental effects on stress corrosion, corrosion of fasteners, and crevice corrosion.

**Damage to Masonry**

For the purposes of this chapter, masonry includes brick/mortar systems, building stone, and concrete. The latter is perhaps the most widely used construction material in the US [44]. Common building stones include granite, slate, sandstones, marble, and limestone. The last three comprise the class "calcareous" or carbonate stone ($CaCO_3$), which is an important building material, especially in Europe. In the United States, marble is used almost exclusively (outdoors) in commercial and institutional buildings and monuments; limestone and sandstones are used regionally in residential construction. Although brick appears (by virtue of its apparent survival in heavily polluted environments) to be largely impervious to air pollution/acid rain damage, mortars using lime (CaO) as a cementing agent may be susceptible to dissolution losses, which can lead to loosening of the sand grains and eventual crumbling. This mode of damage also applies to some sandstones. Thus, in this section damage mechanisms will be discussed generally in the context of $CaCO_3$, followed by a discussion of damage to concrete.

*$CaCO_3$ Damage Mechanisms*

Information on rates of environmental deterioration of stone comes from at least three separate sources: environmental tests on stone samples, time-lapse measurements and photographic comparisons on buildings and monuments, and geological erosion data. In addition, physical property data are available from laboratory experiments. All of these sources concur that the stone types with the highest rates of loss are calcareous stones. In spite of its reputation as "eternal", calcareous stone is not a very durable material, even in a clean atmosphere. Although $CaCO_3$ is only "slightly soluble" in natural water (depending on temperature and atmospheric $CO_2$ concentration), if an equilibrium solution could be realized as a result of rain washing, even in a clean environment the recession rate for 1 m of rain annually would be in the range of about 3–20 mm in a thousand years [94], which is probably unacceptable for a national monument intended to last indefinitely. The solubility of calcite is responsible for the formation of limestone caves and other landform features. The solubility rate increases with rain acidity (the kinetic dissolution rate increases by about a factor of 8 as pH drops from 5 to 4) and with ambient $CO_2$ concentration [95]; both of these factors have increased in modern times.

In addition to damage by wet deposition, calcite is subject to chemical attack from dry deposition. Both $CaSO_4$ and $Ca(NO_3)_2$ are more soluble than $CaCO_3$; thus dry deposition and uptake of $SO_2$, $NO_2$, and $HNO_3$ can lead to calcite damage. As discussed above, rates of deposition will be controlled by both aerodynamic factors (functions of meteorological conditions and material configuration) and surface resistance (functions of surface chemistry). $HNO_3$ deposits virtually without surface resistance: the surface resistance to $SO_2$ depends on surface wetness and chemical buffering in the moisture layer, and surface resistances to $NO_2$ are generally high and may be increased by surface wetness. Braun and Wilson [96] exposed pure $CaSO_4$ to $SO_2$ and found much less uptake than with $CaCO_3$, presumably because the subsequent reaction with $CaCO_3$ provided better buffering of the surface moisture layer. For this reason, stone damage by $SO_2$ depends on rain exposure (to remove corrosion products) or the presence of other pollutants which may affect subsequent $SO_2$ deposition. For example, in urban areas, rain sheltered areas on stone structures often develop black crusts which have been shown to be combinations of carbon and gypsum [24]. A possible explanation is carbon-catalyzed oxidation of dissolved $SO_2$ in the moisture layer, which would facilitate further $SO_2$ deposition. In general, such crust formations have been shown to be quite destructive, since they eventually spall off and damage the stone surface features in the process.

Further complications in the processes of pollution damage to stone are associated with stone porosity and internal structure. As salts penetrate the stone, they may cause cracking due to their higher molar volumes ($CaSO_4$ vs. $CaCO_3$, for example). Increased porosity due to selective dissolution and leaching can lead to frost damage. Loosening of surface material can lead to increased wind erosion damage.

*Thoeretically-Based Calcite Damage Function*

Theoretical calculations presented by Reddy [97] and laboratory physical property data lend themselves to formulation of a theoretical damage function for calcareous stones. The damage function recognizes three sources of material loss:

1. *Dissolution in "clean" rain.* At pH ranges higher than those expected in industrialized regions (pH < 5.5), calcite solubility is a function of atmospheric $CO_2$, temperature, and the physical characteristics of the stone (such as porosity). There may be contributing factors relating loss of large particles to dissolution of the binding matrix (sandstones, for example).

2. *Attack by gaseous air pollutants,* notably $SO_2$ and nitric acid. The deposition velocity is needed in order to estimate flux to the stone surface as a function of atmospheric conditions and stone configuration. Values can range from about 0.1 to 3.0 cm/s. Once deposited, it is expected that all of the acid will react.

3. *Acceleration of dissolution due to rain acidity.* Not only is the equilibrium solubility affected by pH [97], the kinetic rate of dissolution is also affected. Rate constant data are available from the literature [97, 98].

Both the "clean rain effect" and the effect of acid rain are dependent on the rainfall rate, and rain is needed to wash away gypsum deposits from dry deposition in order to reactivate the surface for further dry ($SO_2$) deposits [96], which is tantamount to postulating a rainfall dependence on the $SO_2$ dry deposition velocity. Thus it may be convenient to express stone damage in terms of material loss per meter of incident rainfall.

In equation form, these mechanisms of stone loss could be represented as

$$M/R = A + B\ V_{ds}\ SO_2 + C\ V_{dn}\ HNO_3 + D\ H^+ \tag{5}$$

where M is the stone mass loss or penetration depth, R is the intercepted rainfall, A is the solubility in "clean" rain, B and C are coefficients incorporating both the stoichiometry and the dependence of deposition velocities on rainfall, the deposition velocities will depend on aerodynamic and surface conditions, and D represents the increase in solubility with precipitation acidity.

In order to consider the question of the kinetics of dissolution as rain flows over a stone surface, note that in the pH range 3–5, the calcite dissolution rates at the beginning of the process can be approximated by the expression $0.051\ H^+$ (in units of nanomoles/cm$^2$/s) [98]. In this case, the delivery of $H^+$ to the stone surface depends on the rainfall rate and can be expressed as $27\star10^{-6}\ R\star H^+$ (nanomoles/cm$^2$/s), with the rainfall rate R in mm/h. Even at R = 100 mm/h (a very high rate), the rate of reaction ($0.051\ H^+$) is an order of magnitude faster than the delivery rate ($0.0027\ H^+$), which suggests that delivery of acid is the rate limiting step, regardless of pH (within this range). In addition to the rainfall rate, the surface configuration may affect the flow rate of acid over the surface.

These calcite dissolution rates look reasonable compared to experimental rates of dissolution of spinning marble cylinders [99], but it is not clear that the experiments were performed under the right conditions since there was unlimited supply of acid. With high rainfall rates (the limiting case), these experiments may be applicable. These dissolution rates imply that there will always be an excess of $Ca^+$ over acid, which is consistent with experimental runoff pH values in the range 7–8 [100]. Since the final pH for the dissolution process is outside the range for approximately linear dependence of the dissolution rate on $H^+$, one must resort to Plummer's curve [98] for the final rate (about 0.01–0.1 nanomoles/cm$^2$/s). This results in reduced sensitivity to the initial pH, but even this rate is within the $H^+$ delivery rate at modest rainfall rates. The observed $Ca^{++}$ concentration tends to vary inversely with rainfall amount (per event) [62] which is as expected according to this model.

*Measured Stone Loss Rates*

Whereas metal corrosion rates are usually determined by weight loss measurements after stripping off the corrosion products, material loss rates for stone may be determined in several ways, including gravimetric. Chemical analysis of precipitation runoff gives a measure of the dissolved material. Surface recession may sometimes be measured directly by reference to more durable mineral inclusions or to metal studs or other reference points. Stone deterioration may also be indicated by probing the depth of the surface gypsum layers; such data have not been included in this table since it is not clear how much of this actually constitutes eventual "damage" [101].

These dissolution or material loss rates will be sensitive to the test configuration and perhaps to the residence time of the rain passing over the stone, as well to atmospheric properties. Also, some test protocols filter the precipitation runoff (notably the NAPAP tests [62]) and thus represent dissolved material only, whereas all the other data will include exfoliation losses and losses of other non dissolved material.

Some relevant loss data for different types of stone are given in Table 5; the data and bibliography of Livingston and Baer [102] were helpful in compiling this table. The index of atmospheric pollution sensitivity given in the table is "stone loss (µm) per meter of rain": the average annual rate of loss of the test material divided by the average annual rainfall for that location. The annual rainfall in most locations is in the range 0.5–1.5 m/yr; however, the data of Reddy et al. [62] are based on about 18 rain events with a total accumulation of about 0.3 m.

The stone loss rate values given in Table 5 are quite variable, but some general observations can be made:
- Material losses tend to be higher on horizontal than on vertical surfaces; thus location of the critical building components may be an important consideration.
- Notwithstanding the near horizontal orientation, loss rates from the NAPAP tests are among the lowest, which is consistent with their test protocol involving dissolution losses only (for this particular data set). By implication, non dissolution losses are significant. In one instance, dissolution loss was indicated to be only about 40% of the total [100].
- The differences among stone types can be considerable, approaching a factor of two. It is thus unlikely that results from a single test program could be taken as representative of a broad stone category such as generic "limestone" or "marble".
- Loss rates tend to be higher in high $SO_2$ locations.

These generalizations notwithstanding, there are still some "outlier" stone loss rate values in Table 5, including the data from St. Paul's [103] and from "natural weathering" [104]. Measurements of surface recession on St. Paul's Cathedral showed a high degree of locational variability (factor of 5) and were described as being in a "flow zone." This may also be the case for horizontal natural stone surfaces [104]; both observations are thus consistent with the importance of water delivery rate.

Finally, based on solubility and kinetic data, one would expect higher loss rates in cool climates with frequent low intensity rains. The negative influence of temperature on solubility has been shown in studies of erosion rates of natural limestone formations [104] in which it is commonly assumed that all the dissolved calcium is in fact the result of erosion. It is also possible that the quarrying and finishing processes used in producing commercial building stone change the surface properties in ways that reduce erosion or solubility (once the initial stone dust has been flushed away).

The measurements of Meierding [114] illustrate two different effects of incident water. First, his cross-sectional comparison used communities with low air

**Table 5.** Precipitation based loss rate data for calcareous stones (μm/m of incident rain)

| Ref. | Material | Data source | Orientation | Location | Av'g SO$_2$ | Loss/m. rain |
|---|---|---|---|---|---|---|
| 62 | Vt. marble | Runoff | 30° horiz. | East. US | 8–22 | 7.5 (diss. |
|  | Ind. limest. | Runoff | 30° horiz. | East. US | 8–22 | 6.4 only) |
| 105 | Vt. marble tombstones | Recession (caliper meas.) | vert. | New York, Long Island | ~300 low | 14 8.5 |
| 106 | Pa. marble tombstones | Recession (caliper meas.) | vert. | Philadelphia & vic. | ~200 low | 34 5 |
| 103 | Prtld lmstn. | Recession | horiz. | London | ~140 | 134 |
| 107 | Portland limestone | Runoff | horiz. | London | ~140 | 80 |
| 65 | Monk's Park limestone | Weight loss | vert. | SE England | 13–130 | 50 |
|  | Portland limestone | Weight loss | vert. | SE England | 13–130 | 30 |
| 108 | Baumberg sandstone | Weight loss | vert. | Germany | 7–120 | 24 |
|  | Krensheim limestone | Weight loss | vert. | Germany | 7–120 | 17 |
| 109 | Portland limestone | Weight loss Weight loss | (prism) (prism) | London rural UK | ~140 ~30 | 50 17 |
| 110 | Ga. marble | Wt. loss | 30° horiz. | St. Louis | 18–50 | 4 –22 |
| 104 | in-situ limestone | Recession | horiz. | "temperate zone" | n/a | 140 |
| 111 | Vt. marble limestone | Recession Recession | horiz.(?) n/a | South Bend, IN New York | n/a n/a | 34 29 |
| 112 | limestone | calipers | vert. | Liege, Belg. | n/a | 1.5– 4.7 |
| 100 | "sandy limest" (60% CaCO$_3$) | run-off | vert. | Mechelen, Belg. | n/a | 26.5 (diss. = 11.3) |
| 113 | porous limest. | wt. loss | vert. | Vienna | 70 | 26.5 |
|  | compact limest. | wt. loss | vert. | Vienna | 70 | 20.1 |
|  | marble | wt. loss | vert. | Vienna | 70 | 18 |
|  | quartz | wt. loss | vert. | Vienna | 70 | 4 |
| 114 | marble tombstns. | caliper | vert. | US transect | low | top = 18; mid = 9 |

SO$_2$ in μg/m$^3$; n/a = data not available

pollution, annual temperatures around 10 °C, and tombstones about 100 years old. The effect of annual rainfall was striking, with statistical significance levels at about the p = 0.001 level. In addition, he noted that tombstones exposed to lawn sprinkling had deteriorated due to mechanical weathering (as opposed to dissolution) one to two orders of magnitude more than those without sprinkling in the same locality. He also noted higher losses at lower temperatures, which he attributed to frost action, at least in part. The finding of higher loss rates at the top of the stones is consistent with rapid reactions, such that the dissolution of CaCO$_3$ in "clean rain" is more pronounced at first contact. It is also possible that some reprecipitation of calcite takes place as the solution flows down the stone.

In general, these stone loss rates are much higher than corresponding data for metals, confirming the conventional wisdom that stonework is the most critical material for cultural resources. Although surface roughening is sometimes detectable after a single acidic rain event, because of the stock thickness used, building stone has a service life of hundreds of years. Even the badly eroded balustrade at St. Paul's has not yet been replaced. Statuary and carved stone may be quite another matter, however. Not only do rounded and protruding surfaces tend to have higher localized deposition velocities than flat surfaces, but once eroded, they are more susceptible to mechanical stresses. Furthermore, relatively small material losses may greatly decrease their value as art.

The crude comparisons possible from Table 5 provide no information on the roles of precipitation acidity, chlorides, or nitric acid, nor information on the relative importance of rain vs. $SO_2$ deposition. For this information, we turn to a closer look at several detailed experimental programs.

*Experimental Determination of Carbonate Stone Damage Functions*

As of this writing, no comprehensive damage functions for calcareous stones have been postulated or published. In this section, some of the available test data are compiled in a form consistent with the theoretical damage function discussed above, and statistical analysis is used to estimate the appropriate coefficients. The results are summarized in Table 6. Since in all cases, the statistical analyses used to estimate the damage function coefficients go well beyond those presented by the original authors, some additional details are included with the discussions of each of the studies used.

*Data from the UK.* Jaynes and Cooke [65] exposed small vertically mounted tablets of two types of English limestone at 25 different locations in southeast England. The samples were mounted on freely rotating carousels in order to eliminate preferential orientations with respect to wind and rain patterns. The exposure period was nominally two years and included sets of samples which were shielded from rain but presumably similarly exposed to gaseous air pollution and wind patterns. This paper contains a great deal of detailed information, including chemical analysis of deposits of the stone. However, rain chemistry data were not included, nor composition of particle deposition. Jaynes and Cooke give correlations between measures of stone damage and environmental factors, but do not attempt to develop damage functions per se.

Three of the test sites seemed to be outliers with respect to the others; they had lower stone loss rates. These three were located south of London or on the south coast, where the rain might be expected to be slightly less acid but chloride effects might be expected; there could be differences in other urban factors as well. A dummy variable ("south") was used to distinguish these three sites in the regression analysis, since no quantitative factor could be assigned to represent their locational differences.

The stone damage metric used by Jaynes and Cooke was percentage weight loss. For the purpose of the present analysis, these figures were converted to average surface recession (μm/yr) based on the reported dimensions of the tablets. Rain amounts and $SO_2$ concentration data were also reported. Using a multiple regression approach, the following relationships can be derived from their data:

$$\text{loss } (\mu m/m) = 25.7 + 0.102(SO_2) + 21.2(\text{type}) - 20.3(\text{south}) \qquad (6)$$
$$\phantom{\text{loss } (\mu m/m) = 25.7 + } (0.067) \phantom{(SO_2) + } (2.8) \phantom{(\text{type}) - } (4.6)$$

where rain is in m/yr, $SO_2$ in $\mu g/m^3$, "type"=0 for Portland stone and 1 for Monk's Park stone, and "south"=0 for all but the three sites mentioned above. The standard errors for each regression coefficient are given in parentheses; the overall $R^2$ for the regression was 0.65 with a standard error of estimate of 10 μm/m (25% of the mean). The most statistically significant variables are the dummy variables representing stone type and location; the $SO_2$ variable fails to achieves the traditional 0.05 significance level, although it is not far off this mark. Individual regressions for each stone type were somewhat less successful, perhaps in part because of the fewer degrees of freedom available. For the less susceptible stone, at typical environmental conditions in Northeastern US cities (rain=1 m/yr, $SO_2$=40 $\mu g/m^3$), this damage function would predict an annual limestone loss rate of about 30 μm/yr, or 33 years to lose 1 mm. Under these environmental conditions, most of this loss (86%) results from the rain action alone.

pH values from bulk collectors in London are in the range 4.3–4.4 [115], which seems consistent with wet-only data taken in Leeds at pH=4.2 [116]. These levels are similar to those in Northeastern US cities [7]. At the locations south of London, the damage function would predict a loss rate about 55% of the London loss rate which seems far too large a difference to be attributed to rain acidity alone. Nitric acid or particulate effects may also be important, but in the absence of local data it is not possible to speculate about detailed causes for this large difference.

Jaynes and Cooke also give chemical composition data on scrapings from stone samples which were sheltered from the rain. Sulfate was by far the most prominent ion, with an average contribution of about 16 mg/g. Approximate calculations translated this into an $SO_2$ deposition velocity of about 0.2 cm/s, which is in reasonable agreement with the results of Braun and Wilson [96] on similar stones in London, also using the composition of surface scrapings to infer deposition. Braun and Wilson also accounted for the sulfur content of rain runoff, which was a minor part of the total. This value is also consistent with the sulfur deposition found on the undersides of the stones exposed in the NAPAP tests [118]. However, Spedding [117] found much higher deposition velocities to limestone in humid air (0.8–1.3 cm/s) in chamber tests with turbulent flow, using new stone samples. The difference in these findings may be due to the comparison of varying long term conditions in the field with specified conditions in the laboratory.

A regression of the Jaynes/Cooke data on sulfate content of deposits with respect to $SO_2$ air concentrations found a statistically significant relationship (p=0.001), but also a substantial intercept. This intercept could be interpreted as indicating other sources of sulfur deposition besides $SO_2$, or alternatively that intermediate variables such as wind speed or time of wetness significantly affect the rate of deposition, since an intercept may be interpreted as one symptom of an incomplete regression model. Assuming a stoichiometric relationship between deposited sulfur and lost calcium (forming $CaSO_4$), the regression coefficient for $SO_2$ in the above damage function (Eq. 6) corresponds to a deposition velocity of about 0.3 cm/s.

An alternative damage function was constructed from the sulfate data, assuming that deposited sulfur will eventually be washed away and thus constitutes a source of weight loss. On the basis of stone loss per meter of precipitation, the result was:

$$\text{loss }(\mu\text{m/m}) = 19.2 + 0.67(SO_4^=) - 6.37(\text{south}) + 20.0(\text{type}) \qquad (7)$$
$$\phantom{\text{loss }(\mu\text{m/m}) = 19.2 + }(0.26) \qquad\quad (4.9) \qquad\qquad (3.0)$$

with an $R^2$ of 0.63 and a standard error of 24.5%.

This regression has a statistically significant coefficient for sulfate, indicating that stone weight loss is better related to a measure of sulfur *deposition* than to $SO_2$ air concentration per se. This in turn implies spatial variability in $SO_2$ deposition velocity, which is also implied by the large change in the "south" regression coefficient between Eqs. 6 and 7.

An alternative estimate of the effects of $SO_2$ that does not require specific knowledge of deposition velocity could be obtained by comparing earlier results on Portland stone obtained by Honeyborne and Price [109] with the results of Jaynes and Cooke [65]. The difference in annual stone loss rate for the two time periods for central London is given as $29 - 16 = 13$ µm/yr or about 20 µm/m rain. The difference in annual average $SO_2$ is about 150 to 200 µg/m³; dividing the two increments gives an $SO_2$ coefficient of 0.10–0.13 µm/yr, in reasonable agreement with the damage function results (Eq. 6) derived from spatial variations. If we attempt to relate the implied sulfur deposition from the sheltered samples to the weight loss of the exposed samples, we find that, expressed as weight loss, sulfate only "explains" about 15% of the observed weight loss. The remainder must therefore be dissolved $CaCO_3$, i.e., the "clean rain" effect. Jaynes and Cooke report very small concentrations of nitrate in the surface scrapings of the rain sheltered samples, even in central London. This implies that the dry deposition velocity of $NO_2$ must be substantially lower than that of $SO_2$, and that the air concentrations of $HNO_3$ (which is known to deposit readily) must also be quite low.

Additional data on loss rates of Portland limestone are given by Bawden [64], including both "natural" outdoor exposures and data from an outdoor $SO_2$ fumigation facility near the coast at Littlehampton. Whereas the natural exposure data fell in line with the data of Jaynes and Cooke [65] and of Honeyborne and Price [109], the data from the fumigation facility displayed a much stronger effect of $SO_2$ on weight loss, approximately 3–4 times higher. Bawden attributed this finding to the effects of the marine atmosphere; however, this hypothesis does not appear to be consistent with the data of Jaynes and Cooke, who found *lower* loss rates near the coast, after the annual rainfall rate is accounted for. A possible alternative explanation may lie with the nature of the fumigation facility, which tends to control $SO_2$ concentrations according to a present frequency distribution. Thus the relationship between air concentration and deposition may be quite different in such a facility relative to the natural atmosphere.

*Stone Exposure Data from the Continent.* Luckat [108] exposed small samples of two types of German calcareous stones thought to be susceptible to air pollution damage: Baumberg sandstone and Krensheim limestone. These tablets were mounted at various heights above ground at 20 different locations (1 height per location) in West Germany and exposed for one year. The reporting of these results has been rather fragmentary; $SO_2$ and $Cl^-$ deposition measurements using the IRMA apparatus (similar to a sulfation plate) were reported in Luckat [108 a], together with the sulfur and chloride contents of scrapings taken from the stone samples. The emphasis of this paper was on the high correlations shown between the content of stone deposits and the IRMA readings. Jaynes and Cooke [65] tabulated stone weight loss measurements (in %) for the Luckat experiments, citing [108 b] for the data, and again reporting good correlations between the sulfur deposition data and the stone weight loss. No information was given on the role of rain.

The average ratio of sandstone loss to limestone loss was 1.36, confirming that sandstone can be more susceptible to damage since whole grains of sand may be lost when the calcite binding matrix is dissolved. The high correlation ($R^2 = 0.88$) between weight losses from the two different stone types implies similar loss mechanisms at work and justifies considering the two as a single data pool.

In order to analyze these data in the same context as other experiments, long term average or "normal" rainfall data were acquired for various locations in Germany and estimates were made for Luckat's exposure sites by interpolation. Percent weight loss was converted into average penetration depth assuming both sides of the stone were attacked and using the reported dimensions of the samples ($60 \times 60 \times 2.5$ mm). Multiple regressions were used to develop estimated damage functions. Taken separately, the functions for each stone showed highly significant $SO_2$ effects, weak effects due to rain and to exposure height, and no effect due to chloride deposition. Using a dummy variable for stone type and pooling both types of stone together (in order to increase the number of degrees of freedom) resulted in the following relationship on the basis of stone loss per meter of incident precipitation:

$$\text{loss } (\mu m/m) = 9.7 + 0.28(SO_2) - 0.09(\text{height}) + 7.9(\text{type}) \qquad (8)$$
$$\phantom{\text{loss } (\mu m/m) = 9.7 + }(0.033) \qquad (0.046) \qquad\quad (2.3)$$

where rain is in meters/year, $SO_2$ in mg/m²d, height in meters, and "type" = 0 for limestone and 1 for sandstone. All the coefficients are statistically significant; the $R^2$ was 0.69 and the standard error, 26%. Eq. (8) predicts an annual loss of about 20 µm for the Northeast US and that the rain effects are about three times the $SO_2$ effects, based on an average (US) deposition velocity of 0.3 cm/s.

Figure 8 presents Luckat's stone weight loss data, arranged to show the effects of annual rainfall and $SO_2$ and their interactions. The fact that the regression lines for the two precipitation groupings are not parallel confirms the role of rainfall in modifying the loss due to $SO_2$. Because of likely collinearity between $SO_2$ and precipitation acidity, it is possible that some portion of this effect may be due to acidity (which would also be proportional to rainfall rate). The weight loss intercepts shown represent the "clean rain" effect.

It is difficult to compare the $SO_2$ regression coefficients from Luckat's experiments to the others previously cited because of the different $SO_2$ measurement

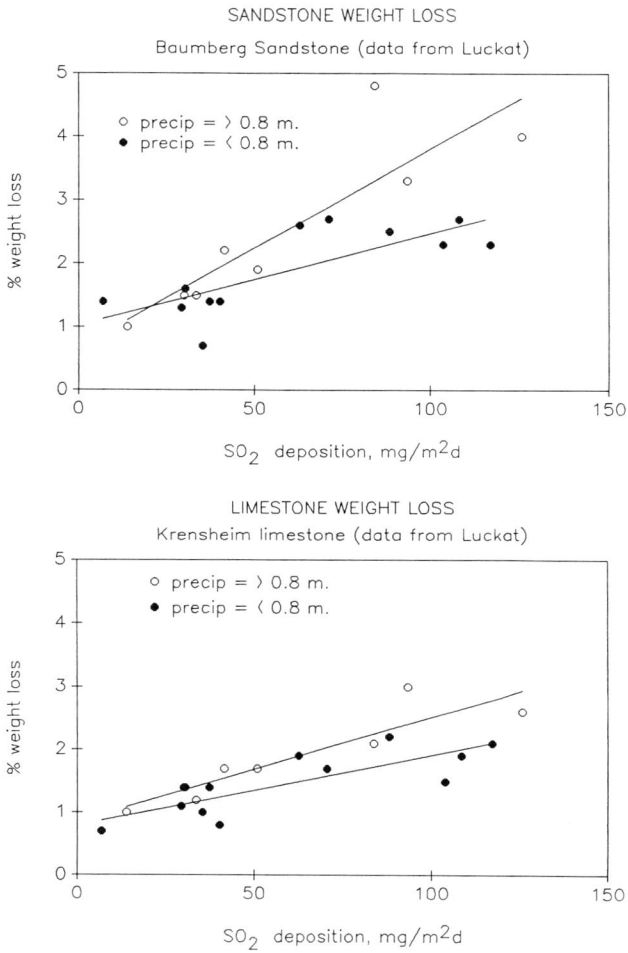

**Fig. 8.** Effects of rainfall and $SO_2$ on stone weight loss. Top: Baumberg sandstone. Bottom: Krensheim limestone. Based on data from Luckat [108]

units used. For a deposition velocity to the IRMA device of about 1 cm/s, units of mg/m²d are nearly equivalent to μg/m³. This would imply a deposition velocity to the stone of about 0.5 cm/s, somewhat higher than others have found [96]. However, Luckat's experiments took no account of rain acidity, particle loadings, or other pollutants, all of which could be positively correlated with $SO_2$ since the highest $SO_2$ readings occurred in the most industrialized area of Germany. If the $SO_2$ effect in the Luckat damage functions is viewed as the sum of all industrial pollutants, then the comparison with other damage functions may be somewhat more reasonable. However, it should also be noted that the Luckat data tend to be more consistent with the fumigation facility data of Bawden [64], for which $SO_2$ concentrations are controlled and thus all other pollutant levels are constant.

In comparing rain effects per se, the frequency of rain should also be taken into account in addition to the annual amount, which may partially explain the higher rain effect seen in the English data [65]. The height coefficient derived from Luckat's data, although a small effect, predicts a slight decrease in stone loss with height above ground, which may be related to atmospheric moisture rising from the ground.

In another series of experiments following Luckat's general design but using the rotating carousel sample mounting arrangement, Weber [113] exposed ten different types of stone samples (porous limestones, compact limestones, quartz-sandstone, and marble) at three locations in Vienna, two on the Cathedral of St. Stephen, and one at a more recent building some distance away. $SO_2$ deposition was measured locally using the IRMA devices. These data show large variability in weight loss among stone types (factor of 10) and due to location on a given building (factor of 2–2.5).

These and several other damage function estimates are presented in Table 6, including a damage function estimated from theoretical data. With $SO_2$ expressed in $\mu g/m^3$ and measured as an air concentration, the $SO_2$ regression coefficients range from 0.02 to 0.51; the 0.02 value results from relatively crude estimates of the $SO_2$ gradient [119], and the 0.51 value from the outdoor fumigation facility appears to be an outlier [64] which may be related to the experimental protocol. The remaining $SO_2$ regression coefficients are rather tightly grouped, with an average value of 0.078 for marble and 0.074 for limestone. By comparison with the theoretical expression, the implied deposition velocity would be 0.41★R ($V_d$ in cm/s, R in m). Expressed as $SO_2$ deposition ($mg/m^2d$), the coefficients range from 0.16 to 0.35, implying that, depending on rainfall, deposition to stone tends to be higher than to an IRMA deposit gauge. The rain acidity coefficients are also variable. The values from the NAPAP runoff experiments (as reanalyzed) correspond closely to the theoretical value; the value obtained from runoff at St. Paul's is much higher, perhaps because of the hydraulics of that particular situation.

On the basis of the theoretical damage function and environmental conditions typical of the Northeast US (1 m precipitation per year, pH=4.2, $SO_2=40$ $\mu g/m^3$, $V_d=0.3$ cm/s), the annual stone loss rate would be 22 mm/1,000 years. Of this amount, about 85% is due to the clean rain effect, 5% is due to acid rain, and the balance (10%) is due to $SO_2$ attack. For the environmental conditions of a decade (or more)★ ago when $SO_2$ was an order of magnitude higher, the annual stone loss rate would have been about 41 mm/1,000 years, of which about half was due to $SO_2$.

*Damage to Concrete and Mortar*

Concrete, or more specifically Portland cement concrete is one of the most widely used materials in the United States and presumably elsewhere [44]. Since $SO_2$ is readily absorbed on cement surfaces [20] it is appropriate to examine the possible consequences for materials damage. This summary follows an inquiry performed for NAPAP by Webster and Kukacka in 1985 [120]. Their analysis identified four possible modes of acid attack on concrete:

**Table 6.** Estimated damage functions for calcareous stones (μm/m. rain)

| Ref. | Material | Recession damage funct. | Statistical methods |
|------|----------|------------------------|---------------------|
| 62 | Vermont marble<br>Indiana limestone | $6.02 + 0.018\,H^+ + 0.085\,SO_2$<br>$4.88 + 0.015\,H^+ + 0.069\,SO_2$ | Time series analysis of 40 runoff events from 3 test sites; dissolution loss only |
| 119 | Vermont marble tombstones | $8.5\ + 0.02\,SO_2$ | Comparison of city and rural sites based on estimated $SO_2$ gradient |
| 96 | Monk's Park limestone | $12.5\ + 0.054\,SO_2$ | Measured $SO_2$ uptake by stone (assumed equal to Ca loss on molar basis) |
| 65 | Monk's Park limestone | $46.9\ + 0.10\,SO_2$ | Cross-sectional analysis of 25 sites, two stone types pooled (2 years) |
| 65 | Portland limestone | $25.7\ + 0.10\,SO_2$ | Cross-sectional analysis of 25 sites, two stone types pooled (2 years) |
| 107 | Portland limestone | $12.5\ + 0.71\,H^+$ | 6 runoff events; no account of $SO_2$ |
| 110 | Georgia marble | $8\ + 0.07\,SO_2$ | Cross-sectional analysis from 9 sites |
| 64 | Portland limestone | $29\ + 0.51\,SO_2$ | $SO_2$ fumigation facility (6 mos. exp.) |
| 113 | porpus limestone<br>compact limestone<br>marble<br>quarzt | $7.7\ + 0.35\,SO_2^*$<br>$11.9\ + 0.16\,SO_2^*$<br>$0.25\,SO_2^*$<br>$4.3$ | Differences between high and low $SO_2$ sites in the same city (Vienna) |
| 108 | Baumburg sandstone<br>Krensheim limestone | $17.6\ + 0.28\,SO_2^* - 0.09\,ht.$<br>$9.7\ + 0.28\,SO_2^* - 0.09\,ht.$ | Cross-sectional regressions on 20 sites, two stone types pooled |
|  | Theoretical calcite (pH 3–5) | $18.8\ + 0.016\,H^+ + 0.18\,V_a\,SO_2/R$<br>($V_a$ in cm/s, rainfall R in m.) | |

$SO_2$ in μg/m$^3$; $SO_2^*$ in mg/m$^2$d, h in a., $H^+$ in meq./m$^2$.
Damage functions based on 1 m. precipitation per year in 100 events.

- dissolving hydrated and unhydrated compounds in the cement paste
- dissolving calcareous aggregates
- physical stresses resulting from crystallization of sulfate and nitrate salts within the pore structure.
- corrosion of reinforcing steel.

The first two modes of attack are direct, but their severity is limited by the generally large masses of concrete materials in relation to the flux of atmospheric acids. This might not be the case in flowing liquid systems. However, the latter two modes are indirect and involve cracking and/or spalling of concrete surfaces due to physical stresses. Freeze-thaw cycles could both enhance the likelihood of these latter modes and the severity of the resulting damage. $SO_2$ has little or no effect on a dry concrete surface, but deposition is enhanced on a wet surface. Diffusion of dry $SO_2$ gas through the porous concrete to reach reinforcing rods is another possibility, but again the severity of this mode of damage will be limited by the low surface concentration of $SO_2$ that would be present on a dry exterior surface. Direct penetration of wet deposition may be of somewhat more concern, especially on horizontal or near horizontal surfaces. Webster and Kukacka's literature review concluded that little qualitative or quantitative data were available on the effects of acid deposition on concrete structures [120]. However, they did conclude that there was valid cause for concern, based largely on anecdotal evidence. A research program was recommended to EPA by a workshop convened to consider the problem, but thus far the topic appears to have been largely neglected.

The need for research on the specific problem of corrosion of reinforcing steel had previously been discussed by Page and Treadaway [121], who pointed out that the steel may lose its protective passivation if the pH of the concrete paste drops from the normal values of above 11 to as low as 9. Acidic substances from the environment that may of concern include $CO_2$, $SO_2$, and chlorides.

A related problem of perhaps more potential concern is the erosion of asbestos cement products by acid rain, releasing asbestos fibers into the environment. Spurny reports that higher ambient concentrations of asbestos fibers were measured in the vicinities of buildings containing asbestos cement products [122]. He also measured emissions of fibers and estimated that 20% became airborne and 80% were washed away. No specific information is available as to the contributions made to this problem by rain acidity, over and above normal weathering.

There is very little information on the rates of deterioration of mortar or brick/mortar systems. Charola and Lazzarini present a largely qualitative discussion of the interactions between bricks and mortar as the system ages in the presence of moisture [123]. Gilardi tested several types of masonry materials (mortar, limestone, sandstone, marble, clay brick and granite) for $SO_2$ absorption [124]. Although these tests do not represent deterioration per se, they indicate the relative susceptibility of the materials to attack by sulfur compounds, which was in the order listed (highest absorption to lowest absorption). These results would appear to confirm the treatment of mortar as a variant of calcareous stone with respect to $SO_2$ attack, but says nothing about susceptibility to flowing water or dissolution by acids. The mortar used by Gilardi had a 3:1 sand/cement ratio.

There is a wide variety of mortar specifications, depending on the application and the compressive strength desired [125]. The ratio of sand to cementitious material ranges from 2.25:1 to 3.5:1; the composition of the cement can vary from no lime up to 80% lime. The strongest mortars appear to have the least lime content [125].

Damage to mortar can only be estimated on a very crude basis, in part because of the lack of pertinent test data and in part because the diversity of mortar types will inevitably result in uncertainties in aggregated damage estimates. Such estimates [126] have been based on lime (CaO) as the cementing agent in mortar that binds the sand aggregate together. As the lime is exposed to the outdoor atmosphere, over time it converts to $CaCO_3$ due to attack by $CO_2$. As the calcite dissolves or converts to gypsum under $SO_2$ attack, the entire mixture crumbles and is lost, so that a multiplicative effect results with respect to the original calcite loss. This "model" is only applicable to those situations in which lime is providing the bulk of the cementing action and obviously breaks down as the CaO fraction approaches zero. For a complete model of mortar deterioration, information on rates of Portland cement loss are needed.

*Conclusions and Major Unknowns Regarding Masonry Deterioration*

The bulk of the evidence points to water as the most important agent of damage for calcareous stone deterioration under present environmental conditions (annual $SO_2 < 100$ μg/m$^3$). Rainfall dissolves calcite at rates depending on the dissolved $CO_2$ content; the additional dissolution resulting from rain acidification is modest at average urban pH levels. Wet $CaCO_3$ surfaces are good absorbers of $SO_2$, but deposition rates will diminish as surfaces become saturated with gypsum. Rain washing can renew these surfaces because of the increased solubility of gypsum relative to calcite. The porosity of the stone becomes an important property with respect to salt penetration, freeze/thaw cycles, crystallization pressures, etc., so that $SO_2$ deposits can influence mechanical damage as well as dissolution.

Major unknowns include the specific roles of nitric acid, organic acids, and atmospheric particles, especially soot. In addition, there has been no comprehensive effort to relate deterioration rates to the physical properties of the various types of calcareous stones, such as porosity or permeability. Finally, configuration effects have not been studied in any detail; both the effects of mortar recessing and of sculptural relief are of interest. The available data imply substantial variations in $SO_2$ deposition velocities; research programs should include $SO_2$ deposit gauges as well as air concentration monitors.

Very little quantitative information is available on air pollution effects on cements and mortars; in view of the wide distribution of these materials and their critical structural use, such research would appear to have a high priority.

## Air Pollution Effects on Paints and Organic Coatings

Although there are many different types of paints and organic coatings, they may be conveniently considered in terms of their functions:
- to beautify the surface (aesthetic)
- to protect the surface (substrate) against deterioration from the environment (protective)
- to make the surface easier to clean (sanitary).

This last function is primarily applicable in specialized indoor settings and will not be considered further in this chapter. The distinction between the first two with respect to this discussion is primarily one of the consequences of failure of the paint coating system and therefore the need for vigilant maintenance.

Other potentially important building components is this group may include bituminous roofing materials, caulking, and sealants. In spite of the importance of these materials to the integrity of a building, virtually no information is available on the effects of environmental exposures on their durability.

## Components of Paint Systems

A paint system consists of several components: the pigment, which imparts the desired color; the film-former, which holds the pigment to the substrate which is to be protected by the paint, an extender or thickener to give the paint body and allow small surface imperfections to be filled; and the vehicle, in which the other elements are suspended or dissolved for ease of application [127]. Each of these elements may vary according to the type of application. The most common applications of paint are probably painted wood for residential housing, painted structural steel of various types, coil-coated metal products such as aluminum siding or gutters, and automotive finishes. In considering the effects of air pollution on each of these applications, it is important to keep in mind the likely mechanisms of paint damage and the attendant consequences, which will differ. "Damage" will be defined as loss of function, as listed above.

## Types of Damage to Paints

The types of adverse effects that air pollution can have on painted surfaces include:

- *Molecular Interactions.* Oxidants can penetrate the molecular structure and create additional polymer cross linking or alternatively chain scission. Such changes can affect the mechanical properties of the paint films and thus their durability, for example because of increased susceptibility to cracking.

- *Application Interactions.* Holbrow [128] showed many years ago that high concentrations of $SO_2$ (at ppm levels) could substantially retard the drying of oil-based paints. It is not known whether this experience would extend to current paint formulations or to more realistic (lower) $SO_2$ concentrations. However, the drying process is extremely important for paint systems, especially with respect to providing adequate coverage at edges and corners. The highest economic penalties associated with paint damage will be those that result in early paint failure, such as peeling or intercoat separation.

- *Permeation through Coatings.* Virtually all coatings are permeable to oxygen and moisture to some degree [129]; oil-based paint films and other polymers are also permeable to $SO_2$ (gas). This raises a number of questions including critical modes of deposition. $SO_2$ deposits most readily to wet surfaces, which would favor moisture penetration (actually dilute sulfurous acid) as the predominant transport mechanism. The consequences of such penetration would obviously depend on the substrate. Williams [130] has shown the accumulation of sulfite at the (wood) substrate paint interface, following diffusion of $SO_2$ through the film. The consequences of such an action are not obvious, since wood has a sulfite component to begin with and is naturally acidic. However, for steel (or zinc) substrates, corrosion could occur as a result of acid penetration. Corrosion of steel often results in delamination of the paint film because of rust and scale formation. Use of a zinc-based primer can be an effective solution to this problem.

- *Leaching of Paint Film Components.* All paint films lose mass or erode over time; this is the preferred failure mode for a properly applied paint system since it will occur typically only after 7–10 years. However, acidic deposition can accelerate the process, depending on the chemical composition of the paint system. Calcium carbonate is perhaps the most vulnerable of common paint components and is used as an extender or filler in interior paints and in some of the cheaper grades of exterior paints. $SO_2$ and/or acidic precipitation can accelerate the loss of $CaCO_3$ by forming the more soluble $CaSO_4$. The resulting erosion or chalking action may also have an ancillary benefit by removing soiling from the surface. Other paint components that may be susceptible to acid leaching include certain esters in oil-based paints, lead pigments, and zinc compounds. The most common white pigment is now rutile $TiO_2$, which is only very slightly soluble in acids. It is not known how susceptible $TiO_2$ is to dissolution under acid precipitation conditions on either vertical or horizontal surfaces.

As can be seen from the above, there are a number of different possible failure modes for paint systems, and the effects of air pollution or acid deposition will likely be different for each. Since many pollutants occur in combination (such as oxidants and acid deposition), combined or interactive effects are also possible.

Intensive research on acid deposition damage to paints is just beginning in the US. The most important type of failure mode/interaction from an economic perspective would be a shift from an erosion/chalking failure mode (7–10 year life) to a delamination (peeling) mode (1–2 year life). Because of the tendency to repaint important structures on a regular basis or within given time periods dictated by weather considerations, small changes in service life of the order of a few weeks or months may be of no practical or economic significance.

### Observations of Environmental Damage to Paints

Actual experimental data on incremental damage to painted surfaces, attributable to environmental factors, are scarce. Most of the early work utilized inert metal substrates, although typically these specimens were mounted approximately vertically rather than on standard ASTM test racks. Thus, the results from these early field programs do not necessarily include the effects of the appropriate substrate. In addition, there are no known data in the literature on environmental effects of repainting over previously weathered painted surfaces. Further, commercial paint formulations tend to change over time and to vary according to regional requirements, such as mildew prevention. Nevertheless, the early data pointed the way towards potential environmental effects on paint system service lives and became the foundation of current research programs.

*Data from Inert Metal Substrates.* Campbell et al. [131] tested four different paint systems for two years at four different US locations. Accelerated laboratory testing was also performed. Several different "damage" metrics were evaluated, including visual ratings, changes in film tensile strength, in gloss, sheen, and surface roughness, and erosion as determined by weight loss. Erosion rates varied from about 0.3 μm/yr to almost 5 μm/yr, confirming that environmental effects do exist. The two paints containing carbonates showed the largest site-site differences with highest erosion rates at the most polluted sites, leading the authors to conclude that the operative mechanism was $SO_2$ attack. The latex house paint used in these experiments contained no carbonates and showed only minimal site-site differences.

These findings were supported by a subsequent test program involving 30-month exposures of latex and oil-based house paints at nine sites in and near St. Louis, MO [132]. Both paints contained silicates rather carbonate extenders and no effects due to $SO_2$ exposure could be demonstrated. It should also be noted that rain amounts and chemistry were not monitored in this study, but the limited geographic extent of the nine sites makes it unlikely that rain chemistry would have differed significantly among sites. However, differences in rain chemistry are a likely confounding factor in the Campbell et al. study [131].

Haynie later combined these results with additional exposure chamber data to synthesize preliminary damage functions for both carbonate- and silicate-containing paints, for use in the NAPAP 1985 Materials Damage Assessment [133]. This damage function was severely criticized, largely on mechanistic grounds [134], and it is also possible that the statistical procedures used resulted in confounding between wet and dry deposition effects. Nevertheless, the results of its application in the 1985 Assessment, showing damage to painted surfaces to be the largest economic component, formed the basis for a greatly expanded paint damage research program by the US EPA, beginning in 1986 [135].

Figure 9 presents some paint erosion data obtained at Research Triangle Park, NC [136], using both oil and latex based paints on inert substrates, mounted vertically facing south. Some of the samples were left exposed to the weather continuously ("uncovered") and others were shielded from ambient rain ("covered"). There are several interesting points to be noted from these data. First, the oil-based paint tends to erode linearly with exposure time, while the latex paint proceeds at $(time)^{0.5}$, similar to metals with passivated surfaces. This implies that at this location the latex paint presents a surface through which pollutants must diffuse in order to effect a loss in mass. Since only the latex paint was affected by rain washing, the operative mechanism may be leaching of inter-

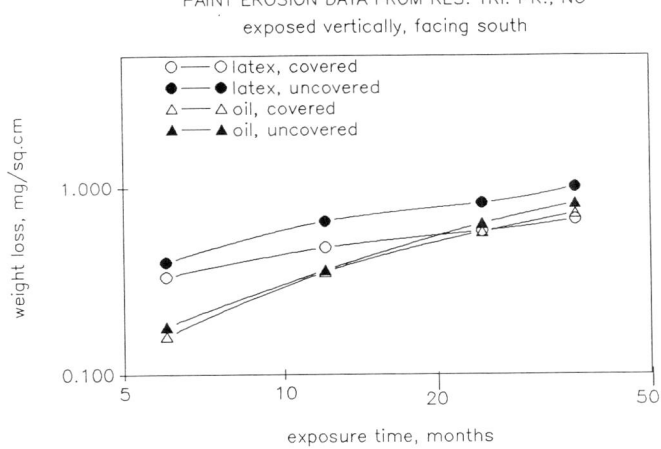

**Fig. 9.** Paint erosion data from Research Triangle Park, NC, based on vertical samples facing South. Data from Ref. 136

nal paint constituents by wet deposition. The linear progression of mass loss from the oil-based paint, unaffected by rain washing, does not show any of the classic symptoms of $SO_2$ attack, and thus may be associated with some other pollutant (nitric acid, ozone) or with loss of volatile components of the paint.

Time series data from the St. Louis tests [132] do not display the behavior shown in Fig. 9. First, the samples were initially exposed in the fall and very little erosion took place during the winter, yielding an S-shaped graph of erosion vs. time. Second, other environmental conditions were changing over time at St. Louis. Finally, the composition of the paints at St. Louis differed from the North Carolina tests; neither contained carbonates and the time histories of both oil and latex paints were similar. This comparison thus suggests that the time histories of paint erosion will be sensitive to both environmental conditions and paint composition.

*Data from Realistic Substrates.* Williams [137] reviewed the available information on damage to painted wood surfaces and noted that virtually all extant research on paint damages ignored the role of the substrate (for example, the Campbell et al. research [131] featured inert metal substrates).

*Wood Substrates.* Preliminary research at the US Forest Products Laboratory showed only minor effects of wet deposition pH, beginning below pH 4.0, on subsequent weathering rates of soaked bare wood specimens of western red cedar [138].

In an informal report to the US Environmental Protection Agency, Starr and Lewis reviewed seven-year exposure data of several formulations (compositions not reported) of acrylic latex exterior paints on two different wood substrates at three US locations: near Dallas; TX; near Chicago, IL; and in suburban New Jersey [139]. These samples received two brushcoats over a primer (which varied). Initial paint thicknesses were not given but may have been as high as 100 µm. During the seven years, no failures occurred due to erosion, flaking or peeling. Mildew and fading were reported to be the two most frequent paint failure modes. Mildew was most frequent in Texas and fading appeared to be most severe at the New Jersey site. It was not possible to draw any conclusions regarding air pollution effects from these data, but they did serve to illustrate the difficulties involved in this type of testing.

The above results are also consistent with measured erosion rates, which are generally less than 5 µm/y. At that rate, even the worst case situation would require around 20 years to completely erode and probably 11 years to begin to "show through" [127, 133]. Since consumers tend to report much shorter repaint intervals [140], it appears that professionally painted test panels do not accurately represent the conditions found on *in situ* structures.

*Metal Substrates.* In contrast to wood, painted structural steel samples have been subjected to long term atmospheric exposure tests. Tests in Sweden of steel painted with a zinc chromate primer and an alkyd top coat showed 20% less service life in an urban atmosphere ($SO_2 = 70$ µg/m$^3$) as compared to a rural site ($SO_2 = 6$ µg/m$^3$) (8 yrs vs. 10 yrs) [141]. Presumably both sites were exposed to nearly the same wet deposition (no data given). Since paint films can be permeable, one mechanism of failure is penetration of $SO_2$ through to the base metal with subsequent rusting and surface expansion, thus delaminating the paint film. With a zinc-based primer, corrosion of the zinc would have to take place first, thus extending the service life of the system.

Coil-coated materials may have a number of advantages leading to longer service lives. First, paint application under factory conditions removes any effects of the outdoor atmosphere on drying or perturbing the raw surface prior to painting (which could be a problem for field painted structural steel, for example). Second, these materials usually feature relatively resistant substrates, such as aluminum, galvanized steel, or zinc-aluminum alloy coatings. Manufacturers of these coated materials now routinely offer 20-year warranties, and some even longer.

The effects of acid deposition on automotive finishes have long been the subject of speculation. At one time, some US manufacturers placed a specific disclaimer in owners' manuals, absolving responsibility for paint spotting due to acid rain. Available information indicates that a number of specialized conditions are necessary for acid rain spotting to occur [142]. First, metallic flake paints seemed most susceptible. Next, according to laboratory simulations, highly concentrated sulfuric acid was required on a hot (120 °F) surface. Such a condition could occur during the daytime drying of acidic dew that had been previously deposited. Last, the problem was only seen on cars with paint 4–6 months old. Because of these limiting conditions, the overall conclusion was that damage to automotive paints is not a major problem. However, acidic particle fallout very near sources can indeed be very harmful and should be considered as a specialized air pollution problem unique to those particular sources.

*Major Unknowns with Respect to Environmental Damage to Paints*

The preceding discussion has indicated many of the difficulties with research on this topic. While there are good reasons to suspect that adverse effects of air pollution exist with respect to paint service lives [143], the basic mechanisms remain to be confirmed through realistic field testing. Thus useful damage functions are not possible at this time. Research on paint damage currently has a high priority on the US NAPAP research agenda, primarily because of the widespread use of this material in the United States.

**Configuration Effects**

Virtually all of the quantitative data on atmospheric damage to materials (especially for metals and paints) have been obtained from small standardized test panels or coupons, frequently mounted on a standard ASTM test rack (about 1 m high, samples inclined 30° to the horizontal, facing south). Some data on deterioration of stone have been obtained from particular building locations and some of the paint coupons have been mounted vertically. It remains a separate task to devise appropriate methods for applying these test results to the generalized prediction of service lives of actual structures or structural components. The specific problems to be considered include:

- aerodynamic conditions differ between standardized test racks and actual buildings. These conditions include both shape and scale factors that affect delivery of airborne pollutants to the surfaces in question.
- surface conditions will differ between test racks and buildings, both due to aerodynamics and thermal effects. Heat transfer from within the structure may affect surface moisture layers.
- surface orientation will affect the amount of rain intercepted and thus the effects of wet deposition.
- the diffusion climatology of cities (where most structures are located) may differ from locations where coupon test data were obtained.

For dry deposition of water soluble gases, the duration of surface moisture films is paramount and may be most affected by thermal factors. In the absence of internal heat transfer, night time radiational cooling is the most important factor and those surfaces with a clear view of the night sky will cool fastest, condense the most atmospheric moisture, and corrode the fastest. For this reason metal roofs are among the most vulnerable building elements.

We will begin this section with consideration of aerodynamic boundary layers, which control the delivery of airborne pollutants to surfaces of various shapes and sizes. Much of this material appeared previously in Lipfert and Wyzga [77]. The section concludes with a discussion of observed configuration effects and of urban spatial gradients in deterioration rates.

**Boundary Layer Concepts**

The concept of a "boundary layer" with respect to the motion of a fluid over a solid body was first expressed by Prandtl in 1904 [144], in which he established that the influence of fluid friction is limited to a very thin layer in the immediate vicinity of the body, outside of which fluid friction may be neglected. Subsequent developments have established the similarity between forced convective heat transfer and fluid friction, and between mass transfer and heat transfer. Deposition of (gaseous) pollutants is an example of mass transfer, and can be described by these same boundary layer concepts.

The atmospheric (or planetary) boundary layer properties primarily reflect the effects of objects on the earth's surface in obstructing the wind flows set up by pressure gradients and other meteorological forces. We must therefore consider not only the details of wind flows around buildings, statues, monuments, etc., but also effects of these objects (usually in large agglomerations) on the atmospheric structure of the wind flow per se.

*The Atmospheric Boundary Layer*

The atmospheric boundary layer can be loosely defined as that portion of the lower atmosphere which manifests the effects of surface features in influencing wind flow. It often extends up to heights of the order of 1 km or to the height of the mixing layer, above which the thermal properties of the atmosphere may effectively insulate it from ground effects. The atmospheric boundary layer is the

carrier for pollutants that affect corrosion. There are several properties of the atmospheric boundary layer of concern here:
- the velocity distribution within the layer
- the temperature distribution within the layer, which will govern the turbulence intensity and hence the dispersion of pollutants as they are released from sources
- the distribution of pollutants within the layer, which will be affected by their release heights as well as by the above two factors.

The velocity distribution is often given by the relation

$$u/u^* = (1/k) \ln(z/z_o + 1) \qquad (9)$$

where u is the local velocity, $u^*$ is the friction velocity given by (shear stress/density)$^{1/2}$, k is the von Karman constant, z is the height above ground, and $z_o$ is the characteristic roughness height of the surface. For grassland, $z_o$ may be of the order of 1 cm; for a suburban neighborhood, perhaps 1 m. Note that $z_o$ is determined from velocity profile measurements and not from the physical size of objects on the ground. Standard National Weather Service (NWS) wind measurements are often referenced to a height of 10 m, in which case the equation above my be used to develop wind speed ratios (u/u(Tu)@10m), cancelling out the $u^*$ factor. The turbulence intensity will be highest near the ground, and can be estimated from (Tu)

$$Tu = k/\ln(z/z_o + 1) \qquad (10)$$

There are two reference heights above ground of interest to this analysis: first, the height at which standard corrosion tests are usually made, about 1 m. Secondly, the appropriate average height for buildings or structures to which these test results may be applied:
- residential buildings, say 3 m.
- fences, 0.5–1 m.
- non residential buildings or structures, 3–30 m.

As an illustration of these wind speed variations, Table 7 presents sample calculations for the three classes of structures and two (extreme) values of $z_o$. We see that the urban-rural variations are the most extreme for smaller objects. Note also that in an urban area with regularly and closely spaced buildings, wind flow

**Table 7.** Average wind velocities and turbulence intensities $u_\infty = 5$ m/s (measured at 10 m)

|  | Local wind velocity | | Local turbulence intens. (%) | |
| --- | --- | --- | --- | --- |
|  | Rural $z_o = 1$ cm | Urban $z_o = 1$ m | Rural $z_o = 1$ cm | Urban $z_o = 1$ m |
| Residence | 4.15 | 2.9 | 7.0 | 29.0 |
| Fence | 3.15 | 1.15 | 9.2 | 71.0 |
| Large building | 5.45 | 6.15 | 5.3 | 13.6 |

patterns will be highly irregular, depending on direction with respect to street orientation, for example. The dramatic increase in turbulence intensity in urban areas is also shown.

The distribution and dispersal of pollutants within the atmospheric boundary layer have been thoroughly discussed elsewhere [35, 37], and will not be elaborated here. However, it should be pointed out that cities create mechanically induced turbulence, and can add heat and moisture to the air (as well as air pollutants). As a result, the air tends to be warmer and less stable in cities, winds tend to be faster at night and slower during the day, and dew tends to form less frequently.

*Boundary Layers on Structures*

All objects immersed in the atmospheric boundary layer perturb its flow in some way, whether they are objects moving through the air or stationary structures exposed to the wind. A new boundary layer is formed on each structure by the passing flow, and the characteristics of this boundary layer govern the transfer of momentum, heat, and mass from the atmosphere to the structure (and vice versa). One of the important characteristics is the physical scale of the object or structure being considered, as well as its shape and surface roughness (texture). This is true not only for isolated objects but for agglomerations (cities, forests), which may in turn have a large influence on the scale of atmospheric turbulence as well as its magnitude.

*Boundary Layer Flux Terms.* Within a boundary layer flow, the surface characteristics differ from those of the unperturbed atmosphere nearby or upstream, giving rise to fluxes:
- *Momentum:* the requirement of zero flow velocity at the surface creates a shear stress or drag on the object due to skin friction ($C_f$);
- *Heat transfer:* if the surface temperature differs from the free stream temperature, heat will flow;
- *Mass transfer:* if the concentration at the surface of some component of the flow differs from the free stream, either because of injection into the boundary layer or because of removal from the stream, mass will flow. Flux to the surface is termed dry deposition, as discussed extensively above.

A well developed theory [144] is available to deal with the calculation of these flux terms for simple situations: flow along a flat plate, around a cylinder or sphere, over an airfoil, etc. Blunt objects such as buildings are generally handled empirically. In addition, these three flux terms have been linked by a relationship (due to O. Reynolds and referred to as "Reynolds' analogy"), which has been verified by numerous classical experiments, usually under conditions which are mathematically tractable.

Our interests here are with removal of pollutants from the air stream; the deposited pollutants then react with the surface and cause corrosion or other damages. This process is referred to as dry deposition, although the presence of a liquid (water) film on the surface is essential for rapid removal of soluble gases such as $SO_2$. The presence of such a film could require a two-layer analysis including phase changes, which is beyond the scope of this preliminary inquiry.

Reynolds' analogy allows estimates to be made of the deposition velocity ratio ($V_d/u_\infty$) based on heat transfer or skin friction tests (or theory), of which the literature abounds. In so doing, one must realize that such a calculation deals only with the delivery of pollutant to the surface, through diffusion.

*Role of Surface Conditions.* As discussed above, if we assume that the pollutant concentration in the boundary layer is zero at the surface (perfect absorption), we have tacitly assumed that the physical chemistry is not limiting, which may only be the case for nitric acid deposition or for $SO_2$ deposition to reactive materials such as zinc or calcareous stones. For $SO_2$ depositing on less reactive materials such as paints or polymers, the surface concentration in the pollutant profile may not be zero, leading to an interaction between physical and chemical processes. Such a situation may occur if the pH in the liquid film drops too low to permit additional $SO_2$ dissolution, as given by Henry's law. Buffering of the film with corrosion products can prevent this from happening (at modest $SO_2$ gas concentrations). This could also be the case for materials such as copper or aluminum which tend to develop protective surface layers over time. However, those materials of most interest are in fact the sensitive ones which either do not build up a protective layer or whose corrosion products are readily soluble at the pH values encountered in precipitation.

However, since the surface resistance is rarely zero, we should expect the deposition velocity derived from boundary layer theory to usually be somewhat larger than observed in practice (neglecting measurement errors). Since the physical chemistry limitations should be independent of size and shape of the object in question, we may use boundary layer calculations to indicate the relative characteristics of different situations.

*Boundary Layer Transition.* Boundary layer properties (governing mass transfer coefficients, for example) are mostly strongly influenced by the transition from laminar to turbulent flow. On a flat plate at low turbulence, such transition occurs naturally at Reynolds' numbers (Re = velocity × density × length/viscosity) between $3.5 \times 10^5$ and $10^6$ [144]. For air flow at 20 °C and 5 m/s, this corresponds to plate lengths between 1.1 and 3.1 m. Free stream turbulence and roughness of the surface can reduce these values under certain conditions. These two parameters can also alter heat transfer and skin friction and thus mass transfer (according to Reynolds' analogy). However, according to stability theory [144], at Reynolds numbers below about $6 \times 10^4$ (about 3 cm in length under the conditions above), transition to turbulent flow cannot begin, since turbulent disturbances in the boundary layer will die out.

Atmospheric turbulence near the earth's surface (Table 7) is generally much higher than found in most wind tunnels, where turbulence levels range up to about 2%. Unfortunately, very few heat or mass transfer tests have been performed under natural outdoor conditions. Surface roughnesses of practical structures of interest may also deviate from laboratory conditions, although boundary layer theory may be used to compute critical roughness sizes and maximum permissible roughnesses, below which the surface is said to be "hydraulically smooth." Table 8 presents some of these values for surfaces of interest for construction materials sensitive to atmospheric corrosion or attack. Since such at-

**Table 8.** Surface roughness data

| | | |
|---|---|---|
| A. | Typical building material surface roughness (mm) | |
| | Smooth stone | 0.005–0.01 |
| | Paint | 0.1 |
| | Galvanized steel | 0.15 |
| | Weathered stone | 1.0 |
| | Corrugated siding | 25 |
| | Carved stone | 150 |
| B. | Admissible roughness, below which surface is "hydraulically smooth:" | 0.33 mm |
| C. | Roughness size creating transition from laminar turbulent (u = 5 m/s): | 3 mm |

tack may enhance surface roughness, it may actually accelerate the process of corrosion. The surface roughnesses associated with smooth stone, galvanized steel, and painted surfaces are "hydraulically smooth," that is they should have no substantial effects on boundary profiles and hence deposition velocity. For roughness elements between 0.33 and 3 mm, transition from laminar to turbulent flow may occur, depending on the Reynolds number.

*Theoretical Relationships.* Reynolds' analogy allows estimates of mass transfer to be derived from tests or theoretical developments for skin friction or heat transfer, as long as the boundary condition of zero surface concentration or resistance is observed. For a two dimensional flat plate this leads to the relationship

$$V_d/u_\infty = 0.5 \, C_f \, (Pr/Sc)^{2/3} \tag{11}$$

where Pr and Sc are the Prandtl and Schmidt numbers, respectively. For the case of $SO_2$ diffusing through air, Eq. (11) becomes

$$V_d/u_\infty = 0.36 \, C_f \tag{12}$$

which provides a convenient means of estimating deposition from data on skin friction. For a laminar flow situation, the theoretical solution for the skin friction coefficient is [144]

$$C_f = 1.328 \, Re_L^{-0.5} \tag{13}$$

which in turn leads to

$$V_d/u_\infty = 0.478 \, Re_L^{-0.5} \tag{14}$$

For the case of flow over a small test coupon, say L = 2 cm, $V_d/u_\infty = 0.00244$; at $u_\infty = 5$ m/s, $V_d = 1.2$ cm/s.

In order to develop relationships for turbulent flow, information about the velocity profile in the boundary layer is required. Experiments have shown that the velocity profile on a flat plate tends to vary as the 1/7 power of distance away from the surface. The following semi-empirical relationship represents skin friction drag on a flat plate for Reynolds numbers between $5 \times 10^5$ and $10^7$ [144]:

$$C_f = 0.074 \, Re_L^{-0.2} \qquad (15)$$

Thus, for a larger plate, say $L = 50$ cm ($Re_L = 10^6$), after substituting into Eq. (12) we find that $V_d/u_\infty = 0.0017$, for a deposition velocity of 0.84 cm/s at $u_\infty = 5$ m/s. The deposition velocities as determined by Eq. (13) and Eq. (15) are not very different, and thus in real outdoor atmospheres it may not matter greatly whether laminar or turbulent flow is assumed. The results of these theoretically-based deposition velocity developments compare reasonably well with semi-empirical formulas for aerodynamic resistance developed by Hicks et al. [3] for flow in the atmospheric boundary layer over a vegetation canopy. Relationships similar to Eqs. (13) and (15) can be worked out for circular cylinders or spheres [77, 144].

*Application of Boundary Layer Theory to Buildings and Structures*

The question of concern here is the application of material deterioration research results to actual buildings and structures, ultimately for the purpose of economic assessment. Use of damage functions such as developed by Lipfert et al. [58] or Haynie [133, 59] implies a direct 1:1 correspondence regardless of size or configuration, in addition to the assumption that the time-of-wetness (presence of liquid film) will be unaffected by size, shape, or surface orientation.

*Rectangular Shapes.* The flow around a building is highly dependent on its situation with respect to neighboring buildings. An isolated (rectangular) building presents a blunt obstacle to the wind: the front face will see stagnation point flow, which will separate around the front corners and reattach at the back corners. The rear face will be in a cavity zone. Flow around the building at a 45° angle will be less chaotic. Since during the course of a year, all of these situations may be expected along each of the building facades, one might try to deduce some sort of average conditions. However, surface maintenance actions may be triggered by the worst case location, such as a corner. In contrast, buildings aligned sufficiently close together along streets will act as a quasi-continuous flat plate, and indeed may channel the wind flow in this way and cause local increases in wind speed. The trailing and leading edges of the block would tend to see somewhat different boundary layer conditions; persistent wind direction would be an important consideration.

We were unable to find in the literature any test data of blunt building-like shapes at sufficiently high Reynolds numbers to simulate real buildings. We did find tests of small square prisms in a wind tunnel (low turbulence) [145] and outdoor tests with natural turbulence for small spheres [146]. The tests on prisms were carried out at various flow angles at Reynolds numbers up to $5.6 \times 10^4$. In contrast, real buildings would have Reynolds numbers greater than $10^6$. The highest local heat transfer (and by analogy, mass transfer) occurred at the rear corners. If extended to a full-scale building, say $Re = 5 \times 10^6$, the deposition velocity ratio $V_d/u_\infty$ would be 0.00048. However, this somewhat low average value should be tempered by the realization that local free stream velocities may be higher, and that the "hot spots" will be higher by about 70–90%.

*Rounded Shapes.* The tests of spheres outdoors yielded average heat transfer values up to 2.2 times higher than in a wind tunnel. This may be due to movement of the flow separation point around the sphere, which would not be appropriate for a less rounded building-like shape. The heat transfer enhancement was most pronounced near the ground, and correlated with turbulence intensity. The heat transfer to spheres at high turbulence was about 70% higher than the average heat transfer to square prisms. Similar results are given by Schlicting [144] for the heat transfer to a circular cylinder in cross flow at varying degrees of turbulence. At the highest turbulence value, the results indicated a factor of 4.8 reduction in $V_d/u_\infty$, corresponding to two orders of magnitude change in Reynolds number. This would yield a full scale building estimate of $V_d/u_\infty = 0.0006$, or 0.30 cm/s for $u_\infty = 5$ m/s. However, as mentioned above, $u_\infty$ might well be substantially higher in such a situation because of the building's effects on the external flow field.

*Flat Plate Model.* Modeled as a smooth flat plate for, say, $Re = 10^7$, the average skin friction coefficient would be 0.003, and thus $V_d/u_\infty = 0.00108$. Turbulence created by window recesses, breaks in the walls, etc., might increase deposition locally. For example, flow in a very rough pipe becomes independent of Reynolds number, but the heat transfer enhancement effects of free stream turbulence are considerably less along a flat plate than for flows around cylinders, etc., where flow separation plays a role. Pedisius et al. [147] showed a heat transfer enhancement on a flat plate of about 20% for a turbulence intensity of about 15%. Drizius et al. [148] showed a heat transfer enhancement of about a factor of 2 for roughness elements of 1.4 mm and $Re_x$ up to about $3 \times 10^6$. For a plate with regularly spaced ribs (similar to corrugation), Veski and Kruus [149] found local heat transfer enhancement to be greatest on the projections. For $Re_x = 10^7$, this gives $V_d/u_\infty = 0.00096$, in good agreement with the preceding. For a 5 m/s wind flow, then, we might expect $SO_2$ deposition velocities of about 0.56–0.9 cm/s on a rough or corrugated building with external free-stream turbulence, but only about 0.56 cm/s on a smooth building. Note that these values are considerably lower than those implied by statistical analysis of outdoor corrosion test results on small plates [58].

## Application of Boundary Layer Theory to Non-Buildings

We have seen how heat transfer and thus dry deposition is reduced on large surfaces, due to the buildup of boundary layer thickness (which reduces the local gradients). However, there are economically important structural objects composed of many elements of small dimension which will show the opposite effect. These include fence wire and fittings, towers made of structural shapes (pipe, angle iron, etc.), flagpoles, columns and the like. Haynie [150] considered different damage functions for different structural elements such as these, but only from the standpoint of their effect of the potential flow in the atmospheric boundary layer. The influence of shape and size act in addition to these effects, and could also change the velocity coefficients developed by Haynie, which were for turbulent flow. Fence wire, for example, as shown below, is more likely to have a laminar boundary layer.

**Table 9.** Deposition velocities to circular cylinders (smooth surface, low turbulence, $u_\infty = 4.5$ m/s)

|  | Diameter (m) | $Re_D$ | $Nu_m$ | Average $V_d$ (cm/s) | Peak $V_d$ (cm/s) |
|---|---|---|---|---|---|
| Fence wire | 0.001 | 301 | 9 | 13.3 | – |
| Fence post | 0.025 | 7520 | 45 | 2.7 | ~4.5 |
| Flag pole | 0.10 | $3 \times 10^4$ | 100 | 1.49 | ~2.5 |
| Structural steel | 0.300 | $9 \times 10^4$ | 220 | 1.09 | ~1.9 |
| Stone column | 1 | $3 \times 10^5$ | 700 | 1.04 | ~1.8 |
| Storage tank | 10 | $3 \times 10^6$ | 4830 | 0.72 | – |

We will consider flow at right angles to an infinite circular cylinder of varying diameter, as shown in Table 9. A dramatic increase in deposition velocity is shown for small diameter objects. This would also apply to isolated portions of a statue, for example.

Schlichting [144] shows data on roughness effects on circular cylinders; for $Re_D$ below about 2,500, there is no effect. For smooth cylinders at low turbulence, there is virtually no Reynolds number effect (or drag) between $Re_D = 1,000$ and $Re_D = 2 \times 10^5$. For large cylinders, with turbulent boundary layers over most of their surface, roughness effects should be similar to those on a flat plate.

### Discussion of Boundary Layer Calculations and Results

*Deposition with Little or No Surface Resistance*

For the few cases for which experimental data for $SO_2$ deposition to zinc are available, good agreement is shown with the boundary layer calculations (Table 10). Unfortunately, experimental data are not directly available for other situations, so inferences must be made by comparing the calculations for similar flow situations.

**Table 10.** Summary of deposition velocity data (cm/s)

|  | Calculated | Measured |
|---|---|---|
| Flat plates |  |  |
| Chamber test [9] | 0.65 | 0.9 |
| Outdoor test racks * | 1.2–3.7 | 1.55–1.75 [58] [+] |
| Large buildings * | 0.5–0.9 | – |
| Circular cylinders * |  |  |
| Fence wire | 13.3 | – |
| Post, columns | 1.0–2.7 | – |
| Storage tanks | 0.72 | – |
| Blunt shapes * |  |  |
| Entire buildings | 0.30 | – |

[+] implied from statistical data analysis  * $u_\infty = 5$ m/s

Larger structures have lower calculated deposition velocities as a result of their larger Reynolds numbers. This effect will be partially countered by higher free-stream velocities for taller structures. Blunt objects will tend to have lower average deposition as a result of their zones of separated flow. This may not pertain to local "hot spots" such as edges or corners, however.

Perhaps the most critical situation will occur for stone objects in the 0.3–1 m diameter range, which could include either statues or columns. A roughened surface due to weathering will increase deposition and hence enhance further surface erosion. Overlaid on all of these results in the tacit assumption that the surface offers no chemical resistance. This appears to be valid for zinc with a water film on the surface; it is less clear for stone, for example, for which the moisture may be trapped below the surface in the pores of the stone. The model as developed here is not valid for essentially inert or less reactive surfaces, such as paint.

## Deposition with Surface Resistance

In the event the surface does offer chemical resistance by virtue of slower reaction rates, resistance to acid attack, etc., the gas concentration will not be zero at the surface and the dry deposition rate will be reduced accordingly. Thus structures made or coated with such materials will show a different relationship between calculated deposition velocities, as presented here, and the actual operational values, which may be governed more by the material properties (i.e., surface resistance) and less by aerodynamic resistance. This would be an important finding to establish, for example by testing over a range of aerodynamic conditions.

## Other Configuration Effects

The above discussion has developed data on the aerodynamic delivery of pollutants in some detail, in part because a suitable body of theory and experimental was available to do so. Real structures can differ from research situations in other important ways, as well. These include effects on surface conditions and the microclimate effects of cities.

*Effects of Material Surface Conditions.* Surface conditions that can influence the deposition and reactivity of air pollutants include temperature (with respect to the outdoor air temperature), orientation with respect to the prevailing wind direction and the sun, and the propensity for developing and retaining surface moisture layers. This last property is perhaps the most important one for dry deposition of $SO_2$. These topics have been discussed in recent comprehensive assessment reports in both the US and Canada [53, 54].

*Orientation Effects.* Several experimental programs have developed data on the effects of orientation of material test samples. Paint samples have been exposed vertically, facing either north of south [131, 132]. In St. Louis, south facing latex paint samples eroded 10–20% faster than north facing samples; for oil-based paints, the south facing samples eroded only slightly faster [132]. This result was also seen in the earlier multi-city study [131], except for the very low pollution city.

In the UK, Shaw exposed small zinc circular cylinders (dry cell casings) at about 100 locations, four samples to a location, one on each corner of an electricity transmission line tower [151]. The variations in mass loss among the four corners at a site was typically quite small, of the order of 0.2% to 1.3%, while the variation between sites exceeded a factor of five. These results seem to demonstrate rather conclusively that local wind effects on a vertical symmetric test sample can be neglected.

*Surface Wetness Effects.* Personal observations at the author's institution indicate that galvanized steel tends to corrode much faster on roofs than on vertical walls of small storage buildings. This probably results both from the increased delivery of wet deposition to roofs and from the increased tendency for dew to form on surfaces with a clear view of the night sky.

Quantitative support for this observation has been reported from tests of an instrumented building in Montreal [152], in which corrosion rates of steel and copper coupons mounted vertically on various wall locations were all lower than corrosion rates of identical samples mounted on standard ASTM racks on the roof. However, in contrast to the circular zinc cans, there were substantial variations among the samples exposed at various wall locations, probably because of both local wind fields around the building and moisture regimes. Sereda [153] had previously shown the existence of large differences in times of wetness in various locations on a building.

It follows from the above that both aerodynamic and surface resistance effects of actual structural configurations must be considered. Shaw's zinc cans may have experienced rather low times of surface wetness (since they were mounted vertically), in which case aerodynamic effects will play a much smaller role. The building in Montreal, on the other hand, experienced both aerodynamic and surface resistance variability.

*Urban Microclimate Effects.* For an assessment of the economic effects of materials damage it is necessary to integrate across one or more metropolitan areas. Questions therefore arise as to the spatial scale and magnitude of intracity variations in corrosion. Obviously, these will depend on the pollution source distribution in the city as well as on climatic factors. For example, Haagenrud et al. [154] showed variations in steel corrosion rate of about a factor of four in an area of $26 \times 31$ km in Norway. However, this area contained a strong $SO_2$ source that contributed most of the spatial variability.

The results of steel corrosion mapping in Melbourne, Australia are perhaps more typical of major metropolitan areas [155]. Samples were exposed for one year at 56 sites covering an area about $50 \times 20$ km. The range in steel corrosion rates was from about 10 to 29 µm/yr. If the marine sites are omitted, the range was from 10 to 20 µm/yr, with one of the locations showing 20 µm/yr apparently near an $SO_2$ source. The authors reported that sites shaded by trees recorded significantly lower corrosion rates. The sulfur and chlorine content of corrosion products were significant predictors of weight loss.

These two studies illustrate the ranges in corrosion rates that may be expected across a metropolitan area. The most important factors appear to be coastal areas and sulfur pollution sources, but other climatic and interference factors contribute to spatial variability as well.

## Concluding Remarks

This chapter has attempted to present a view of the current (ca. 1987) state of knowledge of air pollution effects on materials damage, with emphasis on those materials of economic and cultural importance. Particular attention has been given to atmospheric deposition processes and to damage to actual structures, as opposed to research findings only applicable in a research context. Overall, a detailed understanding is emerging for many materials, painted surfaces being a notable exception. A common theme is the need to analyze and compare materials damage findings from many different research programs and settings, in order to fully consider all the relevant factors. For example, such an approach afforded a better understanding of the various contributing factors for stone damage. In order to achieve useful cause and effect relationships for painted surfaces, it appears that a substantial body of new experimental data will be required, simulating the conditions pertinent to actual consumers' applications.

### Acknowledgments

This chapter reflects the results of many conversations with colleagues and the research results of many individuals and organizations. In many cases, I have added interpretations and data analyses beyond those published by the original authors. In all cases, the results, including any errors or omissions, are my responsibility alone. No endorsement by any governmental or private organization should be inferred.

## References

1. Smith, R.A.: The Beginnings of a Chemical Climatology, Longman, Green and Co., London 1872
2. Nriagu, J.O.: Deteriorative Effects of Sulfur Pollution on Materials. In: Nriagu, J.O. (ed.) Sulfur in the Environment, Part II: Ecological Impacts, Wiley-Interscience, New York 1978
3. Hicks, B.B., Baldochi, D.D., Hosker, R.P., Hutchinson, B.A., Matt, D.R., McMillen, R.T., Satterfield, L.C.: On the Use of Monitored Air Concentrations to Infer Dry Deposition, NOAA Tech. Memo ERL ARL-141, 1985
4. Graedel, T.E., McGill, R.: Environ. Sci. Technol. *20:*1093 (1986)
5. Sinclair, J.D., Psota-Kelty, L.A., Munier, G.B.: Accumulation rates of ionic substances on indoor surfaces. In: Pruppacher, H. (ed.) Precipitation Scavenging, Dry Deposition, and Resuspension, Elsevier 1984
6. Baer, N.S., Banks, P.N.: Int. J. Museum Mgmt. Curatorship *4:*9 (1985)
7. Lipfert, F.W., Dupuis, L.R., Alvarez, L.: Urban and local source effects on precipitation chemistry, presented to the 77th Annual Meeting of the Air Pollution Control Association, San Francisco, June 1984. APCA Paper 84-21.3. Also see Dupuis, L.R.: Analysis of precipitation chemistry data in New York City and surrounding region, Brookhaven National Laboratory Report BNL 38705, 1986

8. Patrinos, A.A.N.: J. APCA *35:*719 (1985)
9. Edney, E.O., Stiles, D.C., Spence, J.W., Haynie, F.H., Wilson, W.E.: Atmospheric Environment *20:*541 (1986)
10. Edney, E.O., Stiles, D.C., Spence, J.W., Haynie, F.H., Wilson, W.E.: In: Baboian, R. (ed.) Materials Degradation Caused by Acid Rain, pp. 172–193, Amer. Chem. Soc., Washington, DC 1986
11. Mikhailovskii, Y.N.: Theoretical and engineering principles of atmospheric corrosion of metals. In: Ailor, W.H. (ed.) Atmospheric Corrosion, Wiley-Interscience, New York 1982
12. Spedding, D.J.: The nature of water associated with atmospheric particulate material. In: Gas-Liquid Chemistry of Natural Waters, BNL 51757, 1984
13. Schwartz, S.E.: Gas-Aqueous Reactions of Sulfur and Nitrogen Oxides in Liquid-Water Clouds, Chap. 4. In: Calvert, J.G. (ed.) $SO_2$, NO, and $NO_2$ Oxidation Mechanisms: Atmospheric Considerations, Butterworth, 1984
14. Adema, E.H., Heeres, P., Hulskotte, J.: On the Dry Deposition of $NH_3$, $SO_2$, and $NO_2$ on Wet Surfaces in a Small Scale Wind Tunnel, proceedings 7th World Clean Air Congress, Sydney, Australia, 1986. Vol. 2, pp. 1–8
15. Pierson, W.R., Brachaczek, W.W., Gorse, Jr., R.A., Japar, S.M., Norbeck, J.M.: J. Geophys. Res. *91:*4083 (1986)
16. Lee, Y.-N., Schwartz, S.E.: J. Geophys. Res. *86,* C-12:11971 (1981)
17. Lipfert, F.W.: J. APCA 39 No. 4 (1989)
18. Huebert, B.J., Robert, C.H.: J. Geophys. Res. *90* D1:2080 (1985)
19. Dollard, G.J., Atkins, D.H.F., Davies, T.J., Healy, C.: Nature *326:*481 (1987)
20. Judeikis, H.S., Stewart, T.B.: Atm. Env. *10:*769 (1976)
21. Judeikis, H.S., Wren, A.G.: Atm. Env. *12:*2315 (1978)
22. Aldaz, L.: J. Geophys. Res. *74:*6943 (1969)
23. McMahon, T.A., Denison, P.J.: Atm. Env. *13:*571 (1979)
24. Camuffo, D., Del Monte, M., Sabbioni, C.: Water, Soil, and Air Pollution *19:*351 (1983)
25. Del Monte, M., Sabbioni, C., Vittori, O.: Atm. Env. *15:*645 (1981)
26. Spence, J.W., Haynie, F.H., Stiles, D.C., Edney, E.O.: A Study of the Effects of Dry and Wet Deposition on Galvanized Steel and Weathering Steel: A Three Year Field Exposure, unpublished manuscript, US Environmental Protection Agency, 1987. Also see Carter, J.P., Lindstrom, P.J., Flinn, D.R., and Cramer, S.D., Materials Performance, July 1987, pp. 25–32
27. Johansson, L.-G.: The corrosion of steel in atmospheres containing small amounts of $SO_2$ and $NO_2$. In: Mansfeld, Haagenrud, Kucera, Haynie, and Sinclair (eds.) The proceedings of the Symposia on Corrosion Effects of Acid Deposition and Corrosion of Electronic Materials, The Electrochemical Society, Vol. 86-6, pp. 407–411 (1986)
28. Ang, C.C., Frank Lipari, Swarin, S.J.: Environ. Sci. Technol. *21:*102 (1987)
29. Baer, N.S., Bernabo, J.C., Lipfert, F.W., Smythe, K.D.: Effects of Acidic Deposition on Materials Degradation: A Synthesis Report. EPRI EA-5424, Electri Power Research Institute, Palo Alto, CA 1987
30. Pierson, W.R., Brachaczek, W.W., Gorse, Jr., R.A., Japar, S.M., Norbeck, J.M., Keeler, G.J.: Atmospheric Acidity Measurements on Allegheny Mountain and the Origins of Ambient Acidity in the Northeastern United States, Atm. Env., *23:*431 (1989)
31. Cadle, S.H., Groblicki, P.J.: The Composition of Dew in an Urban Area. In: Samson, P.J. (ed.) "The Meteorology of Acid Deposition"; APCA Specialty Conference, Hartford, CT, 1983, pp. 17–29
32. Lipfert, F.W., Cohen, S., Dupuis, L.R., Peters, J.: Predictor Equations for Relative Humidity from Relevant Environmental Factors, BNL Report 38957, Brookhaven National Laboratory, Upton, NY 11973, 1985
33. Boynton, R.S.: Chemistry and Technology of Lime and Limestone, 2nd ed., Wiley, New York
34. Sickles, II, J.E., Michie, R.M.: Investigation of the Performance of Sulfation and Nitration Plates, report to US Environmental Protection Agency, Research Triangle Park, NC, May 1984
35. Turner, D.B.: Workbook of Atmospheric Dispersion Estimates, Pub. 999-AP-26, National Air Pollution Control Administration, Cincinnati, Ohio 1969

36. Luckat, S.: Stone Deterioration at the Cologne Cathedral and Other Monuments Due to Action of Air Pollutants, Proc. 4th Int. Clean Air Congress, pp. I-36-8 (1977)
37. Lipfert, F.W.: A Generic Analysis of the Good Engineering Practice Stack Height and Prevention of Significant Deterioration Regulations for New Power Plants, Proceedings of 2nd AMS Joint Conference on Applications of Air Pollution Meteorology, New Orleans, 1980, pp. 719–726
38. Yocom, J.E., McAldin, R.O.: Effects of Air Pollution on Materials and the Economy. In: Stern, A.S. (ed.) Air Pollution, 2nd ed. Also see the 3rd edition (1977); materials damage chapter by Yocom, J.E. and Upham, J.B.
39. Shreir, L.L. (ed.): Corrosion, Newnes-Butterworths, London 1976
40. Barton, K.: Protection Against Atmospheric Corrosion, Wiley, New York 1973
41. Flynn, D.R., Cramer, S.D., Carter, J.P., Lee, P.K., Sherwood, S.I.: Acidic deposition and the corrosion and deterioration of materials in the atmosphere: a bibliography, 1880–1982. EPA-600/3-83-059, US Environmental Protection Agency, 1983
42. North Atlantic Treaty Organization, Committee on Challenges to Modern Society, Impact of Air Pollutants on Materials. Report of Panel 3, NATO/CCMS Pilot Study on Air Pollution Control Strategies and Impact Modeling (1982)
42a. Benarie, M.: Critical review of the available physico-chemical material damage functions of air pollution. In: Environment and Quality of Life, Environment and Consumer Protection Service, Commission fo the European Communities, EUR 6643 EN, 1980
43. Jorg, F., Schmitt, D., Ziegahn, K.-F.: Materialschäden durch Luftverunreinigungen., Ecomed, Landsberg, West Germany 1987
44. Salmon, R.L.: Systems Analysis of the Effects of Air Pollution on Materials – Final Report, report to National Air Pollution Control Administration, Midwest Research Institute, Jan. 1970
45. Fink, F.W., Buttner, F.H., Boyd, W.K.: Technical-Economic Evaluation of Air-Pollution Corrosion Costs on Metals in the US, Report to US Environmental Protection Agency APTD-0654, Battelle Memorial Institute, Feb. 1971
46. Meredith, R.E.: The Cost of Corrosion and the Need for Research, report to US Dept. of Energy, Office of Energy Systems Research, Nov. 1983
47. Schmitt, D.: The Influence of Atmospheric Pollution on the Reliability of Industrial Products, presented to the International Symposium on Product Design Assurance in Engineering, London, June 1985
48. Murray, D.R., Atwater, M.A., Yocom, J.E.: Assessment of the Costs of Materials Damage from Air Pollution in Los Angeles, California, presented to the Annual Meeting of the Air Pollution Control Association, Minneapolis, APCA Paper 86-85.9, June 1986
49. Manning, M.I.: Effects of Structural Materials, CEGB Research, August 1987, pp. 53–61
50. Baboian, R. (ed.): Materials Degradation Caused by Acid Rain, American Chemical Society Symposium Series 318, American Chemical Society (1986)
51. Mansfeld, F., Haagenrud, S., Kucera, V., Haynie, F., Sinclair, J. (eds.): Proceedings of the Symposia on Corrosion Effects of Acid Deposition and Corrosion of Electronic Materials; The Electrochemical Society, Vol. 86-6 (1986), pp. 108–154 (1986)
52. Wyzga, R.E.: J. APCA *37*:679 (1987)
53. Assessment of the State of Knowledge on the Long-range Transport of Air Pollutants and Acid Deposition. Part 6, Effects on Man-made Structures. prepared by P.J. Sereda for Environment Canada, August 1986
54. The National Acid Precipitation Assessment Program, Interim Assessment, The Causes and Effects of Acid Precipitation, Vols. I–IV, Sept. 1987
55. Ailor, W.H.: ASTM Atmospheric Corrosion Testing: 1906–1976. In: Coburn, S.K. (ed.) Atmospheric Factors Affecting the Corrosion of Engineering Metals, ASTM STP 646, Amer. Soc. for Testing and Materials, pp. 129–151 (1978)
56. Haynie, F.H., Lipfert, F.W.: Derivation of Damage Functions for Atmospheric Degradation of Materials, APCA Paper 86-85.3, presented to the 79th Annual Meeting of the Air Pollution Control Association, Minneapolis, June 1986
57. Pourbaix, M.: The Linear Bilogarithmic Law for Atmospheric Corrosion. In: Ailor, W.A. (ed.) Atmospheric Corrosion, pp. 107–121. J. Wiley and Sons, NY, 1982

58. Lipfert, F.W., Benarie, M., Daum, M.L.: Metallic Corrosion Damage Functions for Use in Environmental Assessments. In: Mansfeld, F., Haagenrud, S., Kucera, V., Haynie, F., and Sinclair, J. (eds.) the proceedings of the Symposia on Corrosion Effects of Acid Deposition and Corrosion of Electronic Materials, The Electrochemical Society, Vol. 86-6 (1986), pp. 108–154 (1986)
59. Haynie, F.H.: Environmental Factors Affecting the Corrosion of Galvanized Steel, presented at the ASTM Symposium on Atmospheric Corrosion, Philadelphia, May 1986 (ASTM STP 965, Feb. 1988)
60. Mansfeld, F.: Regional Air Pollution Study – Effects of Airborne Sulfur Pollutants on Materials, EPA-600/4-80-007, US Environmental Protection Agency, Research Triangle Park, NC
61. Atteraas, L., Haagenrud, S.: Atmospheric Corrosion in Norway. In: Ailor, W.H. (ed.) Atmospheric Corrosion, J. Wiley and Sons, NY, pp. 873–891, 1982
62. Reddy, M., Sherwood, S., Doe, B.: Limestone and Marble Dissolution by Acid Rain, 5th International Congress on Deterioration and Conservation of Stone, Lausanne, Switzerland, September 1985
63. Spence, J.W., Haynie, F.H., Edney, E.O., Stiles, D.C.: A Field Experiment to Partition the Effects of Dry and Wet Deposition on Metallic Materials. In: Baboian, R. (ed.) Degradation of Materials due to Acid Rain, American Chemical Society Symposium Series 318, American Chemical Society 1986
64. Bawden, R.J.: Weathering Rates of Fresh Portland Limestone, European Cultural Heritage Newsletter on Research, $2$:10 (1988)
65. Jaynes, S.M., Cooke, R.U.: Atm. Env. $21$:1601 (1987)
66. Lipfert, F.W.: Research Strategy for Environmental Damage to Materials, Brookhaven National Laboratory Report, 1988
67. Vernon, W.H.Y.: Trans. Farad. Soc. $27$:260 (1931)
68. Burdick, L.R., Barkley, J.F.: Effect of Sulfur Compounds in the Air on Various Materials, Bureau of Mines Circular 7064, April 1939
69. Lipfert, F.W., Daum, M.L., Cohen, S.: Methods for Estimating Surface Area Distributions of Common Building Materials, Brookhaven National Laboratory, Upton, NY 11973, 1985
70. Guy, A.G.: Elements of Physical Metallurgy, Addison-Wesley Press, Inc., Cambridge, Mass. 1951
71. US Steel Corp.: The Making, Shaping, and Treating of Steel, H.E. McGannon, ed., 9th edition, 1971
72. see, for example: Weast R.C. (ed.): Handbook of Chemistry and Physics, CRC Press, Inc. 58th ed., 1978, or Dean, J.A. (ed.): Lange's Handbook of Chemistry, 12th ed., McGraw-Hill, New York 1979
73. Spence, J.W., Haynie, F.H.: Theoretical Damage Function for the Effects of Acid Deposition on Galvanized Steel Structures, EPA/600/3–88/027, U.S. Environmental Protection Agency, Research Triangle Park, NC. Sept. 1988
74. Cramer, S.D., Carter, J.P., Linstrom, P.J., Flinn, D.R.: Environmental Effects in the Atmospheric Corrosion of Zinc, EPA/600/D-86/202, US Environmental Protection Agency, Research Triangle Park, NC, 27711, 1986
75. Ailor, W.H. (ed.) Atmospheric Corrosion, Wiley-Interscience, New York 1982
76. Benarie, M., Lipfert, F.W.: Atm. Env. $20$:1947 (1986)
77. Lipfert, F.W., Wyzga, R.E.: Application of Theory to Economic Assessment of Corrosion Damage, in Degradation of Materials due to Acid Rain, Baboian, R. (ed.) American Chemical Society, Washington, DC. 1986, pp. 411–432
78. Lipfert, F.W.: Materials Performance pp. 16–24 (July 1987)
79. Lipfert, F.W.: Acidic Deposition Damage to Materials in the Midwestern States, in Acid Rain: Clouds over the Midwest, Science and Solutions, Conference Proceedings, Chicago, March 7–8, 1986. National Clean Air Fund, Washington, DC
80. Mattsson, E., Holm, R.: Atmospheric Corrosion of Copper and Its Alloys. In: Ailor, W.H. (ed.) Atmospheric Corrosion, Wiley-Interscience, New York, 1982, pp. 365–381
81. Sherwood, S.I.: US National Park Service, personal communication, 1985

82. Evans, U.R.: Corr. Sci. *9:*813 (1969)
83. Mikhailovskii, Y.N., Sokolov, N.A.: Zashchita Mettalov. *21:*214-220 (1985). Translated by Plenum Publishing Corp.
84. Sydberger, T., Vannerberg, N.-G.: Corrosion Sci. *12:*774 (1972)
85. Sowinski, G., Sprowls, D.O.: Weathering of Aluminum Alloys. In: Ailor, W.H. (ed.) Atmospheric Corrosion, Wiley-Interscience, New York 1982
86. Summitt, R., Fink, F.T.: Pacer Lime: An Environmental Corrosion Severity Classification System, AFWAL-TR-80-4102, report to US Air Force Systems Command, Wright-Patterson AFB, Ohio (August 1980)
87. The Aluminum Association: Guidelines for the Use of Aluminum with Food and Chemicals, 5th ed. (1984), Washington, DC, 20006
88. Pourbaix, M.: Understanding and Laboratory Prediction of the Atmospheric Corrosion Behavior of Steels and of Non-Ferrous Metals and Alloys, Annual Report RA.26 to US Army Research, Development and Standardization Group, Belgian Center for Corrosion Study, January 1984
89. Gayle, M., Look, D.W., Waite, J.G.: Metals in America's Historic Buildings, US Dept. of Interior, United States Government Printing Office 1980
90. Callaghan, B.G.: Atmospheric Corrosion Testing in Southern Africa. In: Ailor, W.H. (ed.) Atmospheric Corrosion, Wiley-Interscience, 1982, p. 894
91. Feczko, J.: Terne Roofs Span Centuries, Metal Building Review *23:*38, Feb. 15, 1987
92. Cook, A.R., Smith, R.: Atmospheric Corrosion of Lead and its Alloys. In: Ailor, W.H.: Atmospheric Corrosion, Wiley-Interscience, 1982, p. 397
93. McCaul, C., Goldspiel, S.: Atmospheric Corrosion of Malleable and Cast Irons and Steels. In: Ailor, W.H. (ed.) Atmospheric Corrosion, Wiley-Interscience, 1982, p. 433
94. Wedepohl, K.H. (ed.): Handbook of Geochemistry, Springer-Verlag, New York 1978
95. Langmuir, D.: The Kinetics of Acid Rain Dissolution of Lime Mortar and Portland Cement, draft report to National Park Service, April 1985
96. Braun, R.C., Wilson, M.J.G.: Atm. Env. *4:*371 (1970)
97. Reddy, M.M., Sherwood, S.I., Doe, B.: Modeling limestone dissolution by acid rain, Proc. Research and Design 85, Los Angeles, March 1985. The American Institute of Architects Foundation, Washington, DC, pp. 383–388
98. Plummer, L.N., Parkhurst, D.L., Wrigley, T.M.L.: Critical Review of the Kinetics of Calcite Dissolution and Precipitation. In: Jenne, E.A. (ed.) Chemical Modeling in Aqueous Systems, American Chemical Society Symposium Series 93, 1979, pp. 537–573
99. King, C.V., Liu, C.L.: Amer. Chem. Soc. J. *55:*1928 (1933)
100. Roekens, E., van Grieken, R.: Rates of Air Pollution Induced Surface Recession and Material Loss of a Cathedral in Belgium, submitted to Atm. Envir., 1987
101. Skoulikidis, T.N.: Atmospheric Corrosion of Concrete Reinforcements, Limestones, and Marbles. In: Ailor, W.H. (ed.) Atmospheric Corrosion, Wiley-Interscience, 1982, pp. 807–825
102. Livingston, R.A., Baer, N.S.: Mechanisms of air-pollution induced damage to stone, Proc. 6th World Congress on Air Quality, Paris, 1983. Vol. 3, pp. 33–40
103. Sharp, A.D., Trudgill, S.T., Cooke, R.U., Price, C.A.: Earth Surface Processes and Landforms F *7:*387 (1982)
104. Smith, D.I., Atkinson, T.C.: Process, Landforms and Climate in Limestone regions. In: Derbyshire, E. (ed.) Geomorphology and Climate, Wiley, London 1976
105. Baer, N.S., Berman, S.M.: Marble Tomstone in National Cemeteries as Indicators of Stone Damage: General Methods. APCA Paper 83-5.7, presented to the 76th Annual Meeting of the APCA, June 1983
106. Feddema, J.J., Meierding, T.C.: Atm. Env. *21:*143 (1987)
107. Butlin, R.N., Cooke, R.U., Jaynes, S.M., Sharp, A.S.: Research on Limestone Decay in the United Kingdom, in Vth International Congress on Deterioration and Conservation of Stone, Lausanne, Sept. 1985
108a. Luckat, S.: Staub-Reinhalt. Luft *41:*440 (1981); Also see Luckat, S.: A Quantitative Investigation of the Effect of Air Pollutants on the Destruction of Natural Stone. Environmental Research Project of the German Home Office: in particular, environmental chemicals and the effects of deleterious substances. July 1981 (in German) (108 b)

109. Honeyborne, D.B., Price, C.A.: Air Pollution and the Decay of Limestone, BRE Note 117/77, 1977 (as cited by Jaynes and Cooke [65])
110. Haynie, F.H.: Durability of Building Materials $1$:241 (1982/3)
111. Winkler, E.M.: Stone: Properties, Durability in Man's Environment, Springer-Verlag, New York, 1973, p. 145
112. Kupper, M., Pissart, A.: Vitesse d'erosion en Belgique de calcaires d'age primaire exposes a l'air libre on soumis a l'action de l'eau courante. Abhandlungen der Akademie der Wissenschaften in Göttingen. Math.-Phys. Klasse III $29$:39–50 (1974) (as cited by Jaynes and Cooke [65])
113. Weber, J.: Natural and Artificial Weathering of Austrian Building Stones Due to Air Pollution, presented at the Vth International Conference on Deterioration and Conservation of Stone, Lausanne, Sept. 1985
114. Meierding, T.C.: Phys. Geogr. $2$:1 (1981)
115. Greater London Council: Acid Rain and London, undated report
116. Clarke, A.G., Lambert, D.R.: Local Factors Affecting the Chemistry of Precipitation, presented at the International Conference on Acid Rain, Lisbon, Sept. 1987
117. Spedding, D.J.: Atm. Env. $3$:683, 1969
118. Youngdahl, A.: personal communication, Dec. 1987
119. Husar, R.B., Patterson, D.E., Baer, N.S.: Deterioration of Marble: A Retrospective Analysis of Tombstone Measurements in the New York City Area, EPA/600/3-85/018, US Environmental Protection Agency, Research Triangle Park, NC (1985)
120. Webster, R.P., Kukacka, L.E.: Effects of acid deposition on Portland cement concrete. In: Baboian, R. (ed.) Materials Degradation Caused by Acid Rain, American Chemical Society Symposium 318, pp. 239–249, American Chemical Society 1986
121. Page, C.L., Treadway, K.W.J.: Nature $297$:109 (1982)
122. Spurny, K.R.: On the Weathering and Corrosion of Asbestos Cement Products by Acid Rain, presented to the International Conference on Acid Rain, Lisbon (1987)
123. Charola, A.E., Lazzarini, L.: Deterioration of Brick Masonry Caused by Acid Rain. In: Baboian, R. (ed.) Materials Degradation Caused by Acid Rain, American Chemical Society Symposium 318, pp. 250–258, American Chemical Society (1986)
124. Gilardi, E.F.: Absorption of Atmospheric Sulfur-Dioxide by Clay, Brick, and Other Building Materials, Ph. D Dissertation, Rutgers, 1966. Univ. Microfilms 67-8187
125. Portland Cement Association: Mortars for Masonry Walls, Concrete Information Pamphlet, 1976
126. Lipfert, F.W., Horst, R., Sherwood, S., Lareau, T.: The 1985 Acid Precipitation Materials Damage Assessment: An Overview of Findings. In: Mansfeld, Haagenrud, Kucera, Haynie, and Sinclair (eds.) The proceedings of the Symposia on Corrosion Effects of Acid Deposition and Corrosion of Electronic Materials, The Electrochemical Society, Vol. 86-6, pp. 10–46, 1986
127. Weismantel, G.E. (ed.): Paint Handbook, McGraw-Hill Book Co., New York 1981
128. Holbrow, G.L.: J. Oil Colour Chem. Assoc. $45$:701 (1962)
129. Zankel, K., Ahearn, J.: Materials Damage from Acid Deposition – Phase I: Planning Study, report to the Electric Power Research Institute, Martin Marietta Environmental Systems, Columbia, MD. March 1987. Also see Tator, K.B.: Materials Performance, Nov. 1978, p. 41
130. Williams, R.S.: US Forest Products Laboratory, Madison, WI. personal communication, 1986
131. Campbell, G.G., Schurr, G.G., Slawikowski, D.E.: Final Report on A Study to Evaluate Techniques of Assessing Air Pollution Damage to Paints, The Sherwin-Williams Co. Research Center, Chicago, Il (1972). Also see Campbell, G.G., Schurr, G.G., Slawikowski, D.E., and Spence, J.W.: J. Paint Tech. $46$:59 (1974)
132. Haynie, F.H., Spence, J.W.: Air Pollution Damage to Exterior Household Paints, presented to the 76 Annual Meeting of the Air Pollution Control Association, Atlanta, GA (1983). APCA Paper 83-5.5. Also see J. APCA $34$:941–944 (1984)
133. Haynie, F.H.: Atmospheric Acid Deposition Damage to Paints, Environmental Research Brief, EPA/600/M-85/019 (1986)

134. Livingston, R.A.: Critique of the Carbonate Mass Loss Model for Paint Damage Functions, presented to the 79th Annual Meeting of the Air Pollution Control Association, Minneapolis, MN (1986). APCA Paper 86-85.7. Also see Ref. 52
135. Spence, J.W., Haynie, F.H.: Research Plan for Determining the Effects of Acid Deposition on Exterior Coatings, US Environmental Protection Agency, Research Triangle Park, NC 1987
136. Spence, J.W.: Coatings Exposure Program Within NAPAP, Appendix A in Ref. 135
137. Williams, R.S.: Effects of Acid Rain on Painted Wood Surfaces: Importance of the Substrate. In: Baboian, R. (ed.) Materials Degradation Caused by Acid Rain, American Chemical Society Symposium 318, pp. 310–311, American Chemical Society (1986)
138. Williams, R.S.: Effect of Dilute Acid on the Accelerated Weathering of Wood, J. APCA *38:*148 (1988)
139. Starr, T.L., Lewis, W.S.: Determination of the Film Failure Mode of Exterior Household Paints, Final Report A-4515 to US Environmental Protection Agency, Georgia Tech Research Institute, Atlanta, GA, Sept. 1986
140. Johansen, J.M.: Information from Better Homes and Gardens Consumer Panel, Meredith Publishing Corp., New York 1986
141. Kucera V: Ambio *5:*243 (1976)
142. Spence, J., Haynie, F.: EPA Workshop on Acid Deposition on Painted Surfaces – Final Report, US Environmental Protection Agency, Research Triangle Park, NC, June 1985
143. Amaral, D.A.L., Balson, W.E., Chen, L., Smith, A.E.: Assessing the Costs of Damages to Paints Caused by Acid Deposition, report to US Dept. of Energy, Decision Focus, Inc., Los Altos, CA, May 1987. Also see Smith, A.E., Amaral, D.A.L., and Balson, W.E.: A Probablistic Model for Assessing Damages of Acid Deposition to Painted Surfaces, presented at the International Conference on Acid Rain, Lisbon, Sept. 1987
144. Schlichting, H.: Boundary Layer Theory; 4th ed., McGraw-Hill Book Company, New York 1960
145. Igarashi, T.: Int. J. Heat Mass Transfer *28:*175 (1985)
146. Kowalski, G.J., Mitchell, J.W.: J. Heat Transfer, pp. 649 (1976)
147. Pedisius, A.A., Kazimekas, P.-V., Slanciauskas, A.A.: Heat Transfer-Soviet Research *11:* No. 5 (1979)
148. Drizius, M.R., Bartkus, S.I., Slanciauskas, A.A., Zukauskas, A.A.: Heat Transfer-Soviet Research *10:* No. 3 (1978)
149. Veski, A.Y., Kruus, R.A.: Heat Transfer-Soviet Research *9:* No. 4 (1977)
150. Haynie, F.H.: Theoretical Air Pollution and Climate Effects on Materials Confirmed by Zinc Corrosion Data. In: Sereda, P.J., Litvan, G.G. (eds.) Durability of Building Materials and Components, ASTM STP No. 691, American Society for Testing and Materials: 1980, pp. 157–175
151. Shaw, T.R.: Corrosion Map of the British Isles. In: Coburn, S.K. (ed.) Atmospheric Factors Affecting the Corrosion of Engineering Metals, ASTM STP 646, American Society for Testing and Materials, 1978, pp. 204–215
152. Hechler, J.-J., Boulanger, J., Dufrene, R., Pinon, C.: Metallic Corrosion in the Atmosphere: A Study of Weight Losses, Wetness and Pollutants Distribution Around a Building (Preliminary Results), presented at EPA Workshop on Damage Functions for Real Structures, Raleigh, NC, Dec. 1987
153. Sereda, P.J.: Weather Factors Affecting the Corrosion of Metals, in Corrosion in Natural Environments, ASTM STP 558, American Society for Testing and Materials, pp. 7–22 (1974)
154. Haagenrud, S.E., Henriksen, J.F., Gram, F.: Dose-Response Functions and Corrosion Mapping for a Small Geographic Area. In: Mansfeld, Haagenrud, Kucera, Haynie, and Sinclair (eds.) Proceedings of the Symposia on Corrosion Effects of Acid Deposition and Corrosion of Electronic Materials, The Electrochemical Society, Vol. 86-6, pp. 78–97, 1986
155. King, G.A., Sasnaitis, I., Terrill, S.: Environmental Factors Influencing the Corrosivity of Melbourne's Atmosphere, CSIRO Research Report (Australia), 1985

# Air Pollution Control Equipment

*Heinz Brauer*
Institut für Chemieingenieurtechnik
Technische Universität Berlin
Ernst-Reuter-Platz 3–5, D-1000 Berlin 10

Man and his Environment . . . . . . . . . . . . . . . . . . . . 188
Emission and Air Quality . . . . . . . . . . . . . . . . . . . . 190
    Global Emissions . . . . . . . . . . . . . . . . . . . . . . 190
    Emission Sources . . . . . . . . . . . . . . . . . . . . . . 191
    Emission Parameters . . . . . . . . . . . . . . . . . . . . . 194
    Damages Caused by Pollutant Emissions . . . . . . . . . . . . 196
Analysis of Anthropogenic Activities . . . . . . . . . . . . . . . 197
    Analysis of a Model of Mass and Energy Conversion Plant . . . . 197
        Description of an Industrial Plant . . . . . . . . . . . . 197
        Entrance Stage . . . . . . . . . . . . . . . . . . . . . 198
        Material and Energy Conversion Stages . . . . . . . . . . 199
        Product Stage . . . . . . . . . . . . . . . . . . . . . 199
        Cleaning Stage . . . . . . . . . . . . . . . . . . . . . 199
        Emission Stage . . . . . . . . . . . . . . . . . . . . . 200
    Path of Pollutants and Carriers Through Industrial Plants . . . . . 200
    Analysis of Industrial Plants with Respect to Pollutant Emissions . . 201
    Technical Measures Applicable for the Abatement of Emissions . . . 202
        Process-Specific Measures . . . . . . . . . . . . . . . . 203
        Equipment- and Plant-Specific Measures . . . . . . . . . . 204
        Comprehension of Process-, Equipment- and Plant-Specific
        Measures . . . . . . . . . . . . . . . . . . . . . . . 205
Technology for the Abatement of Particulate Pollutant Emissions . . . 205
    Properties of Particulate Pollutants . . . . . . . . . . . . . . 205
        General Survey on Particulate Matter . . . . . . . . . . . 206
        Particle Size Distribution Curves . . . . . . . . . . . . . 207
        Particle Size and Sedimentation Velocity . . . . . . . . . . 209
    Separation Efficiency of Equipment . . . . . . . . . . . . . . 210
        Fractional Separation Efficiency . . . . . . . . . . . . . 211
        Total Separation Efficiency . . . . . . . . . . . . . . . 212
        Particle Size Distribution at Exit Separator . . . . . . . . 212
    Equipment for Dust Removal . . . . . . . . . . . . . . . . . 213
        Dry Dust Removal Equipment . . . . . . . . . . . . . . 213
        Wet Dust Removal Equipment . . . . . . . . . . . . . . 227
        Selection of Dust Removal Equipment and Cost . . . . . . . 228

Technology for the Abatement of Gaseous Pollutant Emissions . . . . . 229
   Physical Processes and Equipment for the Abatement of Gaseous
   Pollutant Emissions . . . . . . . . . . . . . . . . . . . . . . 230
      Absorption Process and Equipment . . . . . . . . . . . . . . 230
      Adsorption Process and Equipment . . . . . . . . . . . . . . 235
   Chemical Processes and Equipment for the Abatement
   of Gaseous Pollutant Emissions . . . . . . . . . . . . . . . . . 241
      General Description of Chemical Conversion Processes . . . . . 241
      Industrial Application of Thermal Conversion . . . . . . . . . 243
      Industrial Application of Catalytic Conversion . . . . . . . . . 246
   Biological Processes and Equipment for the Abatement
   of Gaseous Pollutant Emissions . . . . . . . . . . . . . . . . . 248
      General Description of the Biological Conversion Processes . . . 249
      Bioabsorption Plants . . . . . . . . . . . . . . . . . . . . . 250
References . . . . . . . . . . . . . . . . . . . . . . . . . . . . . . 252

## Summary

Man is the architect of his environment and has to accept full responsibility for his environment. With respect to this chapter on air pollution control equipment, man is responsible for air pollution and all its effects on man, fauna, flora, and all kinds of buildings. To meet this responsibility, man has to observe the following general rules:
1. Stop the production of pollutants, or
2. reduce the production of pollutants to the lowest possible level.
3. Produce the unavoidable amount of pollutants in such a way, that the properties of the pollutants will guarantee easy separation from the carrier fluid, or effective conversion into harmless products.
4. Develop processes and equipment for effective separation of the pollutants from the carrier fluid or conversion into harmless products.

    Although all aspects of these rules are discussed in this chapter, the main emphasis is put on processes and equipment for separation or conversion of pollutants.

## Man and his Environment

### PROTECT YOUR ENVIRONMENT!

This is a commandment formulated in our times and accepted by all people populating the earth independent of religious and political beliefs. What, however, is the environment?

    Wherever traces of mankind can be found on earth, they give proof of man's activities. Man always endeavored to shape the world according to his needs and wishes. Shaping and reshaping the world, into which man had been born, has been the strategy for survival, for making life more comfortable, for giving life more of a human touch. Man's needs and man's will produced and cultivated a man-made world.

Air Pollution Control Equipment

The author had ample opportunity to become acquainted with living conditions in the african desert Sahara and the brasilian state Amazonas. Wherever he met people, he saw and experienced a world that has been shaped after the will and the ability of these people. Man lives in a world he cultivated. And whenever man has not been able to shape the world according to his needs he has been forced to leave this very part of the world.

The author only once saw a world untouched by human activities. This was when he was flying high over the Hindu Kush on his way from New Delhi to Kabul. The devastation caused by water torrents of the monsoon raging through deep valleys and far into the open country created a picture of cruel strength and deadly lonesomeness. This was a natural world, against which man could not stand up to, which man could not reshape. There were no traces of man, nor of flora and fauna.

Man lives in a world which he shaped and to which he could adapt. In this sense man's world is man-made. Man's environment is the world of his companions. To protect the environment is the command to protect his fellow-men, to accept responsibility for life and needs of his fellow-men in a man-made surrounding. In other words:
Protection of the Environment is a Command for all People:
Accept Responsibility for the Environment.

In a cultivated environment there must be a well balanced composition of all living beings: of man, animals and plants. But this can only be achieved when man respects a natural right for the existence of all living beings and accepts responsibility for the life of his fellow-men, for flora, and for fauna. This, however, does not exclude interference with the life of plants and animals. As a matter of fact, responsibility for his environment forces man into action in favour and against the life of animals and plants.

In order to cultivate his environment, a universally observed set of rules of conduct is required by which man is guided in all his activities:
Man is the architect of his environment.
Considering only a few aspects of the environment, man is forced to take responsibility for the quality of air, water, and soil, which, in a certain sense, cover the basic needs of all living beings. Man has to prevent deterioration of the quality of these elements, he has to prevent all harmful emissions from anthropogenic sources.

In this chapter only air pollution control will be considered in some detail. Special attention is given to processes and equipment which serve to reduce harmful emissions. This chapter is based on a book written by the present author and his Indian colleague: Air Pollution Control Equipment [1].

## Emission and Air Quality

In general, emissions consist of inert and non-inert components. The inert medium is the carrier for the pollutants. Only the non-inert components, that are the pollutants, exert a detrimental effect on the environment, although the concentration of the pollutants in the emission streams is rather low. This is the reason why the costs of pollution control are primarily due to the handling of the inert components.

In the following sections a few remarks will be made on global emissions, on a classification and definition of emission sources, and on emission parameters.

### Global Emissions

Pollutants are emitted by natural and anthropogenic sources. A general description of pollutants and inert components of the atmosphere is given in Volume 1 and Volume 2 of The Handbook of Environmental Chemistry [2, 3].

Figure 1 gives the global emission of some of the more important pollutants and the contribution of natural and anthropogenic sources. According to these data emissions form anthropogenic sources are almost negligible when compared with emissions from natural sources. However, the environment is predominantly endangered by the few percent of global emissions from anthropogenic sources. The differences in emissions from natural and anthropogenic sources and their relative importance can best be explained by the following equation:

$$\dot{M}_E = \dot{m}_E A_E . \tag{1}$$

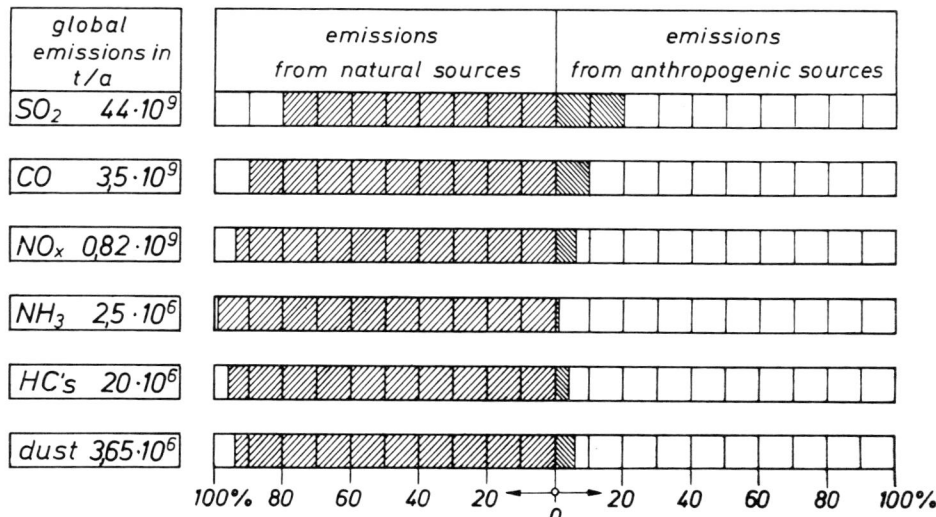

**Fig. 1.** Global emissions from natural and anthropogenic sources

Air Pollution Control Equipment                                                                                          191

$\dot{M}_E$[kg/s] is the mass flow rate of the emitted pollutant, division by the emission area $A_E$[m²] gives the specific mass flow rate $\dot{m}_E$[kg/m²s]. Natural emissions are in most cases characterized by very small specific mass flow rates of the pollutants and by very large emission areas. A typical example for this case is the decomposition of organic matter that leads to the emission of hydrocarbons, sulfur dioxide etc. Typical sources of anthropogenic emissions are chimneys. In this case the emission are $A_E$ is very small, the specific emission flow rate $\dot{m}_E$ is, however, rather large.

It is, first of all, the specific flow rate that makes the pollutant dangerous for the environment. Furthermore, the specific flow rate indicates that air pollution is in many cases a local problem. With a few exceptions pollution of the environment is caused by emissions from anthropogenic sources. It is man who endangers the environment with pollution caused by his activities. It is therefore man's responsibility to reduce pollutions to an acceptable level or stop emitting pollutants altogether.

**Emission Sources**

There are four groups of sources, which may be defined as follows.
1. Point-sources
2. Line-sources/one-dimensional-sources
3. Area-sources/two-dimensional-sources
4. Volume-sources/three-dimensional-sources.

   The best known point-sources are high chimneys of power stations and other industrial plants. Figure 2 shows a photograph of a large power station with three high stacks.

   A typical line-source my be a waste water conduit or a traffic highway. In Fig. 3 a photo shows the traffic on a highway at night.

   Examples for area-sources are the open tanks of large waste water treatment plants and the assemblage of small chimneys on the roofs of houses in residential areas. Figure 4 shows a photo of a densely populated area with chimneys on all roof tops.

   Volume-sources are often found in industrial installations. Figure 5 shows as a typical example a chemical plant. Important technical properties of emission sources are:
1. Locality of emission source
2. Size of emission area
3. Height of emission source
4. Strength of emission flow rate
5. Type of emitted pollutants.

These properties can be easily determined for a point source. As a matter of fact, the point source is the best controlled emission source. All other sources are far more difficult to keep under control. Technology available today is best suited for abatement of emissions from point sources. For other sources the abatement of emissions poses serious technical problems.

The emissions from volume-sources are less than those from any other source under control. This is due to the fact that a volume-source consists of a very large

**Fig. 2.** Photograph of a power station with three high stacks

**Fig. 3.** Night view of highways as an example for line-sources

Air Pollution Control Equipment

**Fig. 4.** View on the roofs of houses in a densely populated area with chimneys on roof tops as an example for two-dimensional sources

**Fig. 5.** Photograph of a chemical plant as example for three-dimensional sources

Fig. 6. Photographic view of a piping distribution station

number of small sources distributed within the volume of the plant. The small sources are for instance flanges and valves. Figure 6 gives a photographic view on a piging distribution station with a great number of flanges and valves. In a case like this emission control becomes a design problem; a plant must be emission proof!

**Emission Parameters**

The abatement of emissions can be achieved by several methods. In this section only the abatement of pollutant emission achieved by a cleaning or separation stage added to the plant will be considered. In the cleaning stage all particulate and gaseous pollutants should be removed as completely as possible from the carrier medium, so that this may be safely emitted into the environment. Since there is no absolutely perfect separation process, a certain amount of the pollutant will be emitted into the environment together with the carrier.

The efficiency of a separation stage will be discussed by means of Fig. 7. The volumetric flow rate of the carrier entering this stage is given by $\dot{V}[m^3/s]$. The me-

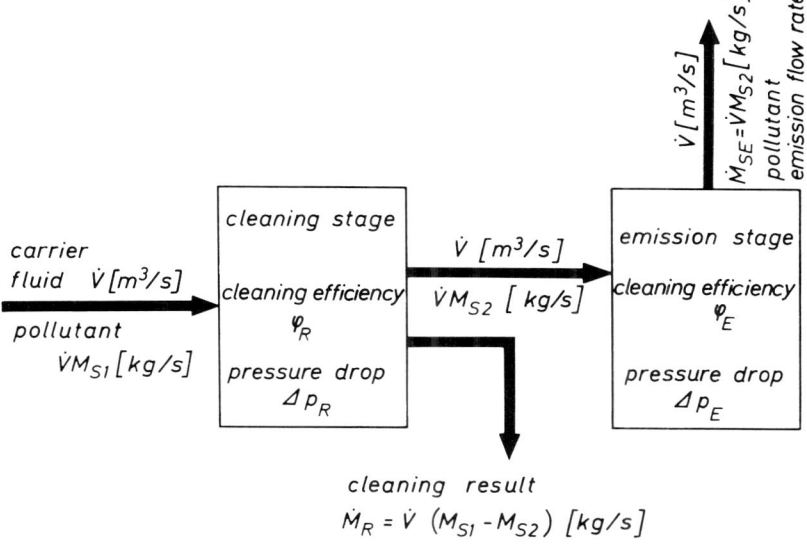

**Fig. 7.** Flow of carrier fluid and pollutant through cleaning and emission stage

dium carries the mass flow rate $\dot{V}M_{s1}$ [kg/s] of the pollutant, with $M_{s1}$ [kg/m³] the mass of pollutant contained in one m³ of carrier fluid, i.e. the pollutant concentration of the carrier at the entrance of the cleaning stage. While passing through the cleaning equipment as much pollutant as possible is separated from the carrier.

At the exit of the separation stage, the pollutant concentration is $M_{s2}$ [kg/m³] and the pollutant mass flow rate is given by $\dot{V}M_{s2}$ [kg/s]. It is assumed that the flow rate $\dot{V}$ [m³/s] of the carrier does not change during flow through the cleaning stage.

The pollutant change in the carrier on its way from entrance to exit of the separation equipment is $M_{s1} - M_{s2}$. Dividing this difference by the value $M_{s1}$ at the entrance, the cleaning or separation efficiency $\varphi$ is obtained:

$$\varphi \equiv \frac{M_{s1} - M_{s2}}{M_{s1}}. \tag{2}$$

This efficiency depends on the type of cleaning equipment, on process conditions, particularly on the flow rate $\dot{V}$ of the carrier, on the concentration $M_{s1}$ of the pollutant, and on the properties of the carrier and pollutant. Consequently, the separation efficiency is a variable value depending on the type of equipment and on operating conditions.

The amount of pollutant $\dot{M}_s$, separated from the carrier while passing through the cleaning stage is given by the equation:

$$\dot{M}_s = \dot{V}(M_{s1} - M_{s2}). \tag{3}$$

This mass of pollutant separated per unit time must be taken care of in some way. In the case of dust separation, $\dot{M}_s$ may be so large, that per day many tons or cubic meters have to be handled. In the simplest case transportation must be available for suitable deposition. In other cases a more or less complicated process for further treatment of the pollutant may be necessary.

After passing the separation stage the waste gas is emitted into the atmosphere. The emission mass flow rate $\dot{M}_E$ consists of the carrier ($\dot{M}_{tE}$) and the pollutant ($\dot{M}_{sE}$):

$$\dot{M}_E = \dot{M}_{tE} + \dot{M}_{sE} . \tag{4}$$

The carrier or transport medium rate is

$$\dot{M}_{tE} = \dot{V}\varrho_t , \tag{5}$$

and the pollutant emission rate is

$$\dot{M}_{sE} = \dot{V}M_{s2} = \dot{V}M_{s1}(1-\varphi_R) . \tag{6}$$

With respect to environmental protection it is the pollutant emission rate $\dot{M}_{sE}$ that is of particular interest. This quantity depends on the flow rate of the transport fluid $\dot{V}$, on the pollutant concentration $M_{s1}$ at the entrance of the cleaning station, and on the separation efficiency $\varphi_R$. According to Eq. (6), the pollutant emission rate $\dot{M}_{sE}$ will increase by a factor of two, when the flow rate $\dot{V}$ is raised by a factor of two.

For environmental pollution, it is the pollutant emission rate $\dot{M}_{sE}$ that counts. The cleaning efficiency $\varphi_R$ or the concentrations $M_{s1}$ and $M_{s2}$ are not very helpful in the interpretation of pollutant emissions.

The gas flow rate $\dot{V}$ should, however, not be underestimated. Is is the flow rate of the transporting medium that is responsible for noise emissions, for size of the cleaning equipment, and for the operating costs.

**Damages Caused by Pollutant Emissions**

Many pollutants have been identified as being directly responsible for health hazards for man. Carcinogenic substances have been put under special control.

The sensitivity of plants is by now a rather well studied phenomenon. Especially in Germany the damage to forests is widely discussed; sulphur dioxide and nitrogen oxides are assumed to cause or encourage such damage.

There is no doubt that sulphur dioxide causes serious damage to most valuable objects of art such as cathedrals, castles, and other documents of our cultural heritage. Figure 8 gives a photographic view of the Porch of the Maidens of the Erechtheum on the Acropolis in Athens [4, 5]. Another example of the destructive action of air pollutants is shown in Fig. 9, which reproduces the head of a bronze statue in Munich [6].

**Fig. 8.** Porch of Maidens of the Erechtheum on the Acropolis in Athens, showing the effects of air pollution on the statues

**Fig. 9.** Head of a bronze statue in Munich showing the effects of air pollution

## Analysis of Anthropogenic Activities

Although there is a multitude of anthropogenic activities which cause harmful emissions, most of these can be considered by means of a simple model which represents a technical system, in which either mass or energy conversion takes place.

### Analysis of a Model of Mass and Energy Conversion Plant

This analysis serves to uncover the production of pollutants and the dispersal of the pollutants into a carrier [7].

*Description of an Industrial Plant*

Figure 10 gives a simplified schematic drawing of a plant for conversion of materials and energy. The plant consists of the following stages:
1. Entrance stage, into which all necessary raw materials and other materials such as air and water required for the process are introduced as a basis for the intended conversion of material and energy.
2. Material and energy conversion stages, in which the processes are carried out that lead to the desired, and to a certain number of undesired products. With

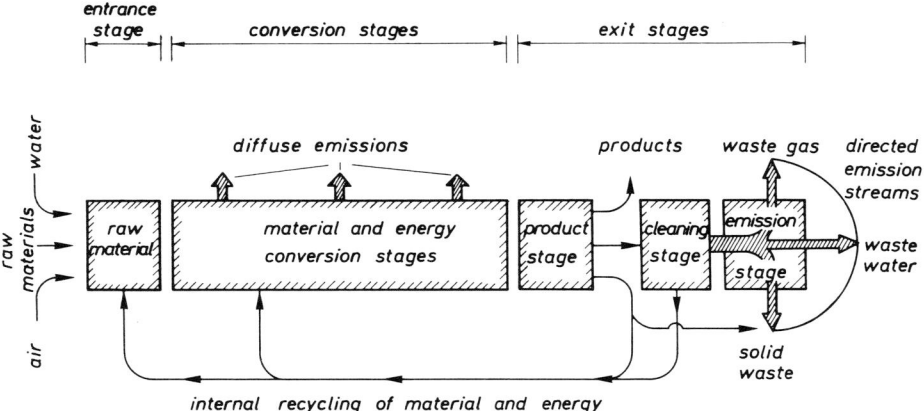

**Fig. 10.** Schematic drawing of an industrial plant for material and energy conversion

transfer into a carrier the "undesired products" may become "pollutants." It is the carrier that will finally carry some fraction of the pollutant into the environment.
3. Product stage, in which the desired products are separated from undesired products and other materials and taken out of the process. At this stage valuable materials are recycled within the internal material and energy cycle.
4. Cleaning stage, in which all materials present in a carrier medium emitted into the environment, are removed from this medium to the necessary degree. Valuable materials are recycled, pollutants are transferred for further treatment or for deposition.
5. Emission stage, which serves to emit the carrier and residual pollutants into the environment under such conditions, that no harm is to be expected.

In the following sections each stage will be discussed with respect of pollutant production, transfer, transport, removal, and finally emission.

## Entrance Stage

The pollutants, species and amount, produced in the conversion stage, depend on the raw materials used for the process. To demonstrate the influence of raw materials on production of pollutants the power station will serve as an example. In the last three decades coal has been partially substituted by oil or gas, with the result, that dust emission decreased. This example may be taken as proof, that the solution of pollution problems starts in the entrance stage, with the selection of raw materials.

But not only the raw materials need consideration. With respect to environment pollution the media that carry the particulate and gaseous pollutants, must be carefully kept under control. Costs for pollutant removal decreases with decreasing carrier flow rate. Although air and water are available at a very low cost, their application should be reduced to the absolute minimum.

## Material and Energy Conversion Stages

All pollutants are produced in material and energy conversion stages. With the raw materials given, pollutants with respect to species, amount, and other properties, depend only on the applied conversion process.

Experience shows that there is no industrial conversion process, producing only the desired products. There will always be a number of undesired products, some of which will be classified as pollutants. Production of pollutants is generally unavoidable. For energy conversion processes imperfections are described by the second law of thermodynamics. According to this law, first formulated by the French scientist Carnot, heat can never be completely converted into any other type of energy. The unavoidability of pollutant production is also one of our basic experiences, although this experience has not yet been expressed by such a simple law as the second law of thermodynamics.

We have to accept material and thermal emissions of technical processes. It is known, however, that there are many industrial processes, which are accompanied by a pollutant production far larger than necessary according to our physico-chemical knowledge available. Material and energy conversion processes offer many opportunities for improvement, i.e. for reduction of pollutant production, and consequently for reduction of pollutant emissions. Processes optimized with respect to low pollutant production rate are the least costly measure that can be taken for reduced environmental pollution.

Another aspect of the conversion process concerns the transfer of the produced pollutants into an inert carrier, as for example air or water. The transfer process is just as important as the production of pollutants. Without the transfer of pollutants into a carrier medium, pollutant removal from the process or the plant would be very easy. It is the dispersion or dilution of the pollutant in the carrier that makes the process of pollutant collection very complicated and expensive. This is the reason why care should be taken to keep the flow rate of the carrier as low as possible.

## Product Stage

In the product stage the desired products are separated from all undesired products and inert media. This separation process should be as efficient as possible to prevent any loss of desired products. The inert media containing the undesired products and a certain amount of desired products are transported to the next process stage.

## Cleaning Stage

In the cleaning stage all particulate and gaseous pollutants, whether these materials are of any value or not, should be removed as completely as possible from the carriers, so that these may be safely let into the environment. There is absolutely no perfect separation process, with the result that always certain amount of pollutant is emitted together with the carrier into the environment.

The efficiency of a cleaning stage is one of the important emission parameters that have been discussed in a previous section. The cleaning efficiency characterizes the reduction of the pollutant concentration of the carrier fluid during flow through the cleaning stage. The cleaning efficiency depends on the type of cleaning equipment, on process conditions, particularly on the flow rate $\dot{V}$ *of the carrier, the concentration* $M_{s1}$ of the pollutant, and the properties of the carrier and the pollutant. Consequently the cleaning efficiency is not a constant typical for the applied cleaning equipment. For simplification it may be said, that the size of the cleaning equipment and the energy requirement depend primarily on the flow rate of the carrier.

The carrier leaving the cleaning station enters the emission stage with the pollutant concentration $M_{s2}$.

*Emission Stage*

The purpose of the emission stage is the emission of the carrier gas and the residual pollutant into the environment. In most cases the emission stage is a chimney. Application of high chimneys will allow pollutant dispersion in a large volume of ambient air, thereby achieving a high degree of dilution, so that the concentration of the pollutant in the air becomes acceptable.

Even when there are no or only negligible traces of pollutants present in the carrier, the emission stage is still required for emission of the carrier. The height of a chimney depends on the degree of dilution of the pollutant, that is to be achieved, while the diameter depends on the flow rate of the carrier fluid.

**Path of Pollutants and Carriers Through Industrial Plants**

An analysis of the path of pollutants and carriers through a production plant will give important information on
1. location and process details of pollutant production,
2. damage caused by pollutants to the plant,
3. reasons for difficulties experienced in the removal of pollutants in the cleaning stage, and
4. measures to be taken for emission proof design of equipment.

The path of the pollutants through material and energy conversion plants may be divided into the following steps:
1. Production of pollutants,
2. transfer of pollutants into the carrier,
3. transport of the pollutants by the carrier through the succeeding stages of the plant,
4. separation of pollutants from the carrier,
5. emission of residual pollutants with the carrier into the environment, and
6. processing or disposition of the removed pollutants.

The path of the pollutant, starting with production and ending with deposition, is closely linked with the path of the carrier. From a technical and economical point of view, the carrier is just as important as the pollutant. The size of cleaning equipment, for example, depends primarily on the flow rate of the carrier. Therefore it is necessary to analyse not only the path of the pollutant but also of the carrier:
1. Introduction and/or production of a carrier,
2. loading of the carrier with pollutants,

Air Pollution Control Equipment 201

3. movement of the carrier with pollutants through the plant,
4. cleaning the carrier from pollutants, and
5. emission of the carrier with residual pollutants.

An analysis of the path pollutant and carrier fluid take through the plant will indicate possibilities for the abatement of harmful emissions.

**Analysis of Industrial Plants with Respect to Pollutant Emissions**

The analysis presented in the preceding section may give the impression that emissions of carrier and pollutants take place only from chimneys. This is unfortunately not the real situation.

Emissions leaving a plant by means of special emission stages will be called controlled emissions. Those emissions, leaving the plant without passing special emission stages, will be called uncontrolled emissions, or diffuse emissions. It is the diffuse emission which forces the plant design engineer to make an emission analysis.

The emission analysis helps to identify all locations of emissions of a plant and to specify the type of pollutants emitted. Figure 11 shows the result of an emission analysis. This example has been published by Baum, Hager, Heiß, and Kellerer [8]. There are basically four types of emissions:
1. Material emissions,
2. thermal emissions,
3. acoustic emissions, and
4. radioactive emissions.

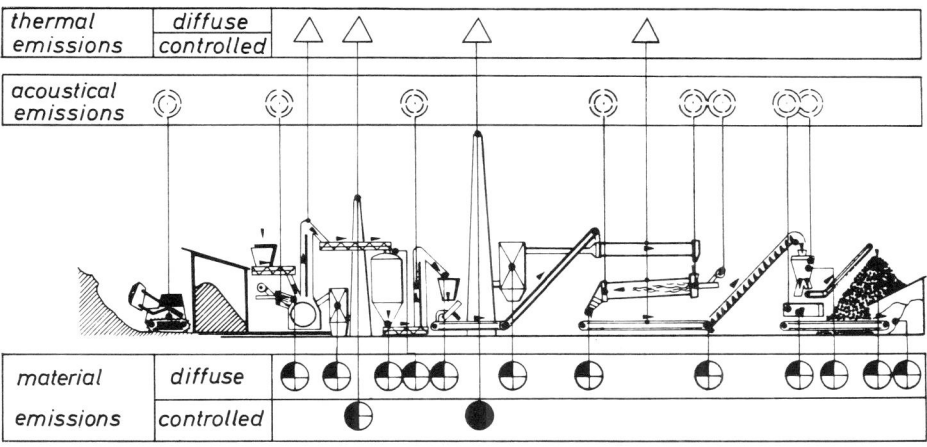

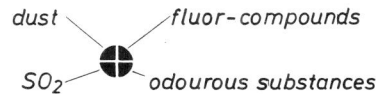

**Fig. 11.** Schematic drawing of an industrial plant with indicated location of thermal, acoustical, and material emissions

In Fig. 11 locations of only the first three types of emissions are identified. For material emissions four typical pollutants are pointed out. It is of course possible to include in such an emission illustration further information on the pollutants and carriers emitted and flow rates.

According to Fig. 11 there are 14 locations with material emissions, but only two of these are controlled, while 12 are uncontrolled emissions. Diffuse or uncontrolled emissions are primarily observed at such locations where elements of a plant are fitted together, as for example, flanges. Great care needs be taken to reduce the number of such locations, thereby reducing the emission potential of a plant. Figure 6 shows a photograph of a section of a plant with a considerable number of flanges and valves.

The emission rate of each of such sources is very small. But the emissions from a large number of small sources may ultimately accumulate to a considerable emission, comparable to that of a controlled source. In the design of ethylene plants the number of flanges and thereby the emission rate has been consequently reduced. This is borne out by the figures given for two plants:

| Year of construction | 1968 | 1971/73 |
|---|---|---|
| Ethylene production capacity | 250 $10^3$ t/a | 400 $10^3$ t/a |
| Number of flanges | 20 $10^3$ | 12 $10^3$ |
| Emission of hydrocarbons | 100% | 31% |

Although the production capacity has been increased appreciably (60%), the number of flanges has been reduced by 40%, and the hydrocarbon emissions by almost 70%.

The emissions from diffuse or uncontrolled sources, which are located in a three-dimensional source field, will first of all affect the people working in the plant. But depending on the strength of the source and on weather conditions people living in the environment of the plant will also be affected by the emissions.

To reduce the emissions, there are in general two methods which can be applied:
1. Process specific measures, and
2. equipment and plant specific measures.

Next to process specific measures, emissions can be successfully attacked by emission specific design methods. The goal of a plant designing engineer should be the uncontrolled-emission proof design.

**Technical Measures Applicable for the Abatement of Emissions**

Generally emissions consist of carriers and pollutants. The abatement of emissions therefore includes abatement of carriers and abatement of pollutants. The discussions in preceding sections revealed the fact that the abatement of emissions may be achieved by process specific means and by equipment and plant specific means. These results of analysis will be taken into account, when in the following sections technical measures for emission abatement are discussed.

# Air Pollution Control Equipment

## Process-Specific Measures

*The process* that takes place in the material and energy conversion stages *depends on*

the raw materials,

the nature of the physical, chemical, and biological conversion processes, and on

the type of equipment applied.

The process should be conducted in such a way, that the following conditions are fulfilled:

1. Production of pollutants should be minimized.
2. If pollutant production is unavoidable, the produced pollutants should have such properties, that removal from the carrier gas can be achieved by simple means.
3. Introduction and production of carriers should be minimized.
4. Transfer of pollutant into the carrier should be minimized.
5. When transfer of pollutants into the carrier is unavoidable, then the amount of pollutant produced should be transferred to as small a volume of carrier as possible.

Plant emissions of pollutants and carriers should be a minimum. This can be achieved with a minimum of cost by minimizing pollutant production, minimizing introduction and production of carriers, and by minimizing pollutant transfer into the carrier.

The rate of pollutant and carrier production depends on the raw materials and on the type of process selected. The condition of minimum emissions should therefore play a decisive role in process selection.

Introduction of carriers such as air and water into the process must be considered under aspects of minimum emissions. As long as air and water were available without cost and could be emitted without further cost after being used and contaminated in the process, care was not taken about these carriers. Since maximum values of the pollutant concentration in the emitted carriers have been fixed, the carriers have to be purified at a relatively high cost. The cost for purification increases with an increasing flow rate of the carrier, because the size of equipment and energy requirement increase. Cost for purification can be substantially decreased, when for a given pollutant production and transfer rate the volumetric flow rate of the carrier can be reduced. It is still often overlooked, that the costs for emission control depend heavily on the carrier. Therefore the importance of carrier media cannot be overestimated.

When pollutant production is unavoidable, the properties of the pollutants should be such that removal from the carrier is made as easy as possible. In dust production for example the particle diameter should be larger than about 10 µm. Particles of this size will not endanger the respiration system, and may be furthermore easily removed in conventional cleaning equipment. If dust particles with a diameter smaller than 10 µm are produced, special measures like agglomeration should be adopted to increase particle size.

Particulate matter has for many technical processes favourable properties not found with matter in the bulk phase. Production of solid materials in the particulate phase increases. For the handling of particulate matter a very important rule applies:

Dust will produce finer dust.

As far as waste dust is concerned, which must be separated from a carrier, this natural process must be counteracted by technical processes with the effect of increasing size.

When the pollutants have been produced and transferred into the carrier gas, they are transported through all sections of the plant to the cleaning stage. This method of handling the pollutants may cause corrosion and erosion, particularly in tube bends, in valves, and in all other elements of equipment in which the flow direction is forcefully changed. The rate of corrosion and erosion is a very strong function of the velocity of the carrier. This is another reason why the flow rate of the carrier should be minimized.

The cost of pollutant removal increases with decreasing pollutant concentration in the carrier and with increasing flow rate of the carrier. The cost is due to the size of the equipment and energy requirement. Size and energy requirement can be substantially reduced by decreasing the flow rate of the carrier and thereby increasing the pollutant concentration, when pollutant production is assumed to be constant.

The final stage of an industrial plant is the emission stage, from which the carrier with the residual pollutants is let into the environment. The cost for the emission stage again depends largely on the flow rate of the carrier.

From this discussion the conclusion can be drawn, that in each stage of an industrial plant process-specific measures can be taken for pollutant emission. In general, the cost for pollutant emission can best be minimized by process-specific means such as minimizing pollutant and carrier production and introduction, and minimizing pollutant transfer into the carrier.

*Equipment- and Plant-Specific Measures*

Equipment- and plant-specific measures can contribute considerably to the reduction of pollutant and carrier emissions. The equipment includes not only those for pollutant removal but also those in which the conversion processes take place.

Any type of equipment should be designed in such a way that in each element of a cross section or volume process conditions are the same. This fundamental condition of equipment design is in many cases not given the necessary attention. Deviation from this rule becomes more serious and more detrimental to pollutant production with increasing size of equipment. Particularly in dead spaces conversion process conditions may be such that pollutant production is increased. Although the general guidelines for equipment design will not be discussed here, it shall be pointed out that shape and size of equipment are important parameters for pollutant production. Equipment-specific measures are of course most important for pollutant removal equipment. Size and shape are important parameters for cleaning efficiency. Existing equipment can still be improved and new equipment must be developed.

# Air Pollution Control Equipment

The analysis of industrial plants with respect to pollutant emission has revealed the fact that diffuse or uncontrolled emissions assume an increasing fraction of the overall emissions. Uncontrolled emissions are observed primarily at flanges and valves, and connections between pieces of equipment, as indicated in Figs. 10 and 11. Emission-proof design of equipment and plants should become a fundamental rule of design, not only for thermal but also for material and other emissions.

*Comprehension of Process-, Equipment-, and Plant-Specific Measures*

In order to reduce emissions from industrial plants the following measures should give a guideline of action:
1. Development of processes and relevant equipment with a minimum of pollutant and carrier production.
2. Development of processes, by which unavoidable pollutants are produced under such conditions, that they may be easily removed from the carrier.
3. Development of processes with a minimum transfer of pollutants into a carrier.
4. Development of cleaning processes and equipment with highest possible cleaning efficiency and lowest possible energy requirement.
5. Development of emission proof equipment and plants in order to avoid all uncontrolled diffuse emissions. Unavoidable emissions should be made controllable.

Measures taken for the abatement of emissions will in general prove to be more effective and less costly when they are taken at the front end of the process. Measures taken at the tail end of the process will in general be the most expensive ones.

## Technology for the Abatement of Particulate Pollutant Emissions

Particulate pollutants may be either dust or drops, and in special cases aerosols. The abatement of the emission of such pollutants will only be considered when the separation of the particulate matter from the gaseous carrier is carried out in special "cleaning equipment." Most of the equipment that has been developed so far, is applicable for dust and drop removal from gas streams. It is only lately that special equipment for drop removal has been successfully developed and applied. The various types of equipment for dust and drop removal depend primarily on the physical properties of the particulate matter. It is therefore necessary to present a brief summary of these properties.

### Properties of Particulate Pollutants

Only the physical porperties of particulate pollutants are important for the type of equipment designed for their removal. With respect to the harm done to the health of man for example, the chemical properties of these pollutants are just as important as the physical ones. These will be dealt with in the following sections. This section will now begin with a general survey on particulate matter.

## General Survey on Particulate Matter

In Fig. 12 various types of particulate matter, which are of some technical importance, are presented in their conventional range of diameter [9]. The diameter of the particles varies over five orders of magnitude, from 0.01 µm to 1,000 µm. Particles with a diameter larger than 5 or 10 µm are sometimes called coarse particles, those smaller than 5 or 10 µm fine particles. It may be assumed, that the fine particles will penetrate the respiration system, particularly the human lung. Furthermore the dust may adsorb gaseous pollutants and carry them into the alveoli of the lung. The gaseous pollutants are thus deposited in the alveoli at a high local concentration. The fine dust is in every respect, i.e. in medical, biological, and technical respects, the most harmful fraction of the particulate matter.

Figure 12 also contains information on methods and instruments, which may be applied in determining particle size. Of particular importance is sedimentation due to gravity and to centrifugal forces. Microscopic methods are also very important, because they not only offer the opportunity of size measurement but also of shape determination. Medical investigations proved that the shape of the particles is of great, possibly of decisive importance for the harm caused by those particles which penetrate the respiration system. According to these investigations fibrous asbestos particles are assumed to cause cancer of the lung. Because of the shape of these particles mechanical damage is done to the lung.

In the lower section of Fig. 12 the range of application of dust removal equipment is indicated.

The amount of particulate matter per unit volume of gas is expressed by the concentration, given in either g of particulate matter per $m^3$ of gas, i.e. in either

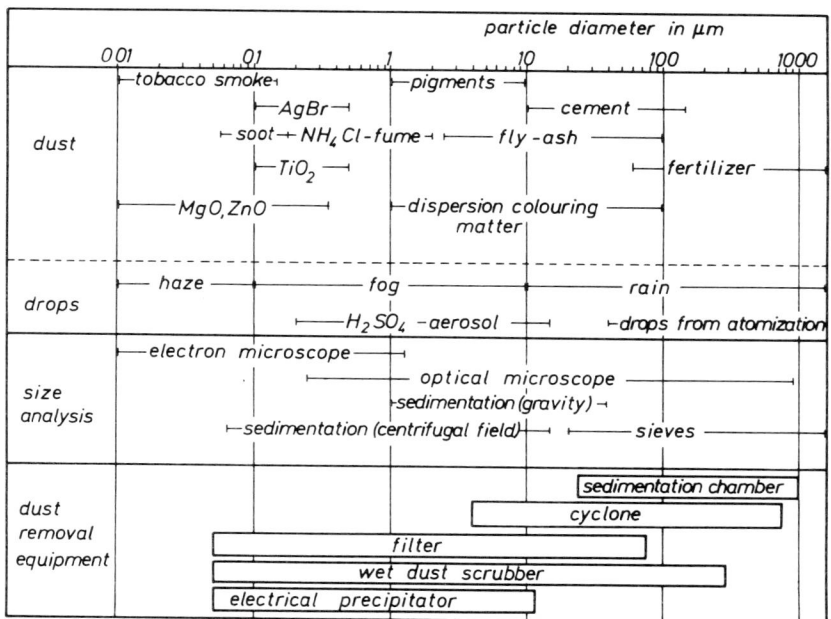

Fig. 12. Typical particulate materials, size analysis procedures, and equipment

**Table 1.** Number n of particles in 1 cm³ of gas

|  | $M_p$ 0.1 g/m³ | $M_p$ 1 g/m³ | $M_p$ 10 g/m³ | $M_p$ 100 g/m³ |
|---|---|---|---|---|
| $d_p = 0.1$ μm | $0.191 \cdot 10^9$ | $1.91 \cdot 10^9$ | $19.1 \cdot 10^9$ | $191 \cdot 10^9$ |
| $d_p = 1.0$ μm | $0.191 \cdot 10^6$ | $1.91 \cdot 10^6$ | $19.1 \cdot 10^6$ | $191 \cdot 10^6$ |
| $d_p = 10.0$ μm | $0.191 \cdot 10^3$ | $1.91 \cdot 10^3$ | $19.1 \cdot 10^3$ | $191 \cdot 10^3$ |

[g/m³] or [mg/m³]. With $M_p$ the mass of the particulate matter and V volume of the gas, the following relation holds for a collective of particles, which all have the same diameter $d_p$:

$$\frac{M_p}{V} = \frac{V_{p1} \varrho_p n}{V} = \frac{d_p^3 \pi \varrho_p n}{6V}, \tag{7}$$

with $V_{p1}$ volume of one particle, $\varrho_p$ density of particles, and n number of particles. From Eq. (7) one obtains for the number of particles contained in gas volume V:

$$\frac{n}{V} = \frac{M_p/V}{\varrho_p d_p^3 \pi/6}. \tag{8}$$

With $\varrho_p = 1$ g/cm³ one obtains for various conditions the results summarized in Table 1.

These figures show that even at very low mass concentration the number of particulates present in only one cubic centimeter or g may be extremely large.

*Particle Size Distribution Curves*

With respect to the separation of particulate matter from the gaseous carrier the size, that is the size distribution, is the most important property of the particles.

There are three particle size distribution curves: residue distribution curve or oversize mass fraction curve, the undersize mass fraction curve, and frequency distribution curve. Oversize and undersize mass fraction curves are cumulative distribution curves. In the case of a sieve process the residue mass $M_{pR}$ [kg] is the mass of that fraction of a mass of particles that remains on the sieve. The undersize mass fraction $M_{pD}$ [kg] is the mass of that fraction of the total mass of particles that falls through the sieve. The fractions $M_{pR}$ and $M_{pD}$ vary with mesh size, i.e. with particle diameter. The total mass $M_p$ [kg] of the particles is given by:

$$M_p = M_{pR} + M_{pD}. \tag{9}$$

Dividing $M_{pR}$ and $M_{pD}$ by the total mass $M_p$ of the particles and multiplying by 100, one obtains the residue or oversize ratio R and the undersize ratio D in %:

$$R \equiv \frac{M_{pR}}{M_p} 100, \quad \text{oversize ratio}, \tag{10}$$

$$D \equiv \frac{M_{pD}}{M_p} 100, \quad \text{undersize ratio}. \tag{11}$$

According to Equation (3) the sum of R and D is given by:

$$R + D = 100. \tag{12}$$

The oversize ratio R and the undersize ratio D are determined as follows:

$$R = \int_{d_p}^{d_{p,max}} dR = 100 - \int_{d_{p,min}}^{d_p} dD, \tag{13}$$

$$D = \int_{d_{p,min}}^{d_p} dD, \tag{14}$$

Figure 13 gives an example of the oversize ratio and the undersize ratio distribution curve, i.e. of R and D, as well as the mass frequency $q_3$ as a function of particle diameter $d_p$. In Fig. 13a the oversize ratio a % is that fraction of the total mass of particles, that has a diameter $d_p$ larger than $d_{pa}$. Accordingly in Fig. 13b the undersize ratio b % = 100–a % ist that fraction of the total mass of particles that has a diameter $d_p$ smaller than $d_{pa}$.

The broken vertical lines cut the diameter coordinates at $d_p'$ and the oversize and undersize ratio curves in the points of inflection, indicated by a double circle symbol.

Differentiating the undersize ratio D with respect to particle diameter $d_p$, the mass frequency $q_3$ is obtained:

$$q_3 = \frac{dD}{dd_p}. \tag{15}$$

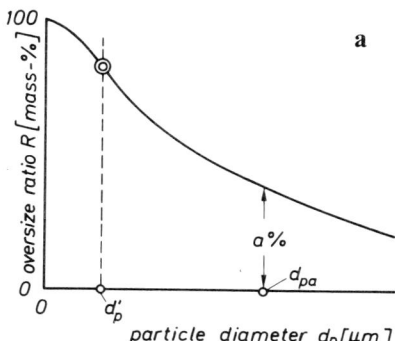

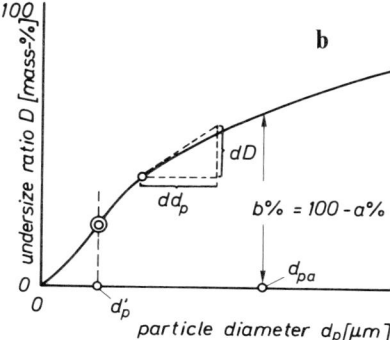

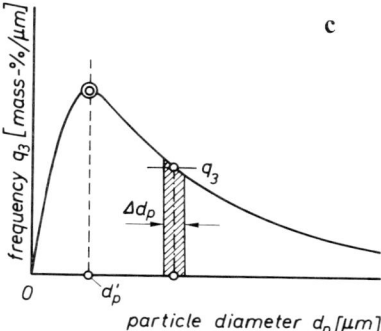

Fig. 13. Particle size distribution curves: a) oversize ratio R, b) undersize ratio D, and c) frequency $q_3$

# Air Pollution Control Equipment

In Fig. 13c the mass frequency is presented over the particle diameter $d_p$. The maximum of the frequency curve occurs at $d'_p$, for which the oversize and undersize ratio curves have a point of inflection. The area $(\Delta d_p q_3)$ indicated in Fig. 13c gives the fraction of the total mass of particles contained in the diameter range $\Delta d_p$.

The shape of the cumulative curves may vary widely, and depends primarily on the process by which the particles have been produced, and on material properties of the particles. Consequently no general equations are available for these curves. If necessary, however, empirical equations can be set up at least for rather simple types of curves. In other cases the curves can be described by a set of discrete values.

Particle size distribution curves must be determined for particle collectives for conditions prevailing at the entrance and at the exit of the cleaning stage. The distribution curves for the emitted particle collective are of particular importance from the medical point of view.

## Particle Size and Sedimentation Velocity

Next to the size the sedimentation velocity is an important property of the particle. There exists a relatively simple relation between both properties, which will be briefly discussed.

Figure 14 gives the relation between particle diameter $d_p$ and sedimentation or settling velocity $w_s$ by means of the following dimensionless parameters:

$$Ar \equiv \frac{d_p^3 g}{v^2}\left(\frac{\varrho_p}{\varrho} - 1\right) \quad \text{Archimedes number}, \tag{16}$$

$$w_s^* \equiv \frac{w_s^3/(gv)}{\varrho_p/\varrho - 1} \quad \text{velocity number}, \tag{17}$$

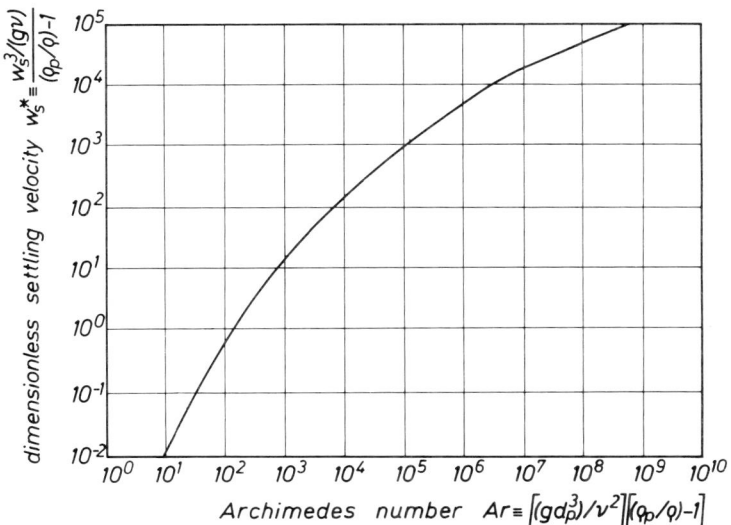

Fig. 14. Dimensionless settling velocity w* over the Archimedes number Ar

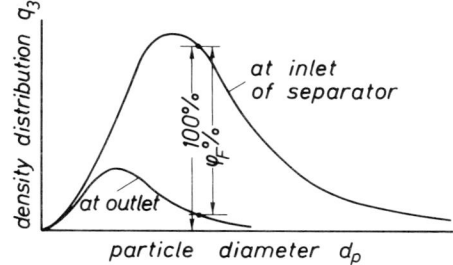

**Fig. 15.** Explanation of the fractional separation efficiency

with $d_p$ particle diameter, g gravitational acceleration, $v$ kinematic viscosity of gas, $\varrho_p$ particle density, $\varrho$ gas density, and $w_s$ settling velocity of particle.

With the diameter $d_p$ known, the Archimedes number Ar may be determined, so that from Fig. 14 the velocity number $w_s^*$ is obtained and with Eq. (17) the desired settling velocity $w_s$. If however the velocity $w_s^*$ is the known parameter, the velocity number $w_s^*$ will lead by means of Fig. 14 to the Archimedes number Ar and hence to the particle diameter $d_p$.

When the particles are smaller than about 40 µm the following relations may be safely applied:

$$w_s = \frac{1}{18} \frac{g d_p^2}{v} (\varrho_p/\varrho - 1), \tag{18}$$

$$d_p = \frac{18 v w_s}{g(\varrho_p/\varrho - 1)}. \tag{19}$$

Velocity $w_s$ and diameter $d_p$ are of great importance for all calculations concerning particle movement in separation equipment.

**Separation Efficiency of Equipment**

The efficiency of dust and drop removal in technical equipment is expressed by the collection or separation efficiency. The most informative quantity is the fractional separation efficiency, which characterizes in the most convincing way the particle removal property of equipment. Less informative is the total separation efficiency, because conditions can be selected in such a way, that for every type of particle removal equipment a total separation efficiency of close to 100% can be achieved.

It should be remembered at this point that for environment protection purposes it is the pollutant emission rate and the mass frequency $q_3$ which should be considered. As pointed out by Eq. (6) in a previous section the pollutant emission rate $M_{s,E}$ is not only a function of the total separation efficiency $\varphi_R$ but also of the volumetric flow rate of the gas stream $\dot{V}$ and the concentration of the pollutant $M_{s1}$ in the gas stream entering the separation equipment. The separation efficiency describes only the quality of separation of the applied equipment.

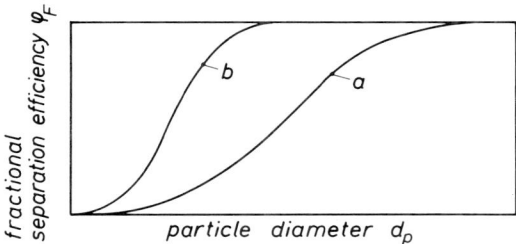

**Fig. 16.** Two curves for the fractional separation efficiency

Closely related to the fractional separation efficiency is the mass frequency $q_3$ or the oversize ratio R of the particulate matter emitted into the environment. Separation efficiency and particle size distribution will be discussed in the following sections.

*Fractional Separation Efficiency*

The fractional separation efficiency characterizes the dust removal property of equipment, and will be denoted by $\varphi_F$. According to Fig. 15, the fractional separation efficiency is that fraction of the total amount of dust contained in a given diameter range which is separated from the gas stream.

With the density distribution or frequency $q_3$ given for the dust at inlet and at exit of the dust removal equipment, the fractional separation efficiency may be expressed by the equation:

$$\varphi_F = (\Delta d_p q_3)_{inlet} - (\Delta d_p q_3)_{exit} . \qquad (20)$$

For $d_p \to 0$ the fractional separation efficiency must also become zero.

In Fig. 16 two types of curves for the fractional separation efficiency $\varphi_F$ are drawn. With increasing particle diameter $d_p$ the efficiency $\varphi_F$ increases also. But there will always be dust fractions that will not be separated. As these fractions contain the fine dust which penetrates the alveoli of the lung, fine dust is dangerous dust. From the two curves prsented in Fig. 16 curve b is the more favourable one because a larger fraction of the fine dust will be separated from the gas stream.

Fractional separation efficiency depends on two groups of parameters pertaining to the
1. type of dust removal equipment (shape, size, physical phenomena of separation) and
2. type of dust material, that is to be separated (size distribution, shape of particles, density).

The properties of the dust may in certain cases be taken into account, when the diameter $d_p$ is substituted by a particle parameter of a more general nature. This has been successfully achieved for example in case of the cyclone. For other types of dust removal equipment such a fractional separation efficiency curve for general application is not yet available. The most important group of parameters is certainly that pertaining to the type of equipment.

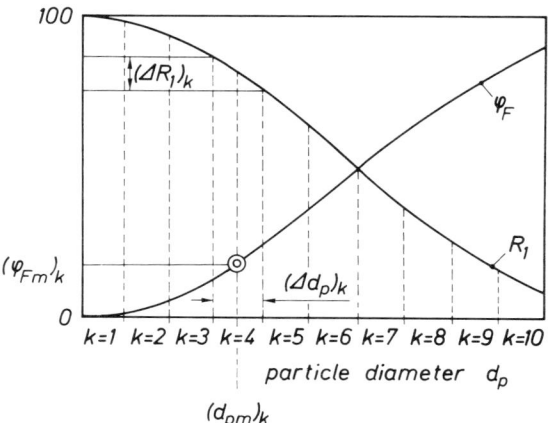

**Fig. 17.** Determination of the total separation efficiency

*Total Separation Efficiency*

The total separation efficiency defined by Eq. (2), will be determined by means of the fractional separation efficiency applying the following equation:

$$\varphi_R = \sum_{k=1}^{n} (\Delta R_1 \varphi_{Fm})_k . \tag{21}$$

The procedure for determination will be explained by means of Fig. 17, in which the curves for the fractional separation efficiency $\varphi_F = f(d_p)$ and the oversize ratio $R_1 = f(d_p)$ of the dust at equipment inlet are presented. In general it will be appropriate and adequate to divide the diameter range in about $k = 10$ equal sections $\Delta d_p$. For section $k = 4$ the quantities $(\Delta d_p)_k$, $(\Delta R_1)_k$, and $(\varphi_{Fm})_k$ are indicated in Fig. 17. $\varphi_{Fm}$ is the mean value of $\varphi_F$ in the section $\Delta d_p$ at $d_{pm}$. For a chosen example the computations have been carried out as demonstrated by Table 2.

*Particle Size Distribution at Exit Separator*

The effect of particulate pollutants on the environment depends strongly on the density distribution or frequency $q_3$ as has been discussed on the basis of Fig. 13c.

**Table 2.** Determination of the total separation efficiency $\varphi_R$

| $(\Delta d_p)_k$ µm | $(d_{pm})_k$ µm | $(R_1)_k$ — | $(\Delta R_1)_k$ — | $(\varphi_{Fm})_k$ | $(\varphi_{Fm}\Delta R_1)_k$ |
|---|---|---|---|---|---|
| 0– 5 | 3.5 | 0.98 | 0.02 | 0.20 | 0.0040 |
| 5–10 | 7.5 | 0.90 | 0.08 | 0.30 | 0.0240 |
| 10–15 | 12.5 | 0.71 | 0.19 | 0.81 | 0.1539 |
| 15–20 | 17.5 | 0.50 | 0.21 | 0.95 | 0.1995 |
| 20–25 | 22.5 | 0.30 | 0.20 | 0.98 | 0.1960 |
| 25–30 | 27.5 | 0.13 | 0.17 | 0.99 | 0.1683 |
| 30–40 | 35.0 | 0 | 0.13 | 1.00 | 0.1300 |

$$\varphi_R = 0.8757 = 88\%$$

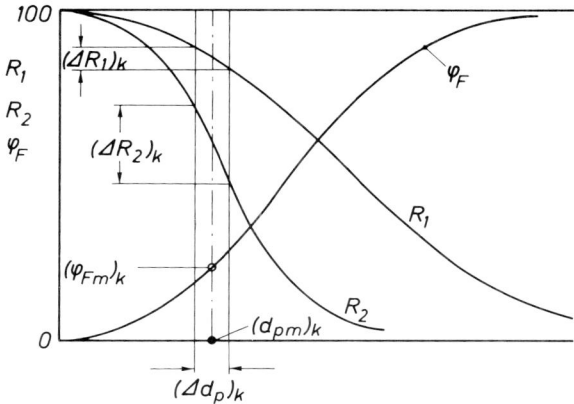

**Fig. 18.** Determination of the oversize distribution $R_2$ of the emitted dust

Instead of the frequency $q_3$ it is sometimes easier to determine the oversize distribution $R_2$ of the dust at the exit of the dust removal equipment. The oversize distribution $R_2$ may be determined by means of the following equation:

$$(\Delta R_2)_k = (\Delta R_1)_k \frac{1-(\varphi_{Fm})_k}{1-\varphi_R}. \tag{22}$$

Application of this equation is illustrated by Fig. 18.

### Equipment for Dust Removal

The dust removal equipment available may be divided into two large groups: Dry dust removal equipment and wet dust removal equipment. For both groups a few representative examples will be discussed.

#### Dry Dust Removal Equipment

General Survey

In dry dust removal equipment dust separation from the carrier gas is achieved directly, i.e. in the dry state, without application of a special dust collection agent, such as water drops for example. The collected dry dust can be directly disposed of. For dry dust removal equipment there is, however, the danger of dust explosion. Therefore, special measures have to be taken for prevention of dust explosion and pressure release in case of explosion.

The separation of dust particles from a gas is due to the action of mainly three forces: Mass or *inertia force, surface* or *adhesion force, and electrical force*. In most of the dry dust separators there is one force, which dominates the other and is therefore the most influential force for the separation process. According to the dominant force there are three groups of dry dust removal equipment:
1. Mass force separator
2. Adhesion force separator
3. Electrical force separator.

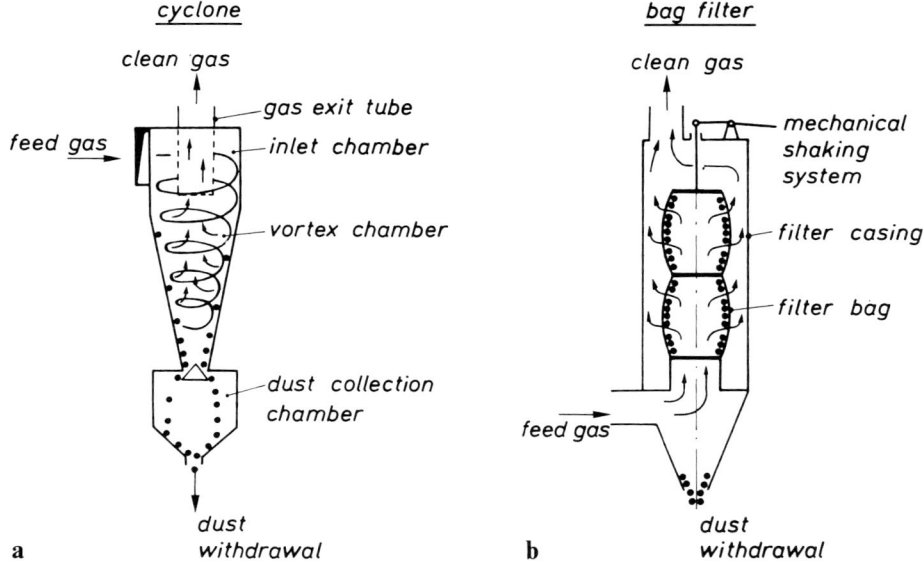

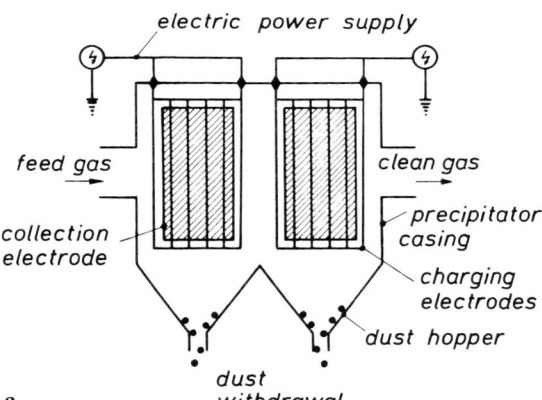

**Fig. 19.** Examples for dry dust removal equipment; a) cyclone, b) bag filter, c) electrical precipitator

A typical mass force separator is the cyclone. Under the action of centrifugal forces the dust particles are separated from the gas phase. Conventional adhesion force separators are bag and pocket filters. A typical electrical force separator is the plate precipitator. Simplified drawings of the three types of dry dust separators are presented in Fig. 19. Design and operation of the three types of dry dust separators will be discussed in special chapters. In order to get a rough idea of the separation properties and of the fields of application, examples for the fractional collection efficiency $\varphi_F$ for the three types of dry dust separators are presented in Fig. 20. These curves are related to a particular dust and to other par-

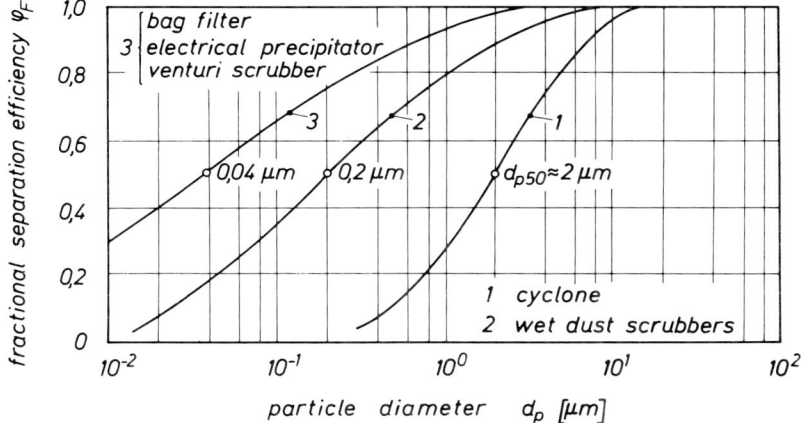

Fig. 20. Examples for fractional collection efficiency curves for dry and wet dust separators

ticular conditions. Therefore the curves are of no general nature but they do show the general trend. The most efficient dry dust separators are fibrous filters and electrical precipitators. The cut diameter $d_{p\,50}$, for which 50% of the dust contained in that diameter section, is about 0.04 µm for fibrous filters and for electrical precipitators and only 2.0 µm for cyclones. The difference in cut diameter illustrates the superiority particularly of fibrous filters over cyclones. It should be mentioned however, that for many technical conditions and applications the cut diameter for fibrous and electrical filters is much larger than given in Fig. 20, so that curve 3 moves closer to curve 1.

For each of the dust separators mentioned there is a VDI-Guideline available with valuable engineering information:
1. Physical fundamentals of separation process,
2. design calculations,
3. examples for various designs,
4. guaranty conditions,
5. operation and maintenance.

Engineers engaged in design and operation of dry dust separation equipment should be in possession of these guidelines:

Mass force separators  VDI-Guideline 3676
Fibrous filters  VDI-Guideline 3677
Electrical precipitators  VDI-Guideline 3678

These VDI-Guidelines (Guidelines of the *Verein Deutscher Ingenieure*) may be obtained from VDI-Verlag GmbH, Düsseldorf, Federal Republic of Germany.

In the following sections further details of design and operation of these three types of dust removal equipment are discussed.

The Cyclone

In Fig. 19a a schematic drawing of a cyclone of conventional design is shown. The cyclone consists of a cylindrical entrance chamber, a conical dust separation or

**Fig. 21.** Photo of an installation consisting of three cyclons arranged in parallel

vortex chamber, an inlet channel, an outlet tube with extension into the cylindrical entrance chamber, a dust collection chamber, and a conical vortex spoiler that separates the separation chamber from the collection chamber, thereby preventing dust reentrainement into the gas. The dust-laden gas enters the cyclone by a tangential inlet channel and leaves the cyclone by the outlet tube. Dust separation takes place in the entrance and in the vortex chamber.

Some of the important properties of the cyclone are as follows.

1. In the cyclone direct dust/gas-separation is possible without introducing any separation agent, like liquid drops. The cyclone is a dry dust separator, that does not generate new pollution problems.
2. The cyclone is designed for a continuous separation process with continuous withdrawal of gas and dust.
3. The technical attractiveness is at least partially due to its simplicity of design. There is no technical sophistication.
4. The physical phenomena of gas/dust-separation in a cyclone are relatively well understood. This is a sound basis for a mathematical description of the process. Pressure drop and separation efficiency can be fairly well determined in advance.
5. The particle cut diameter is in the range of 2 to 5 µm, with a fair chance to reduce it still further.
6. If the cut diameter does not meet the requirements, then the cyclone may be applied as a preseparator.

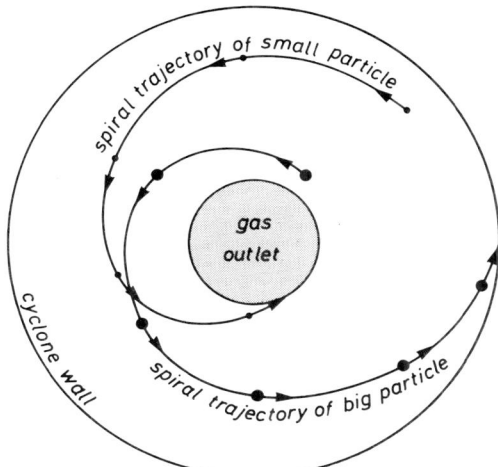

**Fig. 22.** Trajectories of small and large particles in a horizontal plane of a cyclone

7. The cyclone may be applied either as a single unit or with many units operated in parallel.
8. Low cost for investment and for operation.

Figure 21 shows a photograph of a technical installation consisting of three cyclones operated in parallel arrangement. Because of its good properties the cyclone has found wide application in industry, and proved its separation efficiency. With air quality regulations becoming increasingly severe it resulted that the cyclone did not meet all demands, particularly as a fine dust separator. Important contributions to an understanding of fluid flow and particle movement in the cyclone have been delivered by Barth, Muschelknautz, Ogawa, and coworkers [9–20]. A computer program for design and performance computation has been developed and tested by Spilger [21].

Separation of particles from the gas stream in the cyclone is achieved by the combined action of centrifugal and frictional forces. All particles will move on spiral-like paths as shown in Fig. 22. The bigger particles will move along the spiral outwards while the smaller particles will move along the spiral inwards. Outwards moving large particles will be separated from the gas, they will collide with the cyclone wall and move towards the dust collection chamber. Inwards moving small particles will not be separated from the gas stream, on the contrary, they will be carried away by leaving the cyclone through the exit tube.

The separation properties of the cyclone will be discussed by means of the Figs. 23 to 25. The results presentred habe been obtained by the mentioned computer program. Figure 23 gives the size distribution curve for the assumed dust. It is a rather fine dust.

The separation efficiency $\varphi$ is presented over the loading ratio $\mu_e$ at inlett of cyclone in Fig. 24. The loading ratio $\mu_e$ is defined by:

$$\varrho_e \equiv \frac{\dot{M}_{pe}}{\dot{M}} = \frac{\dot{V}M_{pe}}{\dot{V}}, \qquad (23)$$

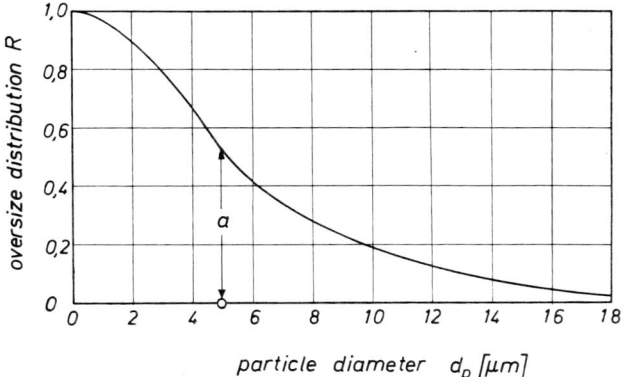

**Fig. 23.** Particle distribution curve used as a basis for cyclone computations

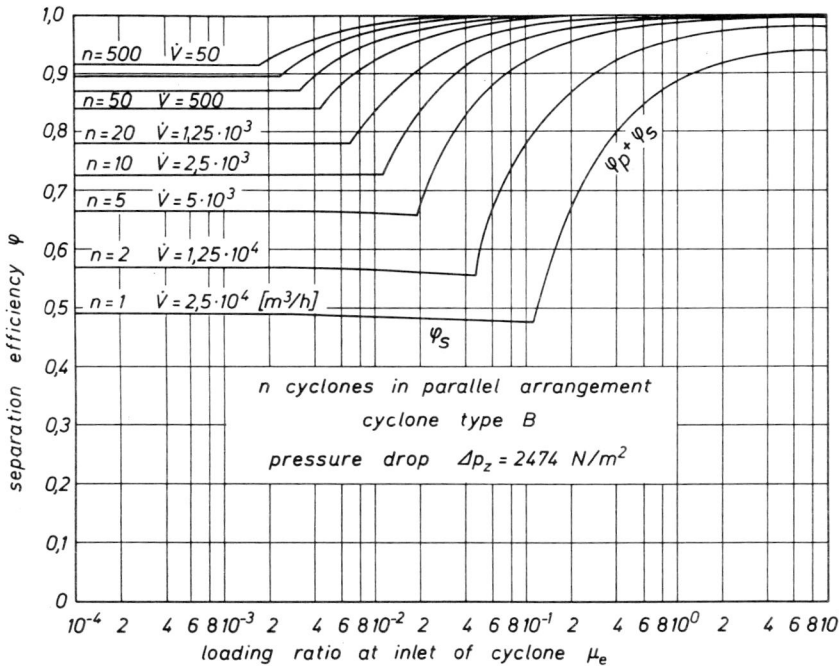

**Fig. 24.** Separation efficiency $\varphi$ for cyclones in parallel arrangement

with $\dot{V}[m^3/s]$ volumetric gas flow rate, $\varrho[kg/m^3]$ gas density, and $M_{pe}[kg/m^3]$, the mass of particles contained in 1 m³ of gas at the inlet of the cyclone. The separation efficiency $\varphi$ has been determined for $n=1$ to $n=500$ cyclones operated in parallel. The gas flow rate is for all cases $2.5 \cdot 10^4$ m³/h. The type B-cyclone is shown in Fig. 25.

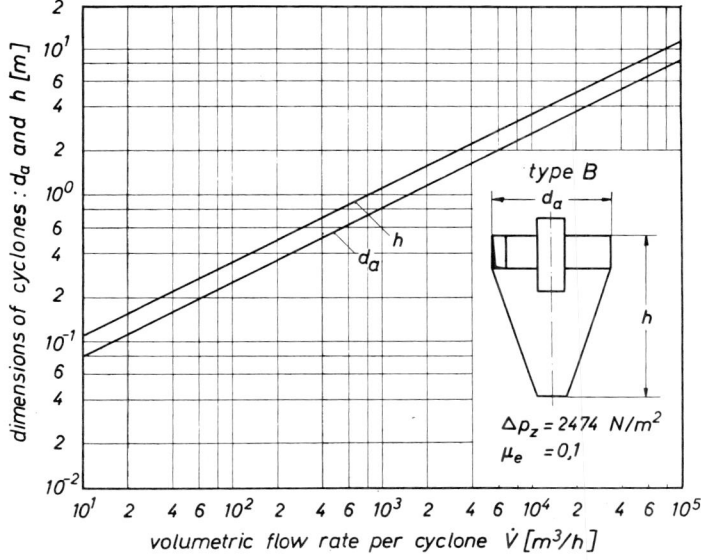

Fig. 25. Dimensions of cyclones over volumetric flow rate of gas

Each efficiency curve consists of two branches. The first branch shows the efficiency to be almost independent of the loading ratio. Due to another separation phenomenon the efficiency curve rises steeply in the second branch of the curve. By arranging small cyclones in parallel extremely efficient dust separation can be obtained.

The dimensions of the cyclone as a function of the volumetric gas flow rate are given in Fig. 25 for a loading ratio $\mu_e = 0.1$.

The Fabric Filter

Fabric materials offer a very effective means for the removal of dust from gas streams. The fabric is arranged in equipment in such a way, that the dust-laden gas has to pass through the small openings of the fabric. Due to different physical phenomena the dust particles impinge on the surface of the fabric material and adhere to it. Figure 26 shows limestone particles adhering to a fiber; the particle size is between 3 and 10 µm, the diameter of the fiber is about 50 µm [23]. A photo of a needle felt that was exposed for a very short time to tobacco smoke is reproduced in Fig. 27.

There are two types of fabric filters, namely bag filters and pocket or envelop filters. Figure 28 gives a schematic drawing of a bag filter. The fabric filter media rests on a metallic frame, that is the support basket. Dust collection may take place in the interior of the fabric filter and on the surface of a very porous layer of dust that builds up on the surface of the fabric filter. Bag filters may be designed as mono- or multi-chamber filters, or as baghouses of rectangular or circular cross section. They are preferably operated under vacuum conditions. In

**Fig. 26.** Limestone particles adhering to a fiber; diameter of the fiber is about 50 μm, while the particle size is of the order of 3 to 10 μm

**Fig. 27.** Structure of needle felt, exposed for a very short time to tobacco smoke

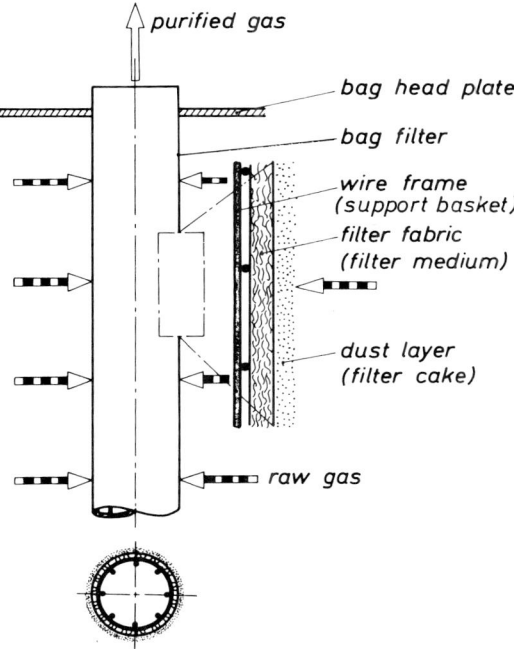

Fig. 28. Arrangement of bag filter and dust collection

order to avoid to high a pressure decrease the dust has to be dislodged and the system cleaned.

Conventional design of a bag filter is given in Fig. 29 [1]. The diameter of a bag filter is of the order of 0.1 to 0.3 m and the height may be up to 10 m. The number of filter bags in small groups may be of the order of a few hundred to a few thousand. The mean relocity $w_\infty$ of the gas normal to the surface of the filter media is very low and *of the order of 2 to 4 cm/s*. Properties of fibers used in filter media are summarized in Table 3. There are two groups of filter media: woven fabrics and non-woven fabrics. Non-woven media are fleeces and felts, especially needle felts.

Dust removal from a gas stream by means of fabric filters is a discontinuous process. After a period of dust collection, which leads to the formation of a layer of dust on the surface of the fabric, a period of cleaning follows in which the dust layer is dislodged by either mechanical or fluid dynamic methods. The dislodging method applied is the most important single parameter affecting the lifetime of bag filters, thus attracting the attention of engineers in the development of low cost filtration equipment and processes.

Fabric filters are primarily designed for:
high collection efficiency,
low pressure drop,
long lifetime, and
minimum cost.

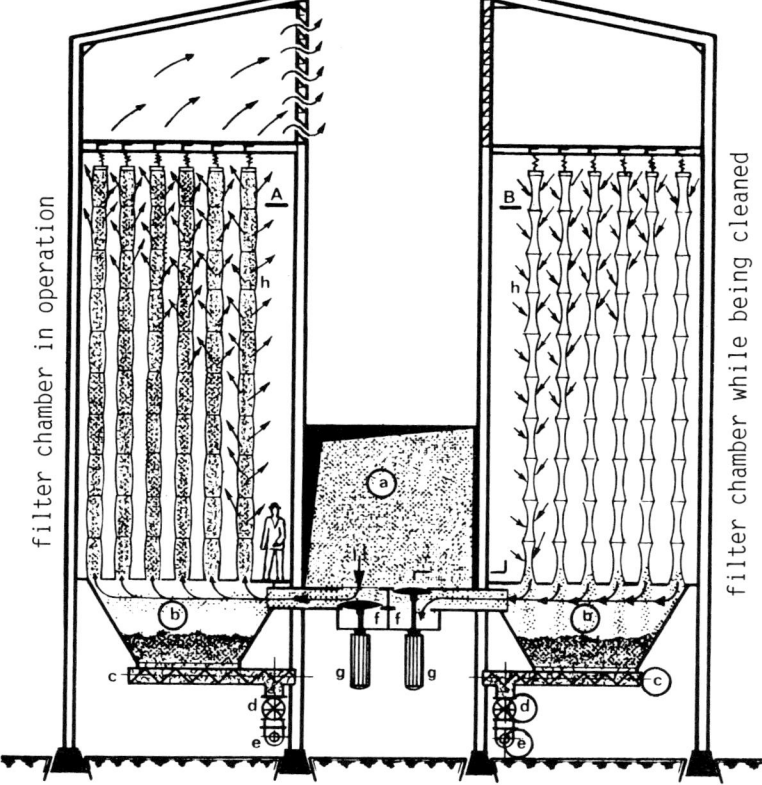

**Fig. 29.** Conventional design of a bag filter. a crude gas pressure duct, b dust hopper, c dust hopper screw conveyor, d bucket wheel lock, e collecting worm conveyor, f flushing gas duct, g cleaning valve, h filter bags

All the parameters depend on the collecting filter surface area $A_c$ and the specific gas flow rate $\dot{v}$, given in m³/(m²s), but quite often in m³/(m²min) or m³/(m²h). The latter dimensions are often preferred because of the figures obtained for the flow rate. A conventional specific flow rate is for instance $\dot{v} = 90[\text{m}^3/(\text{m}^2\text{h})] = 1.5[\text{m}^3/(\text{m}^2\text{min})] = 0.025[\text{m}^3/(\text{m}^2\text{s})]$. The collecting filter surface area $A_c$ may be determined by means of the following equation.

$$A_c = \frac{\dot{V}}{\dot{v}}, \tag{24}$$

where $\dot{V}$ is the volumetric flow rate of the raw gas.

The fabric filter is the most efficient dust removal system available today. Disadvantages are, however, the intermittent operation and the extremely large volume of the filter houses. A rough idea of the fractional separation efficiency curve applicable to bag filters is included in Fig. 20.

Table 3. Properties of some fibers used in filter media

| Fiber | Relative mechanical strength | Maximum usable temperature °C | Relative resistance to attack by | | | Special properties | General applications |
|---|---|---|---|---|---|---|---|
| | | | Acid | Base | Organic solvent | | |
| Cotton | Strong | 80 | Poor | Medium | Good | Low cost | General low temperature dust jobs |
| Wool | Medium | 95 | Medium | Poor | Good | Low cost | Smelters |
| Polyamide (nylon) | Strong | 100 | Medium | Good | Good | Easy to clean | Low temperature, abrasive dust jobs |
| Polyester (dacron) | Strong | 135 | Good | Medium | Good | Easy to clean | Smelters, chemicals, arc furnaces |
| Tetrafluoroethylene | Medium | 260 | Good | Good | Good | Expensive | Chemicals |
| Glass | Strong | 280 | Medium | Medium | Good | Poor resistance to abrasion | Smelters, arc furnaces, carbon black |
| "Nomex" nylon | Strong | 230 | Good | Medium | Good | Poor resistance to moisture | Smelting, arc furnaces |

The Electric Precipitator

Dust particles are removed from a gas stream in electrostatic precipitators due to the action of electrostatic forces. Between a possitive and a negative electrode an electric field is establised, in which the charged dust particles are forced to move in the direction of the collecting electrode, on the surface of which a very porous layer of dust is built up with time. The dust layer has to be removed periodically for instance by rapping of the collecting electrode. The dislodged dust falls into a hopper at the base of the precipitator [1].

Electrostatic precipitators are not only used for the purpose of dust separation but also for the separation of mist from a gas stream. The liquid particles coalesce on the collection electrode, so that a liquid film is established that flows downwards due to gravity forces and drains into a sump at the bottom of the precipitator.

The basis of electrostatic precipitation is Coulomb's law, that relates electric attraction and repulsion of bodies with the distance between the bodies. According to Coulomb's law the strength of the electric forces decreases with the square of the distance. Coulomb (1736 to 1806), was a French scientist who devoted some his scientific investigations to the study of magnetic and electric attractions. The unit of electric charge is named, in his honour, the Coulomb. Industrial application of Coulomb's law in dust and mist separation processes is due to the American engineer Frederick G. Cottrell, who developed around 1910 the "Cottrell precipitator." The work of Cottrell was initiated directly by environmental protection considerations. An interesting summary of investigations on the nature of electrostatic forces and engineering development of electrostatic precipitators was included in the book by White [23].

*Characteristic properties of the electrostatic precipitator are:*
1. Low pressure drop, of the order of 100 to 1,000 [$N/m^2$],
2. high gas capacity, $10^3$ or $10^6$ [$m_n^3/h$] are quite common,
3. low energy demand, about 0.1 to 0.8 [$kWh/1,000\ m^3$],
4. very high collection efficiency in the small diameter range of dust particles, better than 99%.

But there are serious drawbacks in the application of electrostatic precipitators which are primarily related to the electric properties of the dust. However, over many decades the electrostatic precipitator has been the backbone in the field of fine dust collection and has been successfully applied in all relevant branches of industry.

The process of dust separation in an electric precipitator consists of the following steps:
1. Generation of an electrical field,
2. generation of electrical charges,
3. transfer of electrical charge to a dust particle,
4. movement of a charged dust particle in an electrical field to the collection electrode,
5. adhesion of the charged dust particle to the surface of the collection electrode,
6. dislodging of the dust layer from the collection electrode,

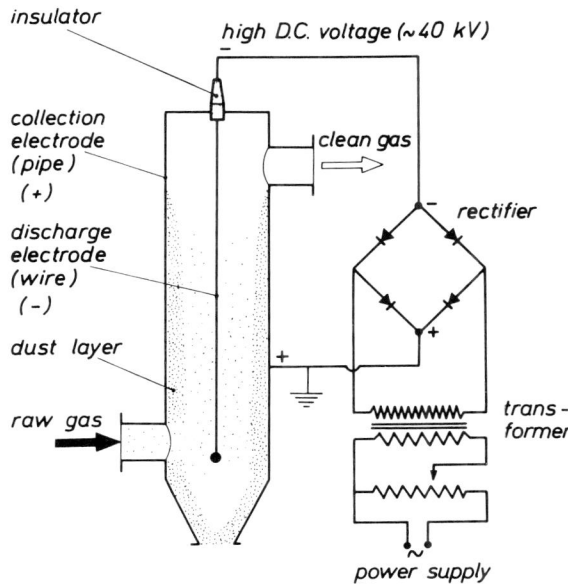

Fig. 30. Schematic representation of a wire and pipe precipitator

7. collection of dislodged dust layer or fragments thereof in a hopper, and
8. removal of the dust from the hopper.

An elementary unit of an electrical precipitator consisting of a discharge and collection electrode must be designed in such a way that at least the first six steps of the precipitation process are carried out effectively.

A very simple elementary unit consists of a wire and a pipe as shown in Fig. 30. The wire, arranged in the axis of the pipe is the discharging electrode and the pipe is the collection electrode. In general the discharging electrode is negative and the collection electrode is positive. The high direct current (dc) voltage applied to the wire electrode is in the range of 20 [kV] to 80 [kV] with a mean value of about 40 [kV]. The voltage depends primarily on the distance between discharging and collection electrodes.

The high voltage applied to the electrodes results in a strong electric field and an electric discharge of ions from the discharging electrode, called the corona. The strength E of the electric field in a wire and pipe system is for ideal conditions given by the following equation.

$$E = \frac{U}{r \ln(r_c/r_d)}, \tag{25}$$

where U is the voltage, r the local radius, $r_d$ the radius of the discharging and $r_c$ of the collection electrodes. The strength of the electric field decreases with increasing distance from the discharge electrode. The electric force $K_e$ acting on a charged particle in an electric field is the Coulomb force, given by:

$$K_e = QE. \tag{26}$$

Q is the electric charge of the particle. The Coulomb force propels the dust particle in the direction of the collection electrode. The collected dust is dislodged either by means of a mechanical rapping system or a liquid sprayed on the electrode. The wire and pipe arrangement is well suited for wet dust precipitation. The more important electrical precipitator makes use of the wire and plate arrangement. The plate is the dust collecting electrode. The collected dust builds up a layer on the plate electrode that must be dislodged by a strong mechanical system that causes the plate to vibrate.

Calculations for the design of electric precipitators are based on the Deutsch formula [24]. From this equation one obtains the dimensions of the precipitator:

$$\frac{A_c}{V} L = \frac{w_z}{w_{pr}} \ln\left(\frac{1}{1-\varphi}\right), \qquad (27)$$

with $A_c$ surface area of collection electrode, V volume of the precipitator element, L length of precipitator in flow direction, $w_z$ mean gas velocity, $w_{pr}$ migration velocity of dust particle, and $\varphi$ total separation efficiency. For a plate precipitator with distance a between two parallel plates and L length of plates, one obtains for one channel:

$$\frac{A_c}{V} L = 2 \frac{L}{a}. \qquad (28)$$

The volumetric gas flow rate $\dot{V}_g$ to be cleaned in the precipitator is related to gas velocity $w_z$ and number n of the elements by the equation:

$$\dot{V}_g = n a h w_z. \qquad (29)$$

In the last equation h is the height of the plates, which is equal to the height of the channels. With a given gas velocity $w_z$ Eq. (29) serves to determine the number n of the channels that have to be arranged in parallel.

Application of Eqs. (27) to (29) requires knowledge of the migration velocity $w_{pr}$, that can only be found by experiment. The migration velocity is a hitherto unknown function of parameters. In general this function may be expressed as follows:

$$\left.\begin{array}{l}\text{migration}\\ \text{velocity}\\ w_{pr}\end{array}\right\} = \text{function} \left\{\begin{array}{l}\text{electric field properties}\\ \text{dust properties}\\ \text{fluid flow properties}\\ \text{geometric properties.}\end{array}\right. \qquad (30)$$

For multizone precipitators mean values of the migration velocity $w_{pr}$ have to be determined for each zone, taking into account that the dust load of the gas and particle size distribution vary substantially. A solution of Eq. (30) has not been obtained so far; it is doubtful whether this will ever be achieved. The relationship between the four groups of parameters and the migration velocity is far too complicated. Therefore solutions for simplified versions only of this equation have been found by experimental techniques. The results belong to the most valuable part of the know-how of electric precipitator manufacturing companies.

# Air Pollution Control Equipment

**Fig. 31.** View on a three-zone, horizontal electric precipitator (Lurgi design)

Figure 31 gives a view of a three-zone electric precipitator of Lurgi design. An example for the fractional efficiency curve for a conventional electric precipitator is given in Fig. 20.

## Wet Dust Removal Equipment

In wet dust removal equipment the dust particles are captured by liquid drops, which are introduced into the gas stream. The gas pollution problem is converted into a liquid pollution problem. This is acceptable, when the volume of the contaminated water is small, and when a waste water treatment plant is available [1].

There is a great variety of wet dust scrubbers available. Only three examples are selected for brief discussion:

1. Packed column scrubber,
2. vortex scrubber, and
3. Venturi scrubber.

In Fig. 32 schematic drawings of these scrubbers are presented. To dedust 1 m$^3$ of gas in these scrubbers about 1 to 3 liters of water are required. Dedusting operation is easily combined with the absorption of gaseous pollutants contained in the gas stream.

In the packed column scrubber the water is distributed over the surface of various types of small packing elements and flowing downwards. The gas moves countercurrently through the porous layer of packing elements, changing its flow direction according to the arrangement of the elements. Because of the inertia forces the particles, particularly the coarser ones, do not follow the gas flow, and impinge on the liquid surface.

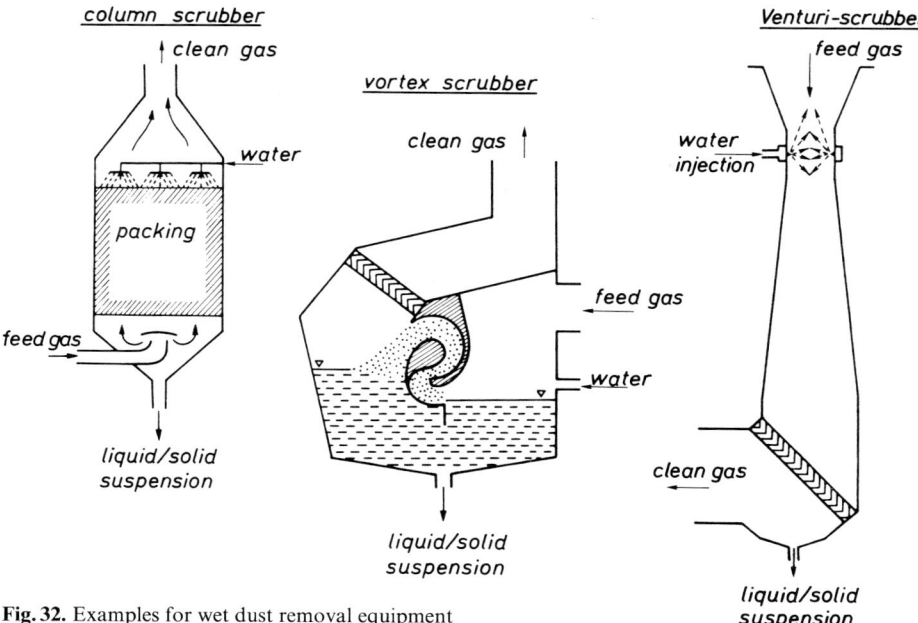

Fig. 32. Examples for wet dust removal equipment

In the vortex scrubber the water drops are produced by the gas stream and carried into the vortex channel. The dust particles are captured in the period of drop generation and drop movement inside the vortex channel. The efficiency of dust collection by drops depends on the relative velocity between drops and dust particles.

The Venturi-scrubber is the most efficient wet dust scrubber. The water is introduced in the throat of the Venturi-scrubber and there dispersed by the high velocity gas stream. The drop in pressure of the gas on its way through the Venturi-scrubber is very high.

A fractional separation efficiency curve for wet dust scrubbers is included in Fig. 20.

Successful application of wet dust scrubbers makes it necessary to reduce the temperature of the gas, in order to prevent undesired evaporation and to separate the drops from the gas after dust collection. The efficiency of wet dust scrubbing depends strongly on the efficiency of drop collection. In general, the drop size is much larger than the dust particle size. Drop separation is therefore generally easier than dust separation but drop separation equipment applied is often badly designed and consequently not very effective.

*Selection of Dust Removal Equipment and Cost*

Selection of dust removal equipment is always related to a specified dust removal problem. Two groups of criteria must be taken into account. These are dust/carrier-gas specific and equipment specific criteria, which are summarized in Table 4 [1].

Air Pollution Control Equipment

**Table 4.** Criteria for the selection of dust removal equipment

| Dust/carrier-gas specific criteria | Equipment specific criteria |
|---|---|
| Dust concentration | Fractional collection efficiency |
| Particle size distribution | Security of operation |
| Particle density | Availability |
| Abrasivity of dust | Size of equipment |
| Gas temperature | Adaptability to varying operating conditions |
| Tendency for agglomeration and incrustation | Pressure drop |
| Chemical reactivity | Maintenance and repairs |
| Inflammability and explosivity | Simplicity of design |
| Toxical properties | Sensitivity to erosion, corrosion, and foam |
| Odorous properties | formation |
| Optical properties | Water requirement |
| Foaming properties | Electric power requirement |
| Wettability | Operating cost |
| Price | Investment cost |

Cost is a very important factor in the selection of equipment. The cost relations given in Table 5 give a rough survey [1].

**Table 5.** Cost relations of dust removal equipment

|  | Investment cost | Operating cost |  |
|---|---|---|---|
| Cyclone | 100% | 100% | |
| Bag filter | 250% | 250% | Dry dust equipment |
| Electrical precipitator | 450% | 150% | |
| Column scrubber | 270% | 260% | Wet dust equipment |
| Venturi scrubber | 220% | 500% | |

More detailed information is included in Table 6 [1].

**Table 6.** Energy requirement and cost of dust removal equipment

|  | Cyclone | Bag filter | Electric precipitator | Wet dust scrubber |
|---|---|---|---|---|
| Cut diameter, $d_{p50}$ [µm] for $\varrho_p = 2.6$ [g/cm$^3$] | 3 –10 | 0.5 | 0.1– 0.8 | 0.1– 1.5 |
| Energy requirement in [kWh/1000 m$^3$] | 0.25– 1.5 | 0.5– 1.5 | 0.3– 1.0 | 0.2– 8.0 |
| Cost of investment in [DM per m$^3$/h] | 1 – 3 | 5 –19 | 9 –27 | 2 –12 |
| Maintenance cost in % of cost of investment | 2% | 12% | 3 – 5% | 7% |

## Technology for the Abatement of Gaseous Pollutant Emissions

Gaseous pollutants may be removed from waste gas streams either by predominantly physical or chemical or microbiological processes. Examples for physical processes are absorption and adsorption, the chemical processes are thermal and catalytic conversion in afterburners, and the microbiological processes are carried out by selected microorganisms in so-called bioreactors. Important gaseous pollutants are sulfur dioxide, nitrogen oxides, carbon oxides, chlorine, and fluorine compounds, hydrogen sulfide, and hydrocarbon compounds. A brief description will follow for the processes mentioned.

**Physical Processes and Equipment for the Abatement of Gaseous Pollutant Emissions**

For matter of simplification absorption and adsorption processes will be considered as predominantly physical processes, although there will always be some kind of a chemical conversion included. Furthermore, purely physical processes will in most cases not reduce the pollutant concentration in the waste gas to the desired extent. Consequently there will be, in most cases, a chemical conversion to follow the physical process. In this chapter however attention is directed to the physical processes.

*Absorption Process and Equipment*

General Description of the Process

Absorption is the process by which a gas is dissolved in a liquid. In most cases absorption is the process of selective removal of a component from a gas mixture. The gas mixture consists of insoluble and soluble components. The soluble component is called the solute; it is transferred from the gas mixture to the liquid. The liquid consists of the solvent and the absorbate. Sometimes the liquid mixture including the absorbate is called the absorbent. Absorption is one of the most advanced techniques applied to the separation of gas mixtures. It is widely used in industry; in the areas of air pollution control it has become one of the important

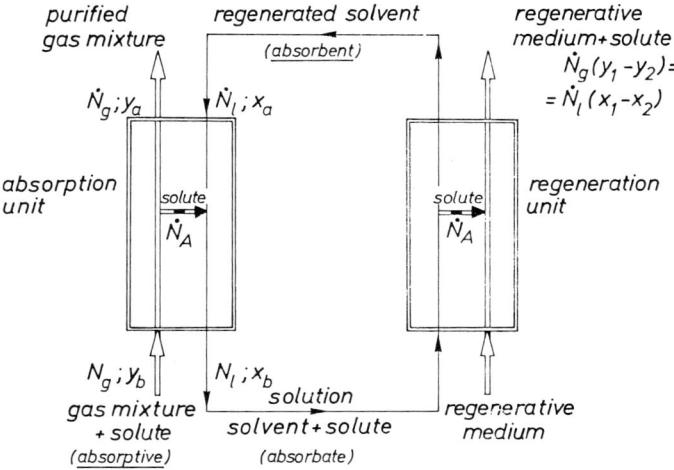

Fig. 33. General layout of an absorption plant

**Table 7.** Typical gaseous pollutants and its sources

| Key-element | Pollutant | Sources |
|---|---|---|
| S | $SO_2$ | Waste gases from power station |
| | $SO_3$ | $H_2SO_4$-production |
| | $H_2S$ | Natural gas; fiber and paper production |
| | COS | Coke oven |
| | $CS_2$ | Fiber industry |
| | Mercaptanes, thiophene | Oil refineries |
| N | NO, $NO_2$, $N_2O_3$, $N_5O_4$, $N_2O$ | Nitric acid production, high-temperature oxidation high-temperature pyrolysis |
| | $N_2O_2$ | Nitrification processes |
| | $NH_3$ | Ammonia production |
| | HCN | Hydrocyanic acid production |
| | Pyridines, xylidines, amines | Coke oven |
| | ClCN | Hardening processes |
| Halogen | HF | Fertilizer production on phosphate basis, aluminium production |
| F | $SiF_4$ | Ceramics, Fertilizer, ferro-silicon production |
| Cl | HCl | Hydrochloric acid production, burning of polyvinyl chloride |
| | $Cl_2$ | Chlorine production |
| | $COCl_2$ | Iso-cyanate processes |
| Br | HBr, Br | Bromine processes |
| Hg | Hg | Alkali-electrolysis |
| C | CO | Fire stations, combustion engines |
| Inorganic | $CO_2$ | Especially power plants |
| C Organic | Hydrocarbons Paraffins, olefins Diolefins, acetylene Aromatics Oxidized hydrocarbons Aldehydes Alcohols Ketones Phenols Halogenated hydrocarbons | Chemical, petrochemical, and allied industries |

processes for the abatement of gaseous pollutants. Table 7 gives a list of pollutants that may be removed from a gas mixture by means of absorption [1].

The general lay-out of an absorption plant is represented in Fig. 33. An absorption plant consists of the absorption unit proper, the regeneration unit, and all necessary auxiliary equipment and machinery. The gas mixture with the gaseous pollutant, the solute or absorptiv, enters the absorption unit at the bottom. Within the absorption unit the gas mixture is brought into contact with the liquid, called the solvent or absorbent. During the time of contact the pollutant is at least partly transferred into the liquid. The purified gas mixture leaves the absorption unit at the top. The liquid that normally enters the absorption unit at the top, leaves the unit at the bottom, carrying with it the pollutant, and from there on is directed to the regeneration unit. Here the absorbent is treated in such a way, that the absorbent may be recycled and the pollutant disposed of appropriately.

Regeneration may be achieved by various processes such as vaporization, rectification, steam stripping, desorption, extraction etc. If necessary a regenerative medium is introduced to extract the pollutant. In most cases energy and mass transfer considerations will lead to the inclusion of heat exchangers in the solvent regeneration cycle. Because of decreasing solubility with increasing temperature and decreasing pressure, the absorption process should be carried out at a relatively low temperature and high pressure. In air pollutant control however, the polluted gases are in most cases available at higher temperatures and low pressure. Therefore the waste gas has to pass through a precooler in order to reduce the temperature. Reducing the temperature by about 15 °C the vapour pressure for many pollutants is reduced the 50% and so is the minimum flow rate of the liquid absorbent. Thus in air pollution control technology the absorption process has to be applied under very unfavourable conditions. This is the reason why all possibilities for the improvement of the absorption process have to be used.

Another important property that has to be observed is the low concentration of the pollutant in the gas mixtures. In many cases the pollutant concentration is so low, that the pollutant has the properties of a tracer that can hardly be traced in the gas mixture. In this case the absorbers have to be improved in such a way, that absorption techniques become tracer techniques.

Although all considerations are concentrated on the absorption unit one has to keep in mind that this is only one component of the absorption plant. In many cases the absorption unit is the smallest element of the plant. This is of special importance when the best suited absorption unit has to be selected for a defined process, and when optimization calculations have to be carried out. In such cases the whole plant has to be taken into account.

## Classification of Absorbers

In an absorber the polluted gas mixture is brought into contact with a liquid absorbent in such a way, that an effective transfer of the pollutant to the absorbent is accomplished. To a certain extent the effectiveness of pollutant absorption depends on the size of the active interfacial area. The available interfacial area may be much larger than the active part in the absorption process. The inactive part of the interfacial area cannot be penetrated by the pollutant molecules. Surface active agents accumulate in the interface and build up a diffusion barrier. This is the reason why in industrial absorbers it is very important to generate a large active interfacial area. This may be accomplished for example by producing wavy liquid films and by the periodic renewal of the interfacial area.

An important property of an absorber is the method of generation of interfacial area. This property serves as a basis for classification. There are three groups of absorbers. The interfacial area is due to the
1. generation of liquid films,
2. generation of jets, and
3. generation of bubbles and drops.

Consequently the three groups of absorbers are:
1. Film absorbers, for example tube bundle and packed columns,
2. jet absorbers, and
3. bubble and drop absorbers, for example tray columns, bubble columns with and without internal circulation and various types of spray columns.

In the following section one of the most efficient spray absorbers will be briefly discussed.

High-Efficiency Absorption Machine

The high-efficiency absorption machine is a new development that has very successfully passed the laboratory stage. The characteristic properties of this absorber, that has been conceived by the author of this chapter [25], may be summarized as follows:
1. Absorption machine with a rotating wheel for liquid dispersion.
2. Periodic redispersion of the liquid.
3. Extremely small volume of the machine.

Figure 34 gives a rough idea of the rotating wheel. The vanes of the wheel have a large number of narrow slots, in which drop formation takes place. In the case of the wheel shown in Fig. 34, the wheel diameter is 0.5 m, the distance between slots is 20 mm, and the width of the slots is 4 mm. In this case the dispersed liquid is redispersed four or five times. The number of redispersals depends on the length

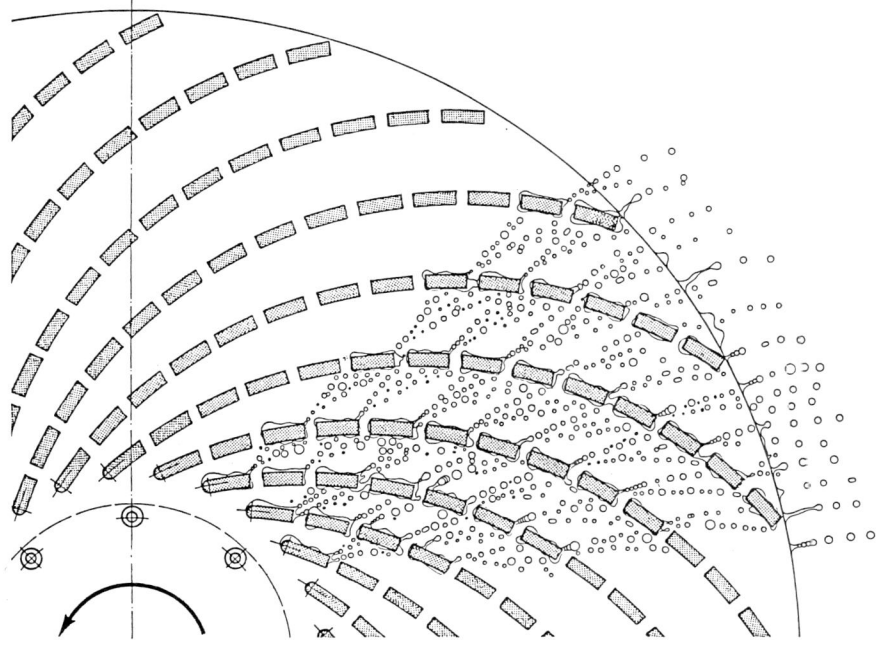

**Fig. 34.** Schematic presentation of liquid dispersion in the rotating wheel of the absorption machine

Fig. 35. Photograph of the rotating wheel with slotted vanes

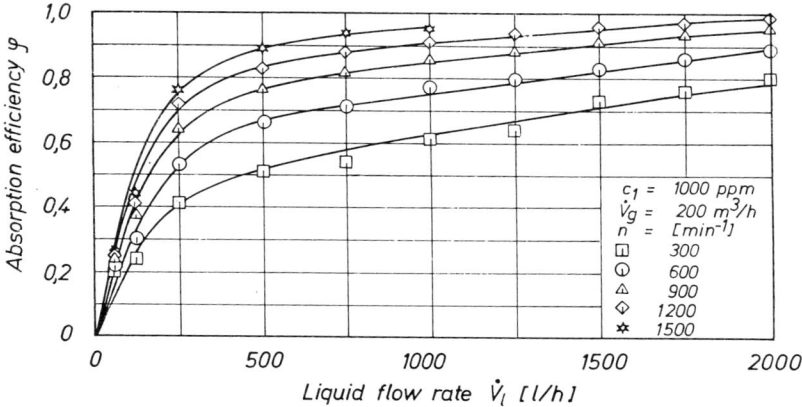

Fig. 36. Absorption efficiency as a function of the liquid flow rate $\dot{V}_1$ for various values of the rotational speed n

of the vanes, that is primarily a function of the diameter of the wheel. A photograph of the wheel is reproduced in Fig. 35.

Some of the experimental results are presented in Figs. 36 to 39. The absorption liquid has been an aqueous solution of 0.05 NaHO with pH = 12.7. The absorbed gaseous component has been sulfur dioxide ($SO_2$) at different concentrations in the air stream. The investigated parameters have been the volumetric flow rate of gas ($\dot{V}_g$) and liquid ($\dot{V}_1$), the $SO_2$-concentration $c_1$, and the rotational speed n [26].

According to Fig. 36 the absorption efficiency increases with increasing liquid flow rate $\dot{V}_1$ and rotational speed n. With $\dot{V}_1$ 1,000 l/h and $\dot{V}_g = 200$ m³/h the liquid/gas-ratio is 5 l/m³. Assuming as rotational speed of n = 1,200 min$^{-1}$, the absorption efficiency is better than 90%.

With increasing gas flow rate $\dot{V}_g$ the absorption efficiency decreases as indicated by Fig. 37. With increasing rotational speed n the absorption efficiency $\varphi$

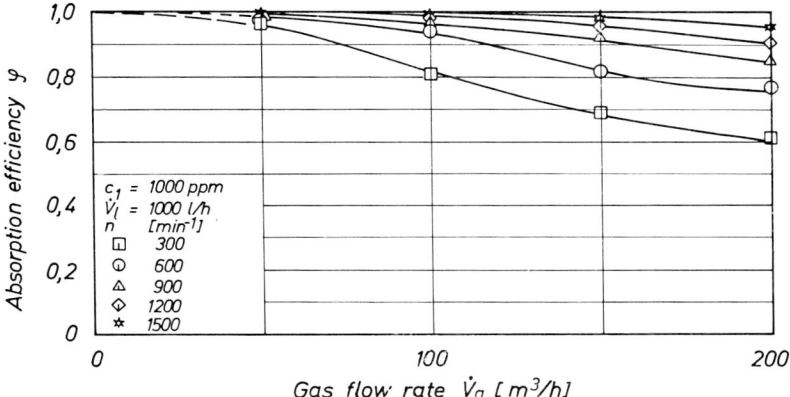

**Fig. 37.** Absorption efficiency $\varphi$ as a function of gas flow rate $\dot{V}_g$ for various values of the rotational speed n

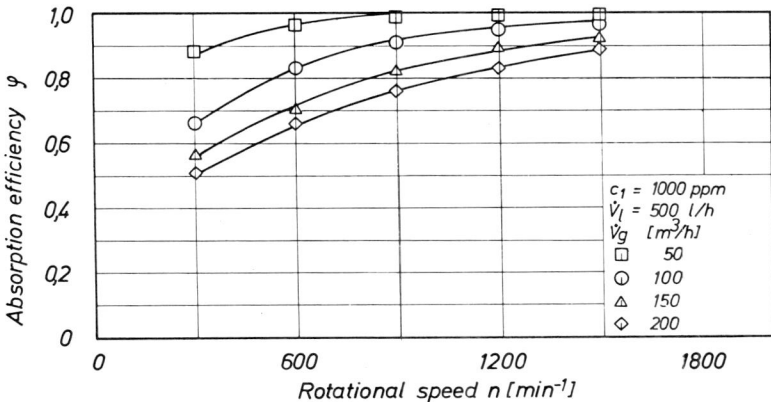

**Fig. 38.** Absorption efficiency $\varphi$ as a function of the rotational speed n for various values of the gas flow rate $\dot{V}_g$

increases according to the results presented in Fig. 38. This effect is of course due to the improvement of the liquid dispersion with increase of rotational speed. An increase of the $SO_2$-concentration $c_1$ will lead to a reduced absorption rate, which is shown in Fig. 39. The absorption efficiency of various types of absorbers is presented in Fig. 40. The new absorption machine is superior to all other absorption systems [27].

*Adsorption Process and Equipment*

General Description of the Process

Adsorption is the process by which residual molecular forces at the surface of solids attract molecules of gases and vapours. In case of air pollution control the relevant gases and vapours are the pollutants that have to be separated from an

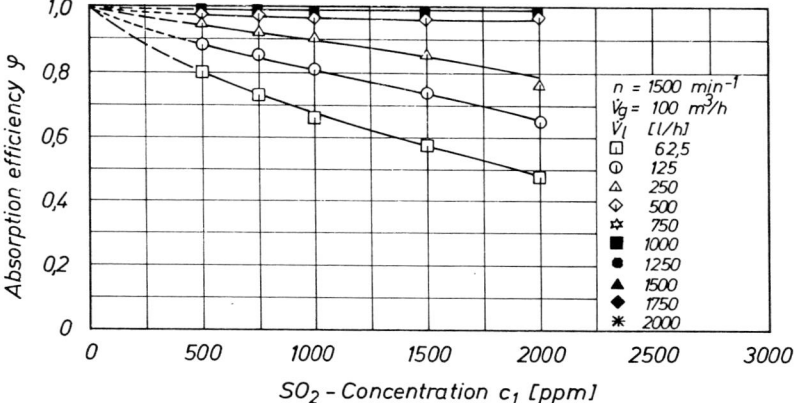

**Fig. 39.** Absorption efficiency $\varphi$ as a function of the $SO_2$-concentration $c_1$ for various values of the liquid flow rate $\dot{V}_1$

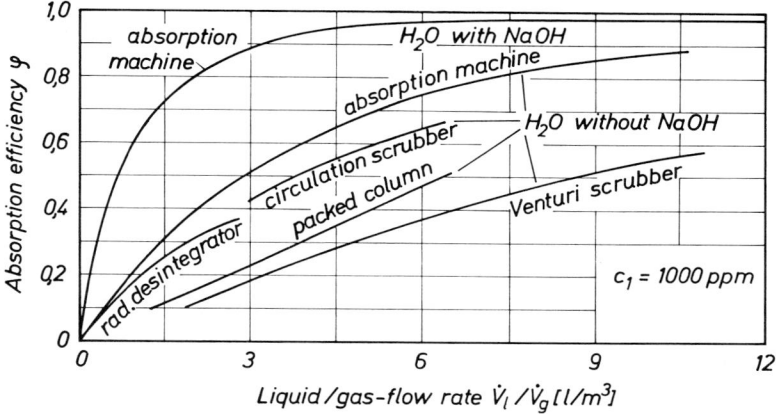

**Fig. 40.** Absorption efficiency $\varphi$ as a function of the liquid/gas-ratio $\dot{V}_l/\dot{V}_g$ for various types of absorbers

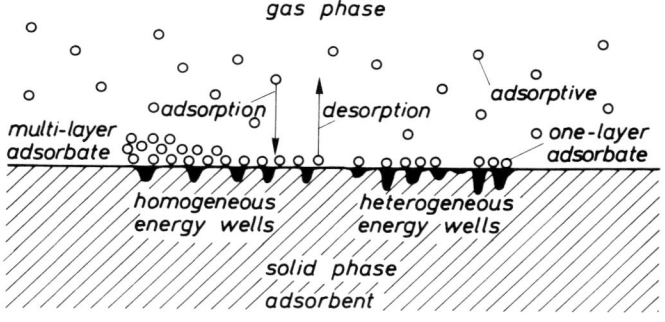

**Fig. 41.** Illustration of notations in adsorption

inert gas stream emitted into the ambient air. Adsorption therefore is a useful means of concentrating the pollutants, facilitating their disposal or recovery [1].

With the notations given in Fig. 41, the adsorbing solid is the adsorbent, while the gas, that is going to be adsorbed is called the adsorptive. Adsorption of gas molecules occurs at active sites of the solid surface. The active sites are called homogeneous, when they all have the same energy potential. With different energy potentials the sites are said to be heterogeneous. Gas molecules that are adsorbed are called adsorbate. The adsorbed phase consists of a thin gaseous layer, that includes the adsorbate, and a thin solid layer, that includes the active sites.

In the process of adsorption the molecule loses kinetic energy, making adsorption an exothermic process. The adsorbate, that is the molecules adhering to the surface of the adsorbent, may be considered to behave like a liquid. The reverse process of adsorption is called desorption. This process is consequently endothermic, heat must be supplied to separate the adsorbate from the adsorbent. In other words, desorption takes place by vapourisation of the adsorbate. The heat of adsorption for organic vapours is several times greater than that of permanent inorganic gases.

In adsorption one must distinguish between physical and chemical bonding of the adsorbate to the adsorbent. Physical adsorption is characterised by low heat of adsorption, that is of the order of 10 to 20 [kJ/mol]. A much stronger bonding exists in chemical adsorption. In this case the heat of adsorption is in general more than four times that for physical adsorption.

The bonding of the adsorbate to the adsorbent is of a chemical nature. The sorption process therefore is often irreversible, and the rate of adsorption may increase with rising temperature. The adsorption capacity is however reduced, because chemisorption is limited to unimolecular coverage of the adsorbate on the active sites.

In air pollution control chemical reactions between various adsorbates may be of importance. Because of the high concentration of the pollutants while they are in the state of adsorbate the reaction rate is considerably enhanced. The adsorbent acts like a catalyst. Sometimes this property of the adsorbent is improved by impregnation with a suitable material. In such a case the adsorption process is irreversible.

The rate of adsorption is proportional to the concentration of the adsorptive in the gas stream, the surface area of the adsorbent, the pore volume of the adsorbent, and other properties of adsorbent and adsorptive. Conventional adsorbents consist of highly porous particles that are arranged in deep layers in suitable vessels. Because of the high porosity the particles have a large internal surface area, compared to which the external surface area is negligibly small. Figure 42 gives an idea of the various pore sizes in a particle:

|  | $d[nm] = 10^{-9} [m]$ |
|---|---|
| Macro-pores | $50 \leq d$ |
| Meso-pores | $2 \leq d \leq 50$ |
| Micro-pores | $1 \leq d \leq 2$ |
| Submicro-pores | $d \leq 1$ |

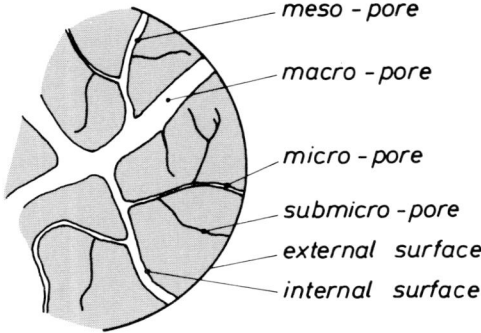

**Fig. 42.** Illustration of porous structure of granular adsorbent

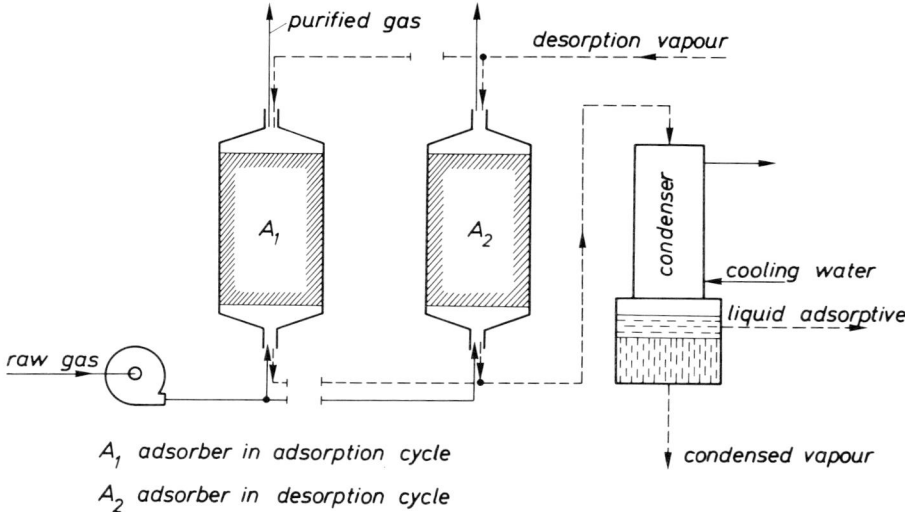

$A_1$ adsorber in adsorption cycle
$A_2$ adsorber in desorption cycle

**Fig. 43.** Schematic drawing of an adsorption unit

The dimension of the pore diameter d is given in nanometer [nm], that is equal to $10^{-9}$ meter [m] and 10 Ångstrøm units [Å]. The diameter of adsorbed molecules is in general of the order of a few Ångstrøm. Capillary condensation therefore takes place in meso-pores. The internal surface area of activated carbon, which is a widely used adsorbent, is of the order of 1,000 [m²/g]. The fraction of the pore volume is of the order of 20 to 40%.

An adsorption plant consists in general of two vessels. One vessel is in the adsorption cycle while the other is in the desorption cycle. Figure 43 gives a schematic drawing of a very simple adsorption plant. Desorption is achieved by the introduction of water vapour. In a condenser the water vapour and the vapourised adsorptive are liquified. Because of the density difference water and adsorptive may be withdrawn through separate outlets.

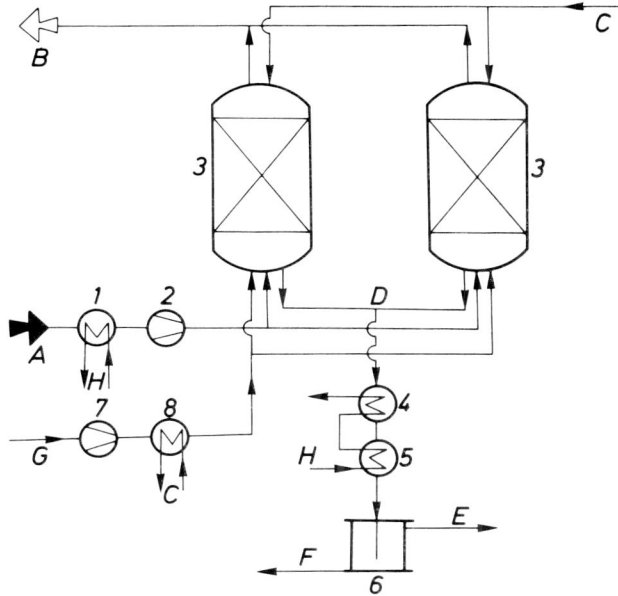

**Fig. 44.** Adsorption plant for removal of organic gases and vapours from waste gases with recovery of solvent. A raw gas, B purified gas, C water vapour, D desorptive and water vapour, E solvent, adsorptive, F condensed water, G fresh air, H cooling water; 1 waste gas cooler, 2 waste gas ventilator, 3 adsorber, 4 condenser, 5 cooler, 6 seperator, 7 fresh air ventilator, 8 fresh air heater

Organic and inorganic compounds may be separated from a gas stream by adsorption. A comprehensive survey on adsorption technology possibilities and future trends has been presented by Brauer [28]. As an example for technical application the adsorption of organic compounds will be considered more closely.

Adsorption and Recovery of Organic Gases and Vapours

Air pollution control processes may often become economic if the organic solvents are recovered from the waste gas and recycled into the production process. Such situations may for instance occur in the following industries [1]:

Printing and paper industries,
fiber industries,
plastic industries,
artificial leather, rubber, and asbestos industries, and
dry cleaning installations.

The adsorption process is carried out depending on the volumetric flow rate of the polluted gas stream in either one or several adsorbers operated in parallel. Figure 44 gives a schematic drawing of an adsorption plant. The adsorbers are of the packed bed type. They are in general operated in a temperature range between 30 and 60 °C. The preferred adsorbent is granular activated carbon with a mean diameter of 2 to 4 [mm].

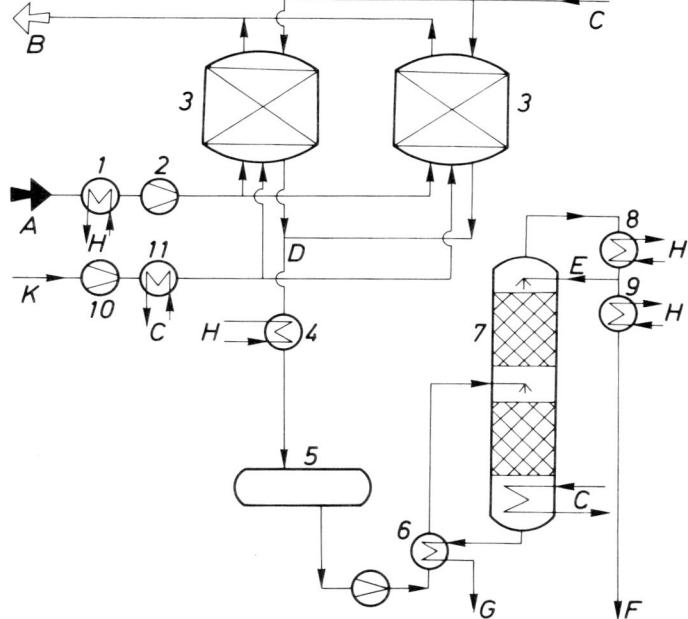

**Fig. 45.** Absorption plant for removal of organic gases and vapours from waste gases with recovery and rectification of solvent. A raw gas, B purified gas, C water vapour, D desorptive and water vapour, E reflux, F solvent, G condensed water, H cooling water, K fresh air; 1 waste gas cooler, 2 waste gas ventilator, 3 adsorber, 4 condenser, 5 storage tank, 6 preheater, 7 rectification column, 8 condenser, 9 product cooler, 10 fresh air ventilator, 11 fresh air heater

When the solvent concentration in the purified gas streams reaches a prescribed maximum value, the operation is switched over to another adsorber. The loaded adsorber goes into regeneration cycle. Water vapour desorbs the solvent from the adsorbent at a temperature of 100 °C or more. The concentration of the adsorptive in the water vapour is much higher than in the original gas stream. In other words the volumetric flow rate of the water vapour is very small compared to that of the gas stream. In a third step the mixture of water vapour and solvent is liquified in a condenser 4 and cooled down to ambient temperature in a cooler 5. In vessel 6 the two fluids are separated if the solvent is not in solution. In case the solvent is in solution with the water, further treatment of the mixture is necessary, as for example in a rectification column shown in Fig. 45.

To ensure continuous operation of the rectification column the water vapour/solvent-mixture is collected in storage tank 5, from which the rectification column receives the feed stream.

Various methods are used for drying and cooling of the adsorbent after the regeneration cycle. If the adsorptive capacity of the wet and hot adsorbent is sufficient for the process, drying and cooling may be achieved by the waste gas itself. In case the adsorptive capacity is not sufficient, drying and cooling are achieved through a fresh air stream.

The length of the adsorption cycle may range to a few hundred hours, while regeneration normally takes a few hours only. In such a situation the adsorption capacity of activated carbon may be improved by operating a two-adsorber system over a certain period of time in series. The regenerated adsorber is the second stage.

Adsorption plants of the types discussed in this section are designed for a volumetric flow rate of the waste gas of several hundred thousand m³/h. The concentration of the solvent in the waste gas varies in general between 5 and 25 [g/m³]. The adsorption plant may be designed and operated for a separation efficency close to 100%. The concentration in the purified gas stream will then meet the requirements of air pollution regulations.

For conventional adsorption plants and operation conditions *per ton of recovered solvent*

| | | |
|---|---|---|
| 3 to | 5 t | of water vapour, |
| 35 to | 50 m³ | of cooling water, |
| 35 to | 250 kWh | of electric energy, and |
| 0.5 to | 1 kg | of activated carbon |

*are required.*

**Chemical Processes and Equipment for the Abatement of Gaseous Pollutant Emissions**

Chemical conversion of organic pollutants contained in gases is the most effective process applicable in air pollution control [1]. The pollutants are primarily gaseous components, but may also be solid and liquid particles dispersed in the gas. Chemical conversion of pollutants may be achieved either by thermal or catalytic processes. Both modes of chemical conversion processes will be briefly discussed.

*General Description of Chemical Conversion Processes*

The wide application of chemical conversion of pollutants is primarily due to the following facts:
1. Approximately complete removal of all organic pollutants independent of the chemical properties.
2. In most cases there will be no waste water or solid waste treatment problem related to the chemical conversion process.

But a high price has to be paid for these advantages. The cost for energy, catalyst, and equipment may be quite appreciable. The conversion of organic pollutants, for example hydrocarbons, is governed by the following overall reaction equation:

$$C_m H_n + \left(m + \frac{n}{4}\right) O_2 \rightarrow \frac{n}{2} H_2O + mCO_2. \tag{31}$$

In this simple case the reaction products are water and carbon dioxide. In most cases however the reactions run through a varying number of intermediate stages and partial reactions, which are not yet well understood. One of the more important intermediate is carbon monoxide, which may subsequently be converted into carbon-dioxide:

$$C + \frac{1}{2}O_2 \rightarrow CO, \tag{32}$$

$$CO + \frac{1}{2}O_2 \rightarrow CO_2. \tag{33}$$

Depending on the composition of the waste gas and the injected fuel there will also be other reaction products like

    $NO_x$            HCl             $SO_2$

*nitrogen oxides, hydrogen chloride, sulfur dioxide,* and others.

The chemical conversion may be either carried out at high temperatures – this process is called thermal conversion – or at low temperatures, applying a catalyst to achieve a high reaction velocity – this process is called atalytic conversion. A catalytic reactor is operated in the low temperature range from 400 to 500 °C, while a thermal reactor is operated in a temperature range from 700 to 1,000 °C or at a temperature beyond that. Both conversion processes are applied in industry, with the thermal conversion as the more important process for the time being.

Chemical conversion of organic pollutants is essentially an oxidation process, that is self-propagating only within a limited range of concentration of the organic pollutant in an oxidant. The lower limit of this range is called the lower explosion limit, while the upper limit is called the upper explosion limit. The explosion range depends on the type of pollutant, composition of the gas mixture, temperature, pressure, ignition temperature, and to a certain extent on the geometry of the vessel [29]. Table 8 gives for a few substances the lower and upper explosion limit. With pollutant concentration outside of this explosion range the oxidation process will be quenched. For technical applications the range of selfpropagating oxidation, that is the explosion range has to be avoided. For matter of safety in technical processes the pollutant concentration has to be outside of this dangerous range.

**Table 8.** Concentration of various substances at the lower and the upper explosion limit for the following conditions: pressure 1 bar, room temperature, ignition energy 10 [J]

| Substance | Lower explosion limit % by volume | Upper explosion limit % by volume |
| --- | --- | --- |
| Ethane | 3.5 | 15.1 |
| Ethylene | 2.7 | 34.0 |
| Carbon monoxide | 12.5 | 74.0 |
| Methane | 4.6 | 14.2 |
| Methanol | 6.4 | 37.0 |
| Pentane | 1.4 | 7.8 |
| Propane | 2.4 | 8.5 |
| Toluene | 1.2 | 7.0 |
| Hydrogen | 4.0 | 76.0 |

There are four types of oxidation processes, for which the characteristic properties will be given.

Type one of oxidation process: The pollutant concentration is about 25% below the lower explosion limit, while the oxygen concentration is above 15% in the gas mixture. In this case there is enough oxygen available for oxidation of the burner fuel that has to be injected in order to achieve reaction temperature. Thermal or catalytic conversion may be applied. The thermal conversion takes place in so-called thermal afterburners at temperatures between 700 and 1,000 °C, or sometimes in furnaces. The catalytic chemical treatment of the organic pollutants is carried out in so-called catalytic afterburners.

Type two of oxidation process: The pollutant concentration is as in the type one process about 25% below the lower explosion limit, the oxygen concentration is however less then 15%. In this case there is not enough oxygen available for oxidation of the injected burner fuel. Supplementary air must be supplied. The process may be carried out either in thermal or in catalytic afterburners. Sometimes the waste gases are treated in a furnace.

Type three of oxidation process: In this case the pollutant concentration is above the upper explosion limit. When there is not enough oxygen in the gas mixture available the conversion process is carried out in flares, either groundlevel or elevated flares. If there is enough oxygen in the gas mixture the conversion process may be carried out in a conventional furnace.

Type four of oxidation process: The pollutant concentration lies within the explosion range. As a safe treatment of such a polluted gas mixture is not possible, it is advised to change the composition in such a way, that either type one or type two of the oxidation process is obtained.

In the following sections an introduction to industrial application of thermal and catalytic conversion processes will be given.

*Industrial Application of Thermal Conversion*

Thermal conversion is generally applied when the concentration of the organic pollutant is below the lower explosion limit. In general, continuous combustion can be achieved only by introduction of supplementary energy. In special cases this is necessary only during start-up operation of the plant that produces the waste gas with the pollutants. The supplementary energy is introduced into the thermal combustor by means of fuel injected by a special burner. The fuel is preferably natural gas, but may also be oil. The heat of the reacted gas is partly transferred to the raw gas in a preheater. This heat transfer process helps to reduce the supplementary energy. The residence time of the polluted gas mixture in the combustion chamber is of the order of 0.5 to 1.0 seconds. Thermal combustion is but one of several types of processes that may be used for purification of waste gases. These processes are absorption, adsorption, and catalytic chemical conversion. If the pollutants are valuable substances that should be recovered, an absorption or adsorption process should be the preferable choice. If recovery is not taken into consideration, the selection will be between thermal and catalytic treatment. In most cases the thermal treatment will be preferred, because certain substances contained in the waste gas mixture will poison the catalyst. Such catalyst poisons are for example chlorine, silicon, phosphorus, selenium, arsenic, and heavy metals. In case it is not required to recover the pollutants, and the pollutants are no catalyst poison, installation and operation costs will subsequently derermine the choice of process and equipment.

The advantages of thermal combustion are besides complete removal of the oganic pollutants with practically no waste water or solid waste problem, exceptionally good adaptability to changes in waste gas flow rate and composition and relatively simple design. Because of the supplementary fuel that has to be injected in most cases the operating costs for thermal combustion are rather high. This may at least in part be compensated by the development of a rather sophisticated heat recovery system, that will of course increase installation cost.

Thermal combustion may be applied for example for treatment of waste gases with predominantly organic pollutants from paint shops, coating shops, oil, fat, and food industries, animal breeding farms, treatment plants for drying and incineration of municipal waste.

More details of application of thermal chemical conversion, giving the production process and the pollutants, are summarized in Table 9 [1].
A typical and rather simple thermal combustion plant is shown in Fig. 46. The plant consists of a thermal combustion chamber (1) with burner (2), the heat exchanger (3), and the stack (4). The raw gas enters the plant at the heat exchanger where it is preheated, from there the raw gas is directed to the combustion chamber, where conversion of the organic pollutants takes place. The purified gas is directed to the heat exchanger where part of its heat is transferred to the raw gas, passes through the stack and into the atmosphere. The thermal combustion carried out in this plant is of the typeone oxidation process, because no supplementary air as oxygen source has to be introduced.

**Table 9.** Application of thermal chemical conversion of pollutants

| Technical process | Pollutants |
|---|---|
| Web offset printing (dryer) | Organic driers (oils), breakdown products of binders, resins, auxiliaries |
| Manufacture of chipboard | Benzene, toluene, aldehydes, phenols |
| Food industry (smoking, roasting: coffee, barley, chicory) | Aldehydes, oleic and fatty acids, sulphur and phosphorus compounds |
| Textile industry (finishing, printing) | Solvents, plasticisers, other textile auxiliaries, dusts, fibres |
| Manufacture of seals, clutch and brake linings | Solvents, phenol, dusts, fibres |
| Floor coverings | Plasticisers, formaldehyde, amines, mercaptans, styrene |
| Impregnation of glass fibre or asbestos insulating mats | Phenol, formaldehyde, carbon monoxide |
| Manufacture of electrodes and bitumen | Hydrocarbons, tar, dust, carbon monoxide |
| Phenolic baths for paint removal | Phenol-cresol |
| Phenolic resin curing chambers | Phenol, formaldehyde |
| Chemical production processes | Phthalic acid and maleic anhydrides, nitric acid, amines, ethylene oxide, etc. |
| Manufacture of laminated paper | Acetone, formaldehyde, cresols, methanol, phenol |

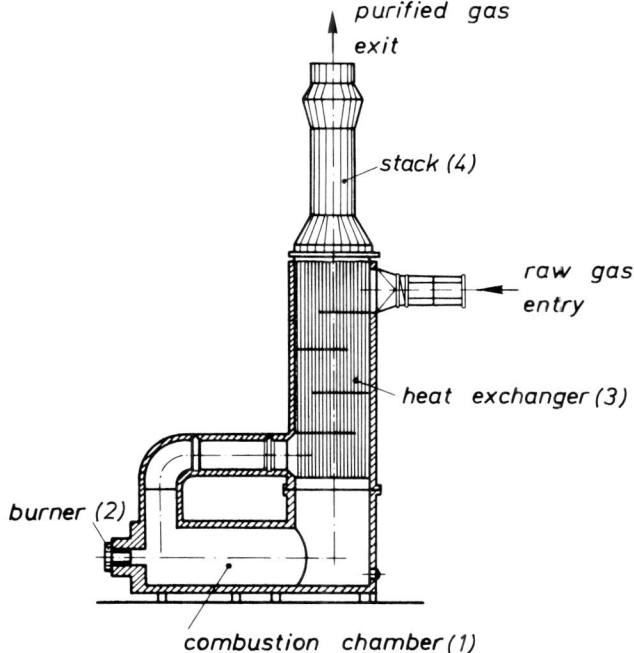

**Fig. 46.** Thermal combustion system

In Fig. 47 a schematic drawing of a plant for thermal chemical conversion of waste gas from a dryer with preheating of the raw gas and the recycled air is shown. The raw gas is taken from the dryer (10) by means of the ventilator (3) to the raw gas preheater (2) and from there to the thermal combustor (1). The fuel injection is regulated by the system (5) using the temperature of the reacted gas as indicator (TIC). The hot reaction gas is taken to the heat exchanger (2), in which part of the heat is transferred to the raw gas. The purified gas, that is still rather hot, is taken to the second heat exchanger (6) where another part of the heat is transferred to the recycled air. The purified gas is then emitted into the environment. From the raw gas leaving the dryer (10) one part is recycled. It is mixed with fresh air and then taken to the heat exchanger (6) for reheating by means of the ventilator (8). The temperature of the recycled gas (TIC) serves to regulate the flow rate of the purified gas through the heat exchanger (6). Starting-up operation is regulated by the system (4).

The chemical conversion of pollutants in combustion chambers is a function of several parameters:
Reaction temperature,
residence time of waste gas in combustion chamber,
concentration of pollutants,
chemical species of pollutants,
concentration of oxygen,
three-dimensional flow in combustion chamber, and
turbulence in flow field.

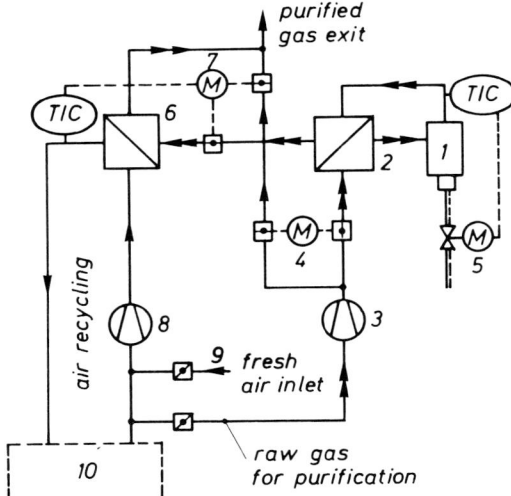

**Fig. 47.** Schematic drawing of a thermal combustion plant with preheating of raw gas and heating of recycled air. 1 combustion chamber with burner, 2 raw gas preheater, 3 raw gas ventilator, 4 bypass regulator for start-up operation, 5 regulator for fuel injection (reaction temperature), 6 heater for recycled air, 7 regulator for temperature of recycled air, 8 ventilator for recycled air, 9 fresh air inlet, 10 dryer

The conversion rate of the pollutants is preferably expressed by the residual carbon concentration – expressed in [mg/m$^3$] – and the carbon monoxide concentration – expressed in [ppm] – as measured in the purified gas stream at the exit of the combustion chamber.

*Industrial Application of Catalytic Conversion*

The industrial process of catalised chemical conversion of pollutants, present in waste gases, consists of the following steps:
1. Collection and transport of waste gas,
2. dust removal from waste gas,
3. heating to reaction temperature,
4. catalytic oxidation,
5. heat recovery from the reaction gases,
6. transport of reaction gases to stack.

Only the second and the fourth step make the difference from the thermal oxidation process. Dust removal from the waste gas may be desirable in order to prevent dust settling on the catalyst surface with the consequence of reduced catalytic activity. Dust removal poses no serious problems in the temperature range of 50 to 150 °C in which the waste gas is in general available. Several methods of dust removal have been discussed in this chapter.

The catalyst is not only sensitive to dust adhesion but also to the adsorption of gaseous compounds on the surface. The activity of the catalyst may be seriously effected by such gaseous "catalyst poisons." It is the unsolved problem of catalyst poisoning that restricts application of catalytic afterburners.

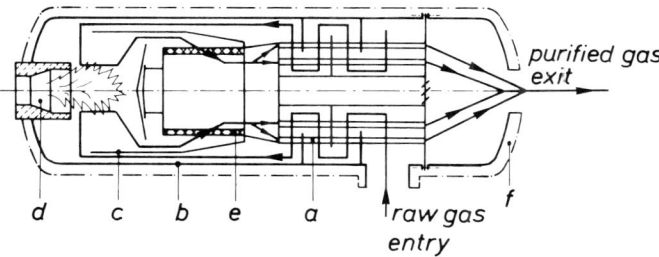

**Fig. 48.** Catalytic afterburner with ring layer of particulate catalyst. a heat exchanger, b outer shell, c inner shell, d burner, e catalytic reactor, f insulation

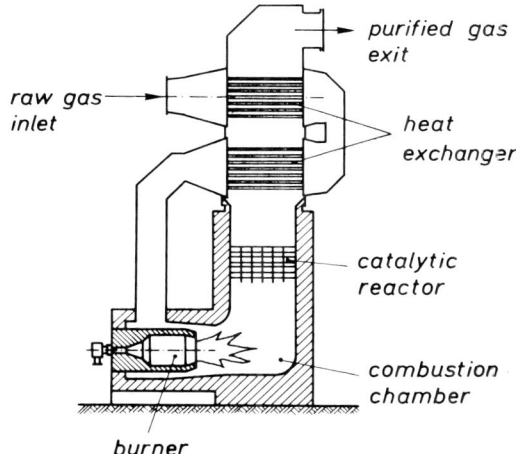

**Fig. 49.** Catalytic afterburner with array of ceramic cylinders coated with active catalytic material

Although the result of catalytic and thermal oxidation is the same, there are fundamental differences in the two processes, and in those parts of the plants in which the oxidation takes place.

In all other steps thermal and catalytic conversion are in accordance with each other. The consequence is that the equipment for both types of processes is rather similar.

Figure 48 shows a very compact type of catalytic conversion system. It consists of a cylindrical shell that contains a burner, catalytic reactor, and heat exchanger. The raw gas enters the system and is preheated in the heat exchanger. Passing through the outer and inner shell of the containment the preheated gas gets into the combustion chamber where it is heated to reaction temperature. The reaction itself takes place in the catalytic reactor, that consists of a ring layer of catalytic active particles. Leaving the reactor the reaction gases flow through the tubes of the heat exchanger where some of the heat is transferred to the raw gas.

Another purification system is shown in Fig. 49. Before entering the combustion chamber the raw gas passes through the tubes of a heat exchanger, where it is preheated. The desired reaction temperature is obtained in the combustion

**Table 10.** Typical applications of catalytic afterburning systems

| Industry | Pollutant, process | Waste gas flow rate $m_n^3/h$ | Inlet temp. °C | Catalyst temp. °C | mg Organic matter per $m_n^3$ of Polluted Gas | Purified Gas |
|---|---|---|---|---|---|---|
| Electrical industry | Solvent (styrene) | 500 | 350 | 370–420 | 1,000–2,800 | 250 |
| Web offset printing | Solvent | 3,500 | 380 | 400 | 500–1,500 | 50 |
| Food industry | Oil gas from automatic frying apparatus | 12,000 | 400 | 400 | 200 | 10 |
| Swimming pool operation | Decomposition of ozone | 500 | 105 | 105 | 2–3 g $O_3/m_n^3$ | |
| Automative industry | Engine exhaust gases from roller type test stand | 900 | 430 | 450 | 700– 900 | 50–70 |
| Food industry | Aldehydes, etc. exhaust gases from smoking chambers | 1,500 | 400 | 420 | 500–1,000 | 100 |

chamber. The oxidation takes place in the catalytic reactor, that consists of an array of cylindrical elements; its surface is coated with catalytic active material. The gas velocity in the empty rectangular vertical channel is of the order of 2 to 5 m/s.

Some characteristic properties of typical catalytic combustion processes applied in industry are summarized in Table 10. Poisoning of catalyst still hinders a wider technical application of this attractive mode of pollutant conversion. In the case of car exhaust gas purification success could only be achieved after regulations imposed restriction on the use of lead as an additive to benzene which is a very strong catalyst poison.

**Biological Processes and Equipment for the Abatement of Gaseous Pollutant Emissions**

In biological waste gas treatment, gaseous pollutants are converted into less harmful or even harmless substances by microorganisms. Biological waste water treatment is an established technology for more than a hundred years. Waste gas treatment by means of microorganisms became known only in 1957, when R. D. Pomeray was granted U.S. Patent 2.793.096 "De-odoring of gas streams by the use of microbiological growth." The attractiveness of biological waste gas treatment is due to the fact that this process – in contrast to conventional chemical conversion – occurs at an ambient temperature.

## General Description of the Biological Conversion Processes

Microorganisms are living chemical micro-reactors in which the conversion process takes place. Microorganisms are available in a great variety. Therefore all kinds of inorganic and organic pollutants can be converted by microorganisms. There will always be a species particularly well suited for the conversion of a special type of pollutant. Futhermore biological evolution will support adaption of microorganisms to pollutants.

The most important pollutants with respect to biological conversion are organic compounds. The microorganisms oxidize the organic compounds, thereby producing cell substance, carbon dioxide and water. This conversion process is of technical importance only when it takes place in a water solution. Pollutants and oxygen must be transferred to the water. Therefore biological waste gas treatment consists of the following two steps:
1. Transfer of organic pollutants from the gas phase into the liquid phase.
2. Microbiological waste water treatment.

Transfer of gaseous organic pollutants into a liquid, preferably into water, is achieved by absorption, and for certain conditions by adsorption. In the case of adsorption, the conversion process takes place in a very thin liquid film, that adheres to the surface of a solid.

Biological waste gas treatment seems to gain rapidly in importance. This is due to the fact, that biological waste gas treatment can be carried out in relatively simple technical equipment. Non-biological waste gas treatment by conventional absorption and adsorption processes turns out to be a rather complicated process, that is particularly due to the necessary regeneration of the liquid and the solid. It is the regeneration process which is carried out by the microorganisms. Biological waste gas treatment requires less complicated technical equipment than conventional processes and is therefore less expensive.

In biological waste gas treatment the pollutants are converted into other, harmless compounds. Recovery of the pollutants is excluded. Therefore biological conversion is restricted to cases of very low pollutant concentration. This is the domain of odourous compounds. As a matter of fact, microbiological waste gas treatment has already found wide application in the degradation of odourous substances.

According to the technical application of biological waste gas treatment in slaughter houses, carcass removal plants, animal breeding plants, compost plants etc. the most important pollutants are ethanol, mercaptane, phenol, cresol, indole, seatological compounds, fatty acids, aldehyde, ketone, carbonic acid, carbonyl, carbon disulfide, ammonia, amine, and many others.

A comprehensive treatment of the fundamentals of biological waste gas treatment and the technical equipment available has been presented in [1, 30, 31].

There are two technical biological treatment plants available: Bioabsorption and biofilter plants. The biofilter plants are primarily applied in animal breeding stations; bioabsorption plants have been successfully introduced into industry. Further discussion will be restricted to bioabsorption plants.

## Bioabsorption Plants

Bioabsorption plants are particularly well suited for the removal of gaseous pollutants with odourous properties. The plant consists primarily of an absorber and a biological waste water treatment plant, in which the absorbing liquid is regenerated by the action of microorganisms.

## General Layout of a Bioabsorption Plant

Figure 50 shows the general layout of a bioabsorbtion plant. Feed gas and water flow countercurrently through the absorber. The odourous substances are absorbed by the water. The purified gas leaves the absorber at the top. The contaminated water leaves the absorber at the bottom, from where it is taken to the bioreactor for regeneration. After regeneration the water is recirculated to the top of the absorber.

Absorption is a physical process, that depends primarily on the fluid flow conditions in the type of absorber selected. It is in general a comparatively fast process. The residence time of the water is therefore of the order of only a few seconds. The biological regeneration of the water is however a comparatively slow process. The residence time of the water in the bioreactor is in the range of a few minutes to about 12 h. Because of the difference in time required for absorption and regeneration, a special bioreactor for regeneration of the water is necessary.

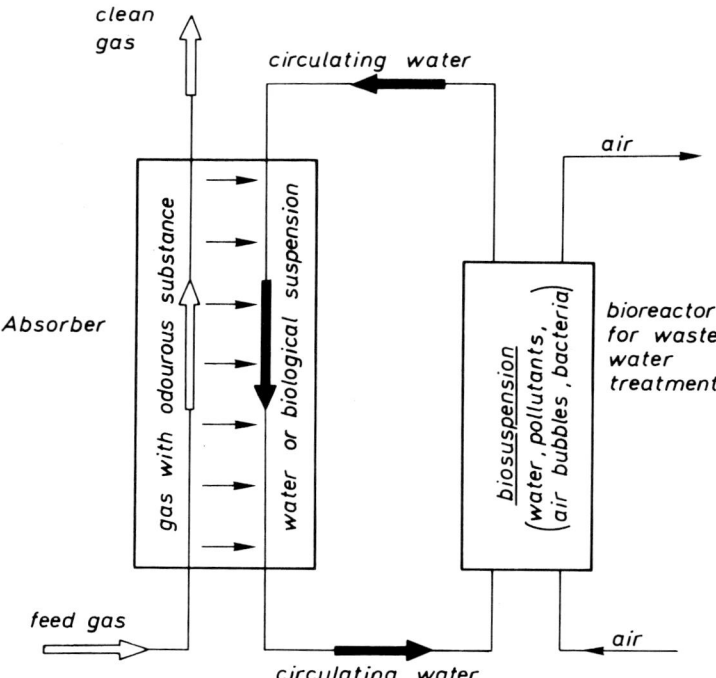

**Fig. 50.** General layout of a bioabsorption plant

If the biological conversion of the absorbed pollutants would take only the same time as the absorption process, no bioreactor would be necessary.

The absorption process has been described in detail in a previous section. These details will not be repeated here. The biological regeneration of the contaminated water will be discussed more closely.

The bioreactor may be an open tank or a closed vessel. In the bioreactor the biochemical conversion of the odourous pollutants takes place. The reaction requires oxygen according to the following equation:

$$\text{odourous pollutants} + \text{oxygen} \xrightarrow{\text{bacteria}} \text{cell material} + CO_2 + H_2O.$$

The oxygen is introduced by means of dispersed air bubbles. The water with bacteria, pollutants, and air bubbles is called the biosuspension. The velocity of the biochemical reaction depends primarily on the velocity of the oxygen transfer. Increasing the oxygen transfer rate, will lead to a smaller size of the bioreactor. According to experimental evidence, the residence time of the waste water in the bioreactor may be decreased to about 15 to 20 min [32]. After the biochemical conversion has taken place, the water is separated from the bacteria and recycled to the absorber. Only in special cases will the biosuspension be circulated. The bacteria rich sludge is recycled to the bioreactor; a certain amount of bacteria will be however withdrawn and otherwise disposed of.

On the way through the bioreactor the air will transfer oxygen into the water and absorb from the water carbon dioxide and a certain fraction of the gaseous pollutants contained in the biosuspension. The contaminated air leaving the bioreactor must be purified if necessary. Because of the small flow rate it may be introduced into the absorber.

### Example for Technical Applications of the Bioabsorption Process

Figure 51 shows the layout of a bioabsorption plant, designed for treatment of the waste gas from a light metal foundry. The waste gas contains amines, phenols, and aldehydes. The plant consists of two bioabsorbers operated in parallel and the necessary auxiliary equipment. In the first stage dust and alkaline constituents of the gas are removed by means of a weak acidic absorbent. In the second stage the gas is contacted directly with the biological suspension. There is one bioreactor for each of the two bioabsorbers. Oxygen is supplied to the biosuspension in the bioreactors by pressurized air. When the bioabsorbers are out of operation the bacteria are supplied with nutrients prepared in a special nutrient vessel.

The operational data of the plant are as follows:

| | |
|---|---|
| Flow rate of gas | $2 \times 60,000 \, m_n^3/h$ |
| Maximum temperature of gas | 35°C |
| Feed gas concentration | 60 to 100 ppm in propane equivalents |
| Clean gas concentration | 5 mg phenol/$m_n^3$ |
| Gas/liquid ratio | $60,000 \, m_n^3/h / 173 \, m^3/h$ |
| Mean residence time of gas in absorber | 9 s |
| Pressure drop of gas in absorber | 400 to 600 N/$m^2$ |
| Specific energy requirement | $2 \cdot 10^{-3} \, kWh/m_n^3$ |
| Equipment material | PE and PVC |
| Specific investment | 8 DM/$(m^3/h)$ |

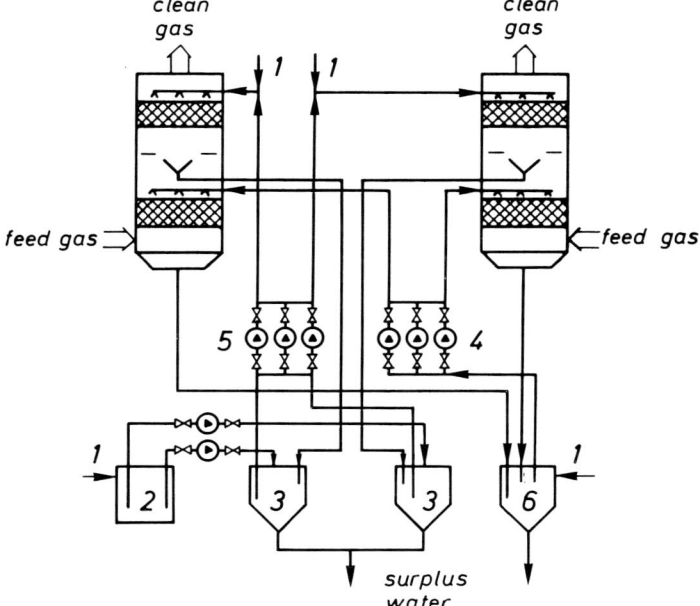

**Fig. 51.** Bioabsorption plant layout for treatment of waste gases from a light metal foundry. 1 fresh water, 2 nutrients, 3 bioreactors, 4 pumps for 1st stage, 5 pumps for 2nd stage, 6 absorbent vessel

A high-efficiency bioabsorption system is described by Brauer [30]. In this system the absorption machine described in a previous section and the reciprocating-jet-bioreactor described in [32] is applied for biological water regeneration.

**References**

1. Brauer, H., Varma, Y.B.G.: Air Pollution Control Equipment; Springer-Verlag, Berlin-Heidelberg-New York 1981
2. Hutzinger, O. (Ed.): Environmental Chemistry, Volume 1, The Natural Environmental and the Biogeochemical Cycles; Springer-Verlag, Berlin-Heidelberg-New York 1980
3. Hutzinger, O. (Ed.): Environmental Chemistry, Volume 3, Anthropogenic Compounds; Springer-Verlag, Berlin-Heidelberg-New York 1980
4. The Acropolis; Rep. of the working group on the preservation of the Acropolis monuments, published by the Hellenic Republic, Ministry of Culture and Science, Athens 1976
5. Green, P.: Parthenon; Verlag Kunstkreis, Luzern/Stuttgart 1975
6. Biesker, H.: Der langsame Tod der Bronzen; VDI-Nachr. 22 (1972)
7. Brauer, H.: Integration von technischen Maßnahmen zur Luftreinhaltung in Stoff- und Energiewandlungsanlagen; VDI-Berichte Nr. 294 (1978) pp. 9/18, VDI-Verlag GmbH, Düsseldorf 1978
8. Baum, F., Hager, J., Heiß, A., Kellerer, L.: Emittenten von luftverunreinigenden Schadstoffen und Lärmanlagen zum Mahlen oder Blähen von Schiefer und Ton; Schriftenreihe Luftreinhaltung, Heft 5, Bayrisches Landesamt für Umweltschutz
9. Muschelknautz, E.: Die Berechnung von Zyklonabscheidern für Gase; Chemie-Ing.-Techn. *44* 1/2, pp. 63/71 (1972)
10. Muschelknautz, E., Rink, N.: Zyklonabscheider; Handbuch des Fortbildungslehrgangs „Mehrphasenströmungen" des VDI-Bildungswerkes

11. Muschelknautz, E.: Hochschulkurs II, Mechanische Verfahrenstechnik; Verfahrenstechnik 6 (1972) 3, pp. I/IV; 5, pp. I/IV; 9, pp. I/IV; 10, pp. I/IV; 12, pp. I/IV and 8 (1974) 1, pp. I/IV
12. Muschelknautz, E., Brunner, K.: Untersuchungen an Zyklonen; Chem.-Ing.-Techn. 39 (1967) 9/10, pp. 531/538
13. Barth, W.: Berechnung und Auslegung von Zyklonabscheidern aufgrund neuerer Untersuchungen; Brennstoff-Wärme-Kraft 8 (1956) 1, pp. 1/9
14. Muschelknautz, E., Krambrock, W.: Aerodynamische Beiwerte des Zyklonabscheiders aufgrund neuer und verbesserter Messungen; Chem.-Ing.-Techn. 42 (1970) 5, pp. 247/255
15. Barth, W., Leineweber, L.: Beurteilung und Auslegung von Zyklonabscheidern; Staub-Reinhaltung Luft 24 (1964) 2, pp. 41/55
16. Ogawa, A.: On the effect of coagulation influencing the total collection efficiency for gas cyclone-dust-collectors; J. College Eng. Nikon Univ. Series A, 19 (1978) 3, pp. 157/166
17. Hikichi, T., Ogawa, A.: On the mechanical balancing particles in a cyclone-dust-collector for the steady and pulsating dust-laden gas flow; J. College Eng. Nikon Univ. Series A, 19 (1978) 3, pp. 167/183
18. Fujita, Y., Ogawa, A.: On the flow pattern in the vortex chamber of the returning flow type; J. College Eng. Nikon Univ. Series A, 19 (1978) 3, pp. 185/197
19. Fujita, Y., Ogawa, A.: The distributions of the static temperature in a turbulent rotational flow; J. College Eng. Nikon Univ. Series A, 20 (1979) 3, pp. 79/95
20. Hikichi, T., Ogawa, A.: Influences of air flow rates flowing into dust-bunkers on re-entrainment of fine solid particles from dust-bunkers of cyclone-dust-collectors; J. College Eng. Nikon Univ. Series A, 20 (1979) 3, pp. 155/162
21. Spilger, R.: Methodik zur Überführung wissenschaftlicher Informationen aus dem Gebiet der Verfahrenstechnik in problemorientierte Rechenprogramme; Dissertation, Technische Universität Berlin, 1978
22. Schönert, K.: Mechanische Verfahrenstechnik, insbesondere Umgang mit feinen Partikeln; Fridericiana – Zeitschrift der Universität Karlsruhe, Heft 21, pp. 12/33
23. White, A.J.: Industrial electrostatic precipitation; Addison-Verlag, Reading, Massachusetts, 1963
24. Deutsch, W.: Bewegung und Ladung der Elektrizitätsträger im Zylinderkondensator; Ann. Phys. 68 (1922) pp. 335/344
25. Brauer, H.: Stoffaustauschmaschinen; Chem.-Ing.-Techn. 58 (1986) 2, pp. 97/107
26. Oevermann, B.: Abscheidung von Schwefeldioxid in einer neuartigen Stoffaustauschmaschine mit alkalischen Absorbentien; Diplom-Arbeit am Institut für Chemieingenieurtechnik, Technische Universität Berlin
27. Kollar, M.: Abscheidung von Schwefeldioxid in einer neuartigen Stoffaustauschmaschine; Diplom-Arbeit am Institut für Chemieingenieurtechnik, Technische Universität Berlin
28. Brauer, H.: Die Adsorptionstechnik – ein Gebiet mit Zukunft; Chem.-Ing.-Techn. 57 (1985) 8, pp. 650/663
29. Bartknecht, H.: Explosionen; Springer-Verlag, Berlin-Heidelberg-New York 1978
30. Brauer, H.: Biologische Abluftreinigung; Chem.-Ing.-Techn. 56 (1984) 4. pp. 279/286
31. Brauer, H.: Von der technischen zur mikrobiologischen Stoffumwandlung; Swiss Biotech 3 (1985) 5, pp. 27/41
32. Brauer, H.: Biological waste water treatment in a reciprocating-jet bioreactor; published in: Fundamentals of Biochemical Engineering, vol. 2 of "Biotechnology," pp. 519/535, VCH Verlagsgesellschaft mbH, Weinheim 1985

# Subject Index

Abatement of Emissions
– Technical Measures  202
Abatement of Gaseous Pollutant Emissions
– Biological Processes  248
– Chemical Processes  241
– Technology for  229
Absorbers
– Classification of  232
Absorption efficiency $\varphi$  235
Absorption plant
– General layout of  230
Absorption Process and Equipment  230
acid deposition
– Sensitivity of various metals to  150
Acidic precipitation
– in Materials Damage  117
Acoustical emissions  201
Adhesion force separator  213
Adsorption
– Illustration of notations in  236
Adsorption plant
– for removal of organic gases  239
Adsorption Process and Equipment  235
Adsorption unit
– Schematic drawing of  238
Aerosol Chemistry
– in Arctic Haze  88
Aerosol size distribution
– in Alaska  86
Air pollutant concentration levels  123
Air Pollution
– and Materials Damage  113
– Control Equipment  187
Air Pollution Effects
– on Specific Materials  138
Air Quality
– Emission  190
Aitken particles  103
Albedo
– over snow and water  101
Aleutian Low  74
Aluminum
– corrosion of  148

– Materials Damage  139
Aluminum*
– and acid deposition  150
American Society for Testing Materials
  (ASTM)  131
Ames test
– of PAN  30
Analysis Methods
– of PAN  17
Analytical techniques
– for SVOC  49
Anthropogenic Activities
– Analysis of  197
Anthropogenic sources
– of arctic haze  94
Archimedes number  209
Arctic Air Mass
– in Arctic Haze  73
Arctic air masses
– total particle number  86
Arctic Air Pollution
– Evolution Sequences for  103
Arctic gas and Aerosol  95
Arctic Haze  69
– major transport pathways of  80
– Seasonal Variability  80
Arctic-Wide Deposition  106
Area of arctic air mass  106
Area-sources  191
Artificial leather, rubber and asbestos
  industries
– Adsorption and Recovery of Organic
    Gases and Vapours in  239
Atmosphere
– semivolatile organic compounds in  39
Atmospheric Boundary Layer  170
Atmospheric deterioration of materials  115
Atmospheric Processes
– and Pollutants  115

Bag filter  214
– Arrangement of  221
– Conventional design of  222

Bag filter 214
– Cost relations of 229
Bare carbon steel
– Materials Damage 139
Benzo(a)pyrene 41
Bioabsorption plant
– General layout of 250
Bioabsorption Plants 250
Bioabsorption Process
– Technical Applications of 251
Biogenics 40
Biological Conversion Processes
– in air pollution abatement 249
Boiling point
– of Peroxyacetyl nitrate 11
Boundary Layer Concepts
– in Materials Damage 170
Boundary Layer Flux Terms 172
Boundary Layer Theory
– Application to Non-Buildings 176
Boundary Layer Transition 173
Bromine
– in Arctic Haze 89
Bronze
– and acid deposition 150
Bubble and drop absorbers 233
Building stone
– Materials Damage 139
Buildings and Structures
– Application of Boundary Layer Theory to 175

$C_2H_4Br_2$ 98
$CaCO_3$
– Damage Mechanisms 151
Calcareous stones
– Estimated damage functions for 161
– Precipitation based loss rates 154
Calcite Damage Function 152
Calcium
– monthly geometric mean in Greenland 83
Carbon
– Light-Absorbing 91
Carbon Steel
– corrosion of 147
– damage function for 148
Carbonate Stone Damage Functions 156
Cast iron[+]
– and acid deposition 150
Catalytic afterburner
– as air pollution control equipment 247
Catalytic afterburning systems
– Typical applications of 248
Catalytic Conversion
– Industrial Application of 246
$CBrClF_2$ 98
$CF_3Br$ 95

$CH_2Br_2$ 98
$CH_2BrCH_2Br$ 98
$CH_2BrCl$ 98
$CH_3Br$ 98
Chamber Studies
– in Materials Damage 135
Chamber tests by Edney 123
Chemical Analysis
– of PAN 23
Chemical Conversion Processes
– in pollutant emissions 241
Chemical Reactions
– of PAN 12
Chemical Tracers
– in Arctic Haze 88
Chemiluminescent Detection
– of PAN 22
Chimney sweeps
– high exposure situations to SVOCs 60
Chlorocarbons 95
Cigarette smokers
– high exposure situations to SVOCs 60
Cleaning stage
– Air Pollution Control Equipment 195
– of an industrial plant 198
CO
– emissions 190
Coal gasification
– high exposure situations to SVOCs 60
Coal tar
– high exposure situations to SVOCs 60
Coal-fired boilers
– source of SVOCs 55
Coatings
– Permeation through 165
Cold traps
– as SVOC collection techniques 48
Column scrubber 228
– Cost relations of 229
Combustion sources
– of SVOCs 42
Computer model calculation
– of PAN 16
Concrete
– Damage to 162
– Materials Damage 139
Control equipment
– for air pollution 187
Copper
– and acid deposition 150
– corrosion of 146
– damage function of 147
– Materials Damage 139
Corrosion of Metals
– from Air Pollution 140
Cotton
– as filter medium 223

# Subject Index

Cyanide concentrations
– in Arctic Haze  82
Cyclone  214, 215
– Cost relations of  229
– Trajectories of particles in  217
Cyclone computations
– Particle distribution curve as basis for  218
Cyclone tracks  74

Damage Function Development
– Experimental Methods for  135
Damage Functions
– Requirements for  132
Deposition
– Experimental Data on  104
– in Arctic Haze  102
Deposition velocities
– to circular cylinders  177
Deposition velocity data
– Summary of  177
Dichotomous sampler  44
Diffuse removal
– of Particles  102
Dilution
– as SVOC collection techniques  48
Distribution Curves
– Particle Size  207
Dry cleaning installations
– Adsorption and Recovery of Organic Gases and Vapours in  239
Dry deposition  104, 119
– in Materials Damage  115
– wet deposition  115
Dry Deposition Velocities
– Typical Values of $SO_2$  120
Dry Dust Removal Equipment  213
Dust
– emissions  190
Dust Removal
– Equipment for  213
Dust removal equipment  206
– Criteria for the selection of  229
– Selection of  228

Economic Studies
– of Materials Damage  128
Electric Precipitator  224
– three-zone, horizontal  227
Electrical force separator  213
Electrical precipitator  214
– Cost relations of  229
Electrochemical corrosion  140
Electromotive series  140
Emission Parameters  194
Emission stage
– Air Pollution Control Equipment  195
– of an industrial plant  198

Environmental Fate
– of SVOCs  57
Equipment for Abatement of Gaseous Pollutants
– Physical Processes  230
Essential oils  103
Exposure Assessment
– of SVOCs  61

F-113  95
Fabric Filter  219
Fiber industries
– Adsorption and Recovery of Organic Gases and Vapours in  239
Fibers used in filter media  223
Field Exposure Studies
– in Materials Damage  136
Field Studies
– for SVOCs  52
Film absorbers  233
Filters
– as SVOC collection media  45
Florosil
– for collection of SVOCs  48
Flow rate of gas
– Dimensions of cyclones  219
Fractional collection efficiency
– for dry and wet dust separators  215
Fractional separation efficiency  210
Free radicals  2

Galvanized steel
– Estimated service life of  146
– Materials Damage  139
Gas Chromatography
– of PANs  19
Gas-Phase Synthesis
– of PAN  6
Gaseous pollutants and its sources  231
Gaseous tracers of arctic haze
– Spatial variation of  97
Gases
– in Materials Damage  116
Glass
– as filter medium  223
– Nomex nylon
Global Emissions  190
Global Troposphere
– Importance of Peroxyacyl Nitrates in  24
Granular adsorbent
– Illustration of porous structure of  238
Graphitic carbon concentration
– in air  93
Greenland Ice Sheet
– Scavenging ratio, for elements  105

$H_2S$
– concentration levels  123

Hazard Evaluation
- of SVOCs  60
Hazardous waste incinerators
- source of SVOCs  55
HCHO*
- concentration levels  123
HCl
- concentration levels  123
Health Assessment
- of SVOCs  59
Heat transfer  172
Henry's Law constant  120
High-Efficiency Absorption Machine  233
High-volume sampling
- for SVOCs  44
$HNO_3$
- concentration levels  123
Hospital incinerators
- source of SVOCs  55
Hydrogen chloride
- chemical abatement methods  242
Hydroperoxyl radical  6
Hygroscopicity
- of Haze Particulates  87

Icelandic Low  74
Indoor air pollution
- in Materials Damage  117
Inert Metal Substrates  166
Inertial removal
- of Particles  102
Infrared spectrum
- of peroxyacetyl nitrate (PAN)  9
Isopleths
- precipitation pH and hydrogen ion deposition  118

Jet absorbers  233

Kraft paper mill boilers
- source of SVOCs  55

Lead
- and acid deposition  150
- Materials Damage  139
Lead pollution
- in Greenland snow  107
Light Absorbing Carbon
- in Arctic Haze  91
Light scattering coefficient
- in the Norwegian Arctic  85
Limestone particles  220
Limestone weight loss
- Effects of rainfall and $SO_2$  160
Line-sources  191
Liquid adsorbents
- as SVOC collection media  45
Liquid-Phase Synthesis
- of PAN  7

Long range plume transport
- of PAN  25
Long-Path Infrared Spectroscopy
- of PANs  17
Los Angeles  3
Luminol  23

Macro-pores  237
Manganese
- monthly geometric mean in Greenland  83
Manufacturing
- source of SVOCs  42
Masonry
- Damage to  151
Mass force separator  213
Mass Spectrometry
- of Pan  24
Mass transfer  172
Material emissions  201
Materials Damage
- and Air Pollution  113
- Damage Function Models  135
- Damage Functions  144
- Semi-Controlled Field Experiments  138
- Temporal Rate Considerations  134
MAVS  44
Mean concentration of sulfur in the atmosphere  106
Measurement Techniques
- for SVOCs  43
Meso-pores  237
Meteorological Considerations
- in Arctic Haze  73
Meteorological Factors
- in Materials Damage  125
Micro-pores  237
Momentum  172
Monochloroperoxyacetyl nitrate  3
Mortar
- Damage to  162
- Materials Damage  139
Motor vehicles
- source of SVOCs  55
Municipal solid waste incinerators
- source of SVOCs  55
Mutagenicity
- of PAN  30

NAPAP Interim Assessment  130
Native stone types
- NATO/CCMS report  127
Natural Weathering  131
$NH_3$
- emissions  190
Nickel
- Materials Damage  139
Nighttime Chemistry
- of PAN  27

# Subject Index

Nitrate radical 6
Nitro-PAH 50
Nitrogen Oxide Cycle 24
Nitrogen oxides
– chemical abatement methods 242
NO
– concentration levels 123
$NO_2$
– concentration levels 123
– dry Deposition 119
$NO_x$
– emissions 190
Non Methane Hydrocarbon Cycles 28
Nonferrous Metals
– corrosion of 149
Nuclear Magnetic Resonance
– of Pan 24

$O_3$
– concentration levels 123
OH radicals 95
Organic Coatings
– Air Pollution Effects on 164
Organic Nitrates
– Relation to PAN 28
Oversize distribution
– Determination of 213
Oxidizing gases
– in Materials Damage 116
Ozone 4

Pacific air mass
– total particle number 86
PAH-quinones 50
Paint erosion data
– from Research Triangle Park, NC 167
Paint Film Components
– Leaching of 165
Painted steel
– Materials Damage 139
Painted wood
– Materials Damage 139
Paints
– Air Pollution Effects on 164
– Environmental Damage to 166
– Types of Damage to 165
PAN
– atmospheric lifetime 25
– Chemical Properties and Stability 11
– Frequency distributions 26
– gas chromatographic analysis of 20
– tropospheric half-lives 15
PANalyzers 21
PANs
– Toxicity 29
Particle Removal in the Arctic 102
Particle Size Distribution
– at Exit Separator 212

Particle size distribution curves 208
Particles 116
– concentration levels 123
– Dry Deposition of 121
– removal from the atmosphere 102
Particulate materials
– size analysis procedures 206
Particulate Matter
– General Survey on 206
Particulate Pollutant Emissions
– Technology for the Abatement of 205
Particulate Pollutants
– Properties of 205
Particulate trains 47
Path of Pollutants
– Through Industrial Plants 200
PCDD/PCDFs
– combustion variables 55
Peracids
– Relation to PAN 28
Peroxides
– Relation to PAN 28
Peroxyacetyl nitrate 3
Peroxyacyl Nitrates
– Formation of 5
Peroxybenzoyl nitrate 3
Peroxybutryl nitrate 3
Peroxyproprionyl nitrate 3
Petro 40
Phase Distribution
– of SVOCs 57
Photochemical smog 2
Photolysis
– of PAN 12
Photostationary state 5
Physical Properties
– of PAN 8
PICs 40
Plant-Specific Measures
– reduction of pollutant emissions 204
Plastic industries
– Adsorption and Recovery of Organic Gases and Vapours in 239
Plastics
– Materials Damage 139
Point-sources 191
Pollutant Air Concentrations
– in Materials Damage 122
Pollutant Delivery Processes
– Atmospheric 117
Pollutant emission rate 196
Pollutant Emissions
– Analysis of Industrial Plants 201
– Damages Caused by 196
Pollutant Interactions
– in Materials Damage 121
Pollutant source types 123

Pollutants
- Classification of 116
Polluted Air
- in Arctic Haze 77
Polyamide
- as filter medium 223
Polyester
- as filter medium 223
Polyurethane foam (PUF) 45
Portland limestone 158
Precipitation scavenging 104
Printing and paper industries
- Adsorption and Recovery of Organic Gases and Vapours in 239

Quantitative metals corrosion data
- NATO/CCMS report 127

Radiation balance
- atmospheric aerosols 100
Radiation Budget
- in Arctic Haze 99
Raman spectra
- of particles collected in Alaska 92
Raman spectrum
- of peroxyacetyl nitrates 10
Reactivities
- of SVOCs 58
Reviews of Air Pollution Damage to Materials 126
Risk approximation
- of SVOCs 61

Sampling Program (AGASP) 95
Sandstone weight loss
- Effects of rainfall and $SO_2$ 160
SAPRC 44
Satellite Sensing
- in Arctic Haze 101
Scatter diagram
- of aerosol optical depths 100
Scattering albedo
- for aerosol 100
Scavenging ratio 104
Secondary copper smelters
- source of SVOCs 55
Sedimentation Velocity
- Particle Size and 209
Semivolatile Organic Compounds 39
Separation Efficiency
- air pollution control equipment 212
- Fractional 211
Separation Efficiency of Equipment 210
Separation efficiency $\varphi$
- for cyclones 218
Size analysis 206
Size Distribution
- of Arctic Haze 85

Size-selective inlet/filter systems 45
Smog 2
Smog aerosols 40
Smog chamber studies
- of PANs 21
Snow
- albedo over 101
$SO_2$
- action on zinc corrosion 143
- Annual emission 79
- dry Deposition 119
- emissions 190
$SO_2$ air concentrations
- Population-weighted 137
$SO_2$ dry deposition velocities 121
$SO_2$ oxidation
- Photochemical 91
- Size-Dependent Chemistry of 91
$SO_2^*$
- concentration levels 123
$SO_4$
- concentration levels 123
Soiling
- in Materials Damage 116
Solar radiation at Fairbanks 74
Solid adsorbents
- as SVOC collection media 45
Source Sampling
- of SVOCs 47
Source Studies
- of SVOCs 54
South Pole
- temperature inversions over 76
Spectral Properties
- of PAN 8
Stainless steel[+]
- and acid deposition 150
Statistical Analysis
- in Materials Damage 136
Steel[+]
- and acid deposition 150
Stone Loss Rates 153
Structure of PAN 16
Structures
- Boundary Layers on 172
Submicro-pores 237
Sulfate
- Annual variation of 82
- in arctic atmospheric aerosols 84
- in Arctic Haze 89
- nonmarine $SO^{2-}$ 90
- pollution-derived $SO_2$ 90
Sulfate in the atmosphere
- variation of 81
Sulfur
- effects on various materials 127
- Mass budget for the Arctic Basin 106

## Subject Index

Sulfur dioxide
- chemical abatement methods 242
Sulfur oxides effects
- NATO/CCMS report 127
Sulphur
- monthly geometric mean in Greenland 83
Surface pressure
- Average values of 78
Surface roughness data 174
Surface Wetness
- in Materials Damage 124
Surface Wetness Effects 179
SVOC blowoff 46
SVOC field studies 52
SVOC see Semivolatile Organic Compounds 39
SVOCs in the atmosphere
- primary sources 42
Synthesis
- of PAN 7

Temperature
- in Materials Damage 125
Tenax-GC
- for collection of SVOCs 48
Terpenoids 103
Tetrafluoroethylene
- as filter medium 223
Thermal combustion plant
- Schematic drawing of 246
Thermal combustion system 245
Thermal Conversion
- Industrial Application of 243
Thermal decomposition of PAN 14
Thermal decomposition reaction
- of PAN 12
Thermal emissions 201
Time Series Analysis
- in Materials Damage 137
Toxicity
- of PAN 29
Trace gas concentrations
- in arctic haze and clean air 96
Trace Gases
- and Arctic Haze 92
Triple point
- of Peroxyacetyl nitrate 11
Troposphere
- formation of Peroxyacyl Nitrates 5
Tropospheric half-life
- of PAN 15
Tunnels
- as SVOC collection techniques 48
Turbulence intensities
- Rural and Urban 171

Ultraviolet-visible absorption
- of peroxyacetyl nitrate (PAN) 8
Unimolecular Decompositions
- of PAN 12
Urban Microclimate Effects 179
Urban Oxidant Burden
- of PAN 29

Vapor pressure
- of Peroxyacetyl nitrate 11
Vapour/particle distribution
- of SVOCs 40
Velocity number 209
Venturi scrubber
- Cost relations of 229
Venturi-scrubber 228
Vertical Morphology
- in Arctic Haze 84
Volatility
- of Haze Particulates 87
Volume-sources 191
Vortex scrubber 228
vs. Pollution-Induced Damage 131

Water
- albedo over 101
Water nucleation 102
Weathering steel
- Materials Damage 139
Wet deposition 104, 117
Wet Dust Removal Equipment 227
Wet Scavenging 106
Wind
- in Materials Damage 125
Wind tunnel 120
Wind velocities
- Rural and Urban 171
Wire and pipe precipitator 225
Wood Substrates
- weathering rates 168
Wood-fired boilers
- source of SVOCs 55
Wool
- as filter medium 223
Wrought iron[+]
- and acid deposition 150

XAD-2
- for collection of SVOCs 48

Zinc
- and acid deposition 150
- Corrosion of 142
- damage function of 145
- monthly geometric mean in Greenland 83
Zinc corrosion 118

## *Environmental Toxin Series*

Editors:
S. Safe, O. Hutzinger

Volume 1

## Polychlorinated Biphenyls (PCBs): Mammalian and Environmental Toxicology

With contributions by numerous experts

1987. 33 figures, 35 tables. X, 125 pages.
ISBN 3-540-15550-3

**Contents:** *S. Safe, L. Safe, M. Mullin:* Polychlorinated Biphenyls: Environmental Occurrence and Analysis. – *L. G. Hansen:* Environmental Toxicology of Polychlorinated Biphenyls. – *A. Parkinson, S. Safe:* Mammalian Biologic and Toxic Effects of PCBs. – *M. A. Hayes:* Carcinogenic and Mutagenic Effects of PCBs. – *I. G. Sipes, R. G. Schnellmann:* Biotransformation of PCBs: Metabolic Pathways and Mechanisms. – *R. J. Lutz, R. L. Dedrick:* Physiologic Pharmacokinetic Modeling of Polychlorinated Biphenyls. – *S. Safe:* PCBs and Human Health. – Subject Index.

Volume 2

## Cadmium

3rd IUPAC Cadmium Workshop, Jülich, Federal Republic of Germany, August 1985

With contributions by M. Stoeppler and M. Piscator

1988. 57 figures, 79 tables. ISBN 3-540-15551-1

**Contents:** Toxicity, Carcinogenicity, Animal Experiments. – Epidemiology. – Cadmium in the Environment. – Methodology and Quality Assessment.

The concern about environmental toxins is ever increasing, as is the need for sound scientific information. The Environmental Toxin Series is dedicated to the publication of comprehensive reviews and monographs on compounds or classes of chemicals which are of importance in environmental toxicology. The series is designed to serve as a background of information for scientific investigation as well as risk analysis and political decision making. The main aim of the series is to describe in as complete a way as possible all potentially hazardous chemicals from the point of view of chemistry, ecology, toxicology, risk analysis and regulatory implications. From time to time conference proceedings on important and urgent topics will be included in the series. We thank the members of the editorial board for their enthusiastic support.

Springer-Verlag Berlin
Heidelberg New York London
Paris Tokyo Hong Kong

# The Handbook of Environmental Chemistry

Editor: O. Hutzinger

Volume 2

# Reactions and Processes

**Part D**

With contributions by P. B. Barraclough, N. O. Crossland, R. Herrmann, W. Mabey, C. M. Menzie, T. Mill, P. B. Tinker, M. Waldichuk, C. J. M. Wolff

1988. 47 figures, 55 tables. XI, 210 pages.
ISBN 3-540-15547-3

**Contents:** *R. Herrmann:* Hydrology. – *N. O. Crossland, C. J. M. Wolff:* Outdoor Ponds: Their Construction, Management, and Use in Experimental Ecotoxicology. – *T. Mill, W. Mabey:* Hydrolysis of Organic Chemicals. – *M. Waldichuk:* Exchange of Pollutants and Other Substances Between the Atmosphere and the Oceans. – *P. B. Tinker, P. B. Barraclough:* Root-Soil Interactions. – *C. M. Menzie:* Reaction Types in the Environment.

An important purpose of **The Handbook of Environmental Chemistry** is to aid the understanding of distribution and chemical reaction processes which occur in the environment. Volume 2, Part D of this series is dedicated to a broad description of chemical reactions of pollutants in environmental compartments, to root-soil interactions and to hydrology.

Springer-Verlag Berlin Heidelberg New York London Paris Tokyo Hong Kong